Lecture Notes in Physics

Edited by H. Araki, Kyoto, J. Ehlers, München, K. Hepp, Zürich
R. Kippenhahn, München, H.A. Weidenmüller, Heidelberg
J. Wess, Karlsruhe and J. Zittartz, Köln

Managing Editor: W. Beiglböck

316

Wolfgang Kundt (Ed.)

Supernova Shells and Their Birth Events

Proceedings of a Workshop
Held at the Physikzentrum Bad Honnef
March 7–11, 1988

Springer-Verlag
Berlin Heidelberg GmbH

Editor

Wolfgang Kundt
Institut für Astrophysik der Universität Bonn
Auf dem Hügel 71, D-5300 Bonn, FRG

ISBN 978-3-662-13647-8 ISBN 978-3-540-46001-5 (eBook)
DOI 10.1007/978-3-540-46001-5

Originally published by Springer-Verlag Berlin Heidelberg New York in 1988
Softcover reprint of the hardcover 1st edition 1988

2158/3140-543210 – Printed on acid-free paper

PREFACE

A joint breakfast at the I.I.Sc. guesthouse in Bangalore with Werner Buckel in the fall of 1985 was the birth event of what has now materialized, at Bad Honnef, as an international workshop on supernova explosions and their shell remnants, supported by the Volkswagenwerk foundation.

At various stages during the organization, Richard Wielebinski's backup and advice had a stabilizing influence, together with his secretary, Gabriele Breuer, and photographic laboratory. During the meeting itself, Norbert Junkes and Reinhold Schaaf took care of everything that would have distracted me from the scientific agenda, and Joachim Debrus from DPG conducted the technicalities of the house in his reliable and friendly manner. To all of them go my thanks.

The subjects of supernova remnants and supernova explosions have an old tradition; however, increasingly better resolution - both spatially and spectrally - keeps them young. In addition, SN 1987 A in the LMC had its independent impact on the workshop, and besides, it became clear right from the first evening that there was enough mutual disagreement among theoreticians to keep the meeting interesting.

Some of these disagreements have survived the workshop and appeared in the printed contributions. As I consider them both important and entertaining, I have highlighted them in an epilogue.

There was also the spontaneous suggestion by Wojtek Markiewicz to record and edit all questions, comments and answers. For him alone, this was an impossible task due to the unexpected flood of comments. Still, he compiled a long list of contributions. A careful editor would have made sure that all questions and answers were secured and proof-read by their authors. I didn't. I left the responsibility to the speakers!

Bonn, July 1988

Wolfgang Kundt

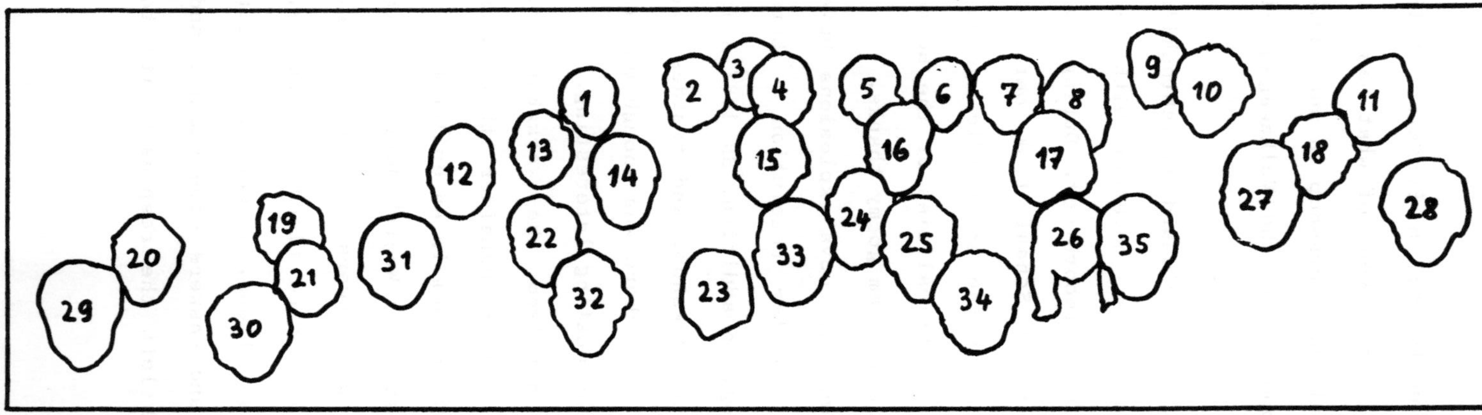

1. Lattimer, Jim
2. Fink, Henner
3. Schaaf, Reinhold
4. Preite-Martinez, Andrea
5. Bandiera, Rino
6. Greidanus, Harm
7. Van den Bergh, Sidney
8. Innes, Davina
9. Fürst, Ernst
10. Brinkmann, Wolfgang
11. Markiewicz, Wojtek
12. Gaskell, Martin
13. Dwek, Eli
14. Biermann, Peter
15. Reich, Wolfgang
16. Arendt, Rick
17. Fischer, Daniel
18. Srinivasan, G.
19. Green, Dave
20. Falle, Sam
21. Strom, Richard
22. Claas, Jacques
23. Reich, Patricia
24. Becker, Bob
25. Braun, Robert
26. Hüttemeister, Susanne
27. Milne, Doug
28. Junkes, Norbert
29. Da Costa, Antônio
30. Meyer, Hinrich
31. Schaefer, Brad
32. Panagia, Nino
33. Bartel, Norbert
34. Kundt, Wolfgang
35. Priester, Wolfgang

CONTENTS

SUPERNOVA SHELLS, GENERAL

SUPERNOVA SHELLS, SPECIAL OBJECTS

INTERPRETATION OF SUPERNOVA SHELLS

W. Kundt
Institut für Astrophysik der Universität Bonn, Auf dem Hügel 71, D5300 Bonn, FRG

Abstract: Our knowledge about Supernova Shells (SNSs) is steadily increasing concerning their number, morphologies and spectra. But there is no general agreement on how to model them. Here I want to comment on their velocity fields, ages, sizes and fine structure. In particular, SN explosions may resemble splinter bombs more closely than pressure bombs; i.e. a blast wave description may be inadequate for several purposes.

1. SNS Kinematics

The motion of matter inside SNSs is often called 'turbulent', in particular by those who like to convert a certain fraction of their kinetic energy into high-energy cosmic rays. If SN explosions resemble splinter bombs - a property which will be derived below from simple assumptions - very little turbulence is expected. In principle, the kinematics can be derived by comparing maps at different epochs (to yield proper motions) and by measuring Doppler shifts.

Figure 1 shows the well-known proper motions inside the Crab nebula and figure 2 those of the optical knots of Cas A. Both velocity fields satisfy $v(\vec{x}) \sim \vec{x}$ within the errors, i.e. correspond to homologous expansions. At greater accuracy, the filaments in the Crab are 8% post-accelerated whereas those of the main shell of Cas A are 6% decelerated w.r.t. those at $\lesssim$ twice the radial distance (whose expansion age agrees with John Flamsteed's SN detection in 1680, see Ashworth (1980). Figure 2 does not contain the velocity vectors of the outlying fast-moving flocculi determined by Fesen et al. (1987) which confirm the described pattern.

The velocity field of the fast-moving knots and flocculi of Cas A is in marked contrast to that of the quasi-stationary flocculi and also to that of the radio features shown in figures 3 and 4. The latter looks quite turbulent indeed. Whereas all three velocity fields were measured in similar positions, their magnitudes differ by factors of 10 and ~ 3 respectively. Apparently we deal with three different components which I like to identify with a) the ejecta, b) the circumstellar medium (CSM) and c) the relativistic component (which may consist of pair plasma). In pressure equilibrium, the mass density of a relativistic gas is $6kT/mc^2 = 10^{-8.2}\ T_4$ times lower than that of a thermal component (of temperature

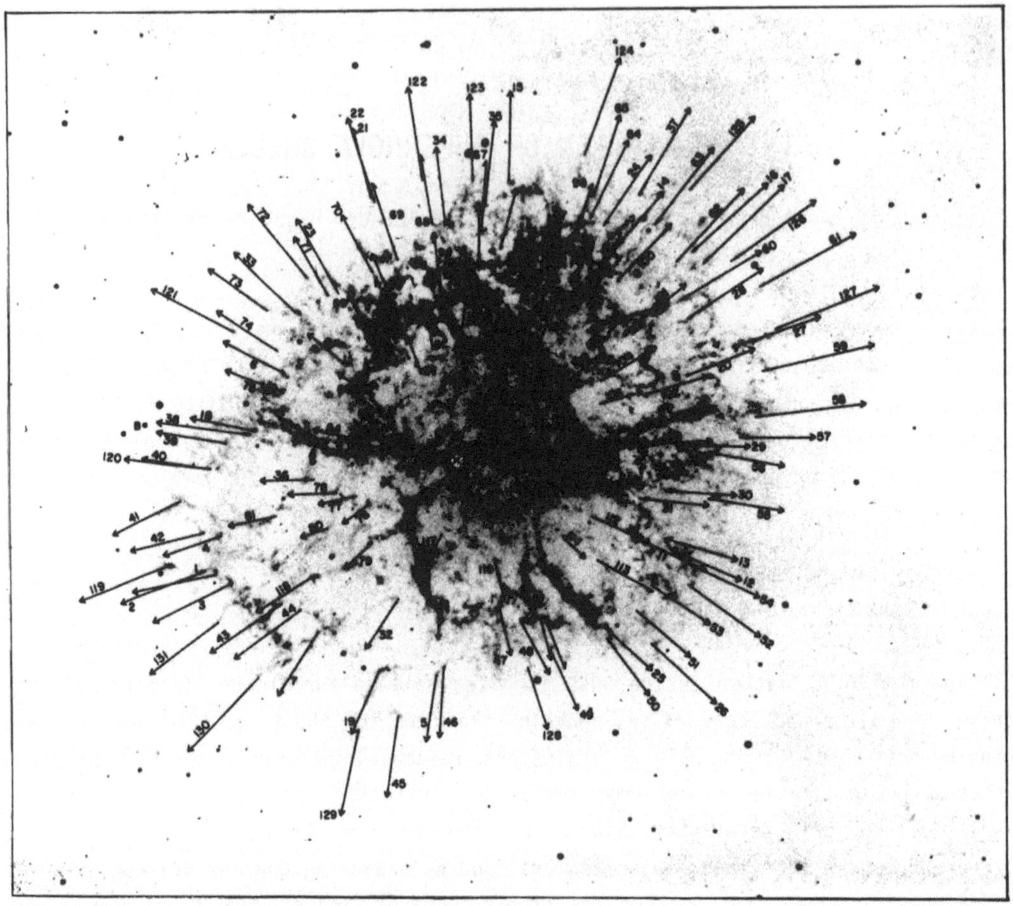

Figure 1: Proper motions for 132 Hα bright spots in the Crab nebula within 270 yr, taken from Trimble (1968). The divergence center is marked. Any noticeable turbulence of the velocity field may be due to difficulties in the identification, i.e. spurious.

$T = 10^4$ K); the motion of the relativistic component therefore must not be confused with the motion of the SN ejecta (cf. Weiler & Sramek, 1988).

Slow-moving components have not only been observed in Cas A but also in Kepler, 3C 58 and MSH 15-52 (Bandiera, this volume; Kirshner & Fesen, 1978; Van den Bergh & Kamper, 1984). In the case of Cas A, the slow component has an average velocity of (165±15) Kms^{-1} towards north-northeast and towards us, inferred from its displaced divergence center and blueshift (Kundt, 1980).

The measured velocity fields therefore show quasi-homologous expansion of the ejecta plus a quasi-reposing CSM which can have a different center-of-mass velocity.

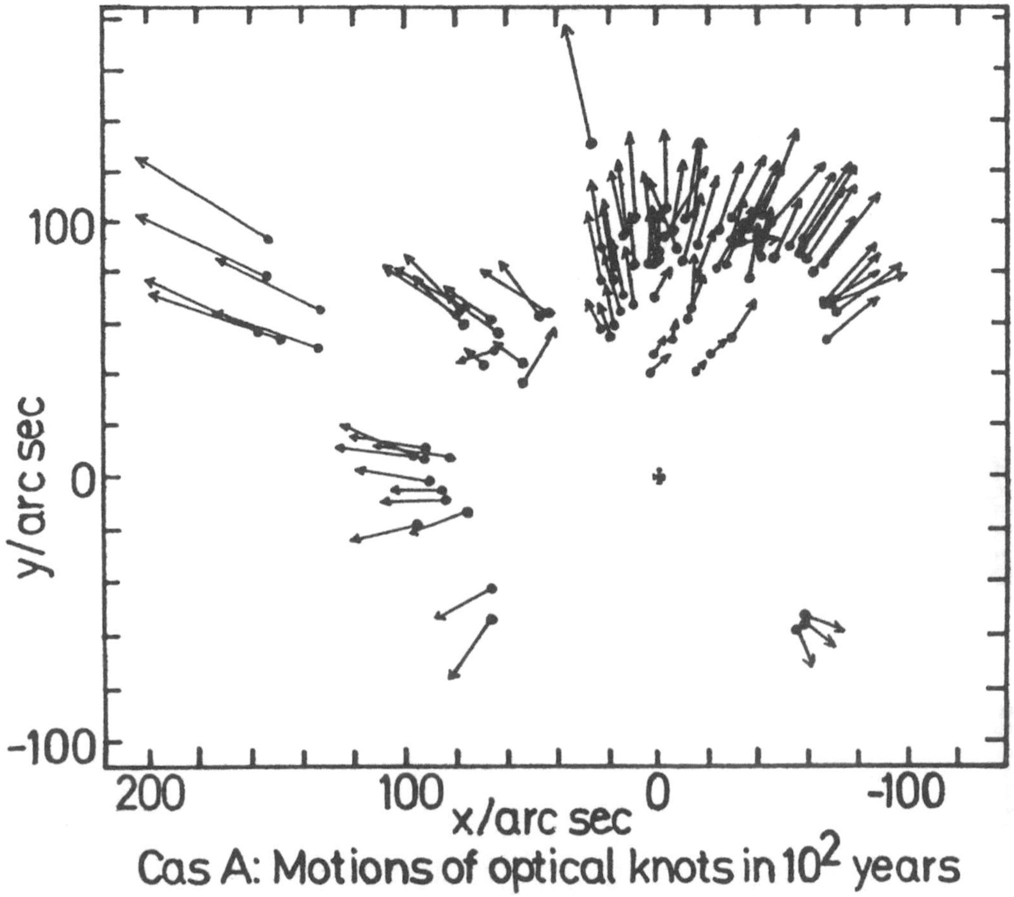

Figure 2: Analog of figure 1 for the (fast) optical knots of Cas A, after Kamper & Van den Bergh (1976). Shown are their proper motions within 10^2 years. The main shell is 6% decelerated w.r.t. the outermost knots. Missing in this plot are the proper motions of the fast optical flocculi found by Fesen et al. (1987); their positions have up to twice the distance (of the main shell) from the divergence center.

2. SNS Ages

Age estimates of SNSs are difficult as long as we do not know the precise deceleration law. Even so, Milne's (1987) table 1 has $1.8 \cdot 10^4$ yrs as the oldest entry. A maximum age of $\approx 10^4$ yr has also been inferred by Van den Bergh (1983) from the ratio of all (considered) SNSs to the (corresponding) historical ones. Thirdly, all (4) pulsars of (spindown) age $P/2\dot{P} \leq 10^4$ yr are found inside SNSs, whereas none of the $\geq 2 \cdot 10^4$ yr old ones are seen surrounded by a shell.

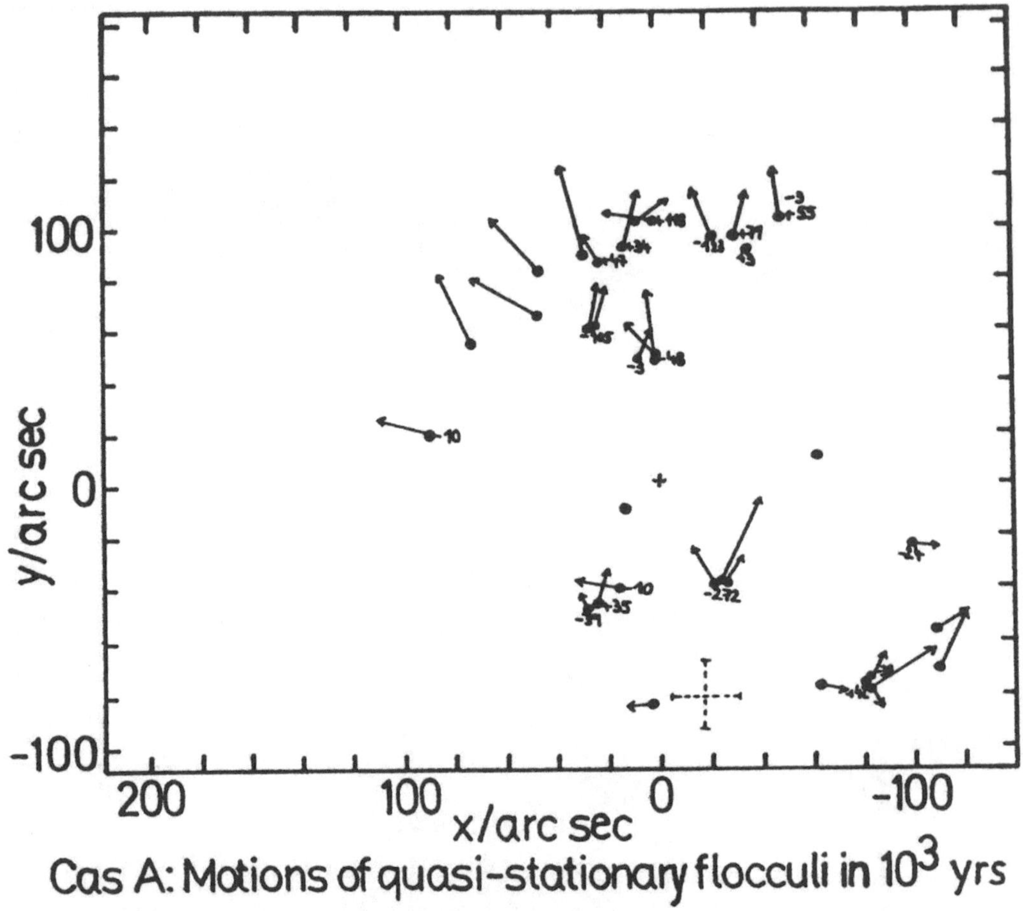

Figure 3: Proper motions of the quasi-stationary flocculi of Cas A, after Kamper & Van den Bergh (1976), copied from Kundt (1980). Shown are their displacements within 10^3 years. Note the large offset of the divergence center (to the South).

In spite of this evidence, one often finds 10^4 yr-old SNSs called 'middle-aged' in the literature and proposed shell ages of $\geq 10^5$ yr. It may therefore be worthwhile to sharpen the above estimates. To begin with, how reliable is the pulsar spindown age? Lyne et al. (1982) have compared it with the kinematic age $(z-z_*)/\dot{z}$ (which is to first order distance independent). Unfortunately, they do not distinguish between the warped molecular-cloud layer and the so-called Galactic plane. Secondly, they assume that pulsars are born in the Galactic plane even though their own data (PSRs 0531+21 and 0611+22) and other young pulsars argue counter. I have therefore drawn their data with (different) error bars in figure 5 and compared with the law of an exponentially decaying transverse magnetic dipole moment of e-folding time $\tau = 10^{5.8}$ yr, cf. Kundt (1988). The common conclusion is

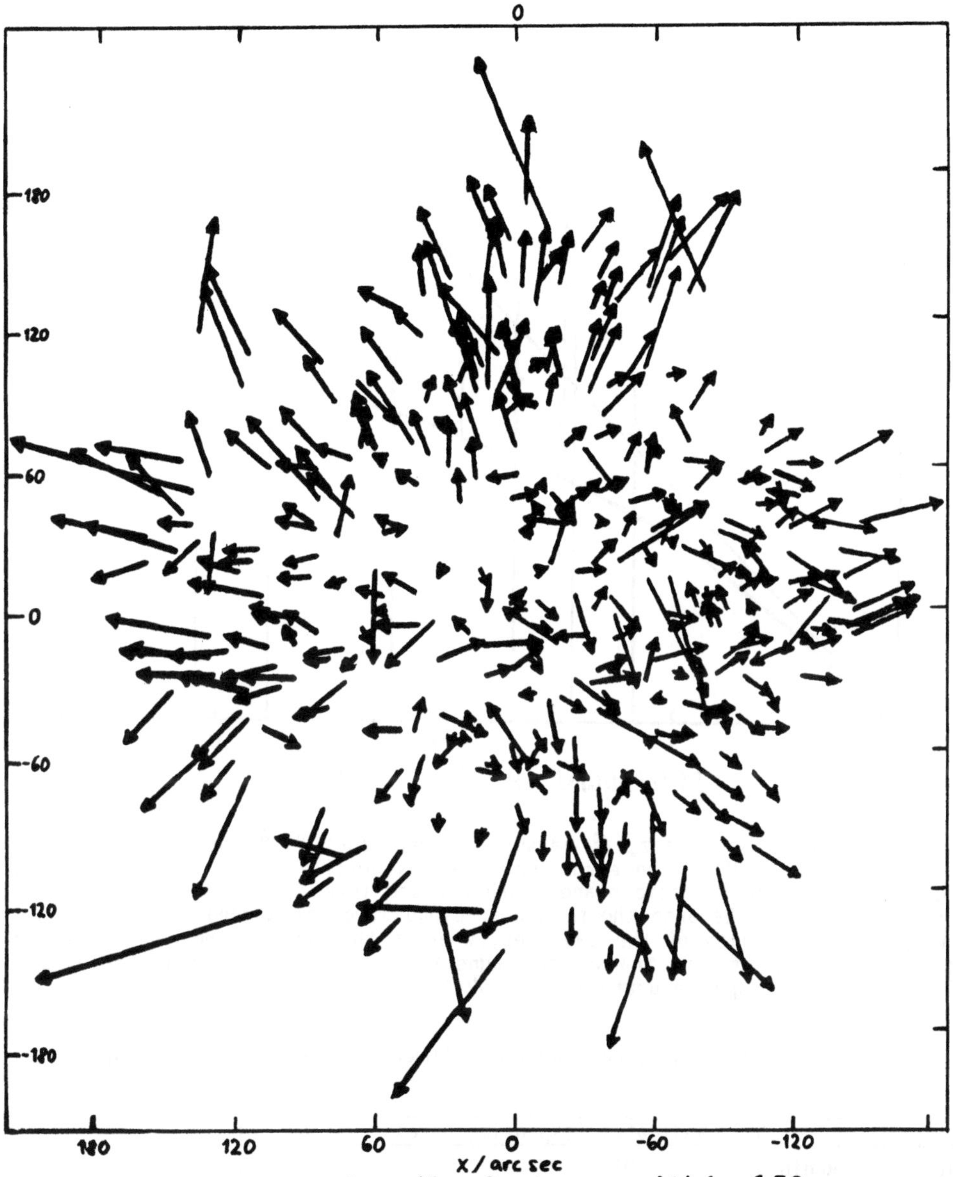

Figure 4: Proper motions of 342 radio features in Cas A, after Tuffs (1986) (his 1983 version has been copied). Their displacements within 150 years are shown. Obviously, these expanding motions contain a random component of comparable amplitude.

that the spindown age $P/2\dot{P}$ overestimates the true age t for $t \gtrsim 3 \cdot 10^6$ yr but tends to be a reliable age estimator for younger values. Of course, since $P/2\dot{P}$ is expected to be non-zero at birth, it can overestimate a pulsar's age even for young values, but only in rare cases (on statistical grounds).

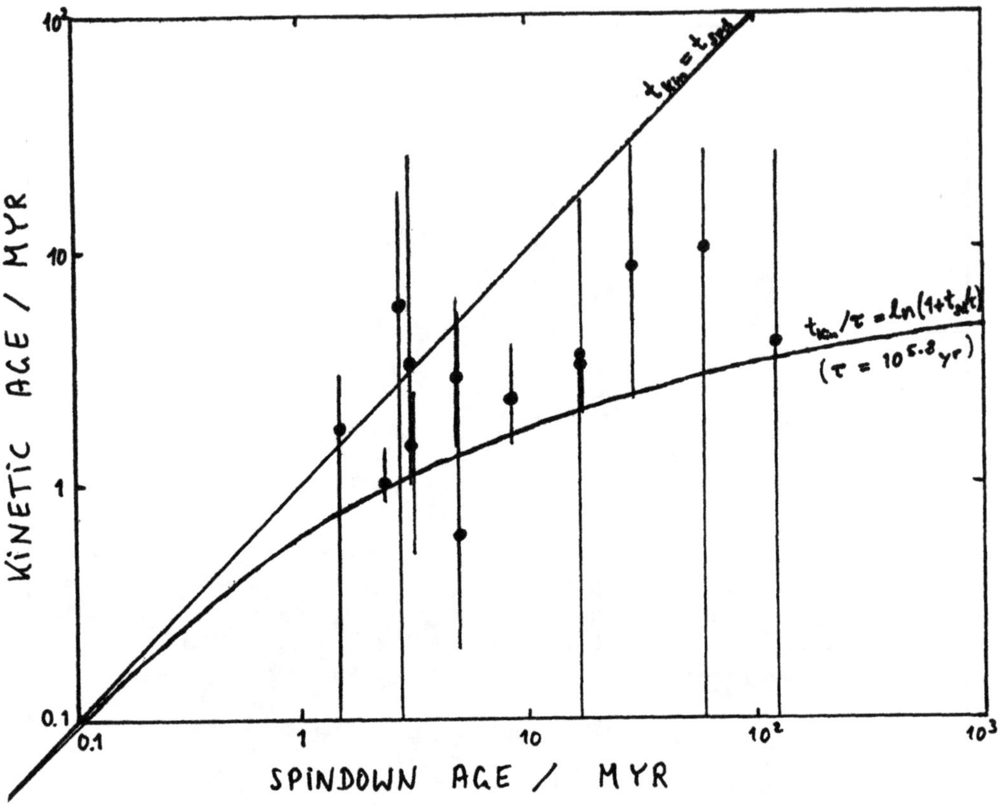

Figure 5: Kinetic ages $(z-z_\star)/\dot{z}$ versus spindown ages $P/2\dot{P}$ for 13 pulsars, taken from Lyne et al. (1982). The error bars are essentially due to an uncertain birth height $z_\star$ above the galactic plane, both because (even) young pulsars have a $z_\star$ of order 200 pc and because the molecular-cloud layer is warped. The straight line corresponds to equality of the two ages, whereas the curved line corresponds to an exponential decay of the transverse magnetic dipole moment (with e-folding time $10^{5.8}$ yr). Kinetic ages saturate around $10^{6.7}$ yr.

With this method at hand, we find the following age groups of well-known SNSs:

Age/yr	SNS candidates (explosion date)
10^2	Radio sources in M82
$10^{2.5}$	Cas A (1680), Kepler (1604), Tycho (1572)
10^3	Crab (1054), SN 1006, 0540-69.3 in LMC, MSH 15-52
10^4	Vela, Cygnus Loop, W50
10^5	CTB 80

Here the first line is a tentative interpretation of the rapidly fading, marginally resolved radio sources in M82 (cf. Bartel et al., 1987) whose age is estimated from the fading timescale. The ages of the 'historical' SNSs are known from European and

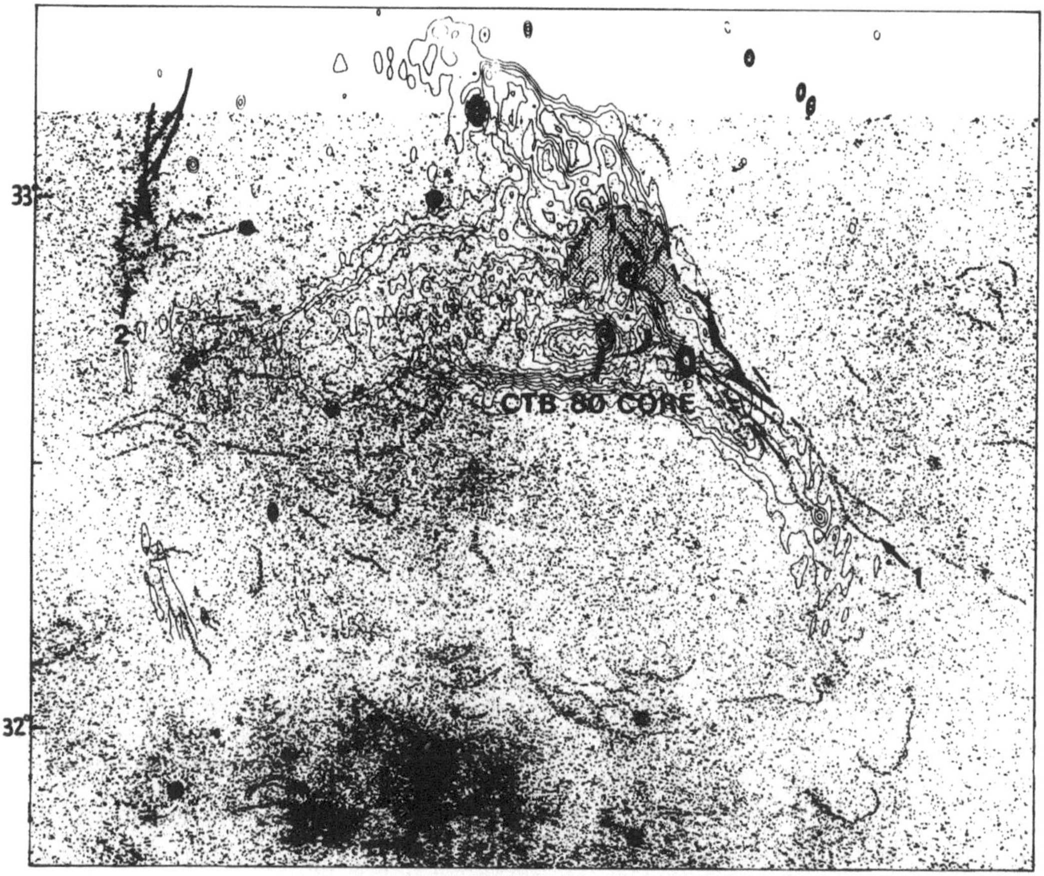

Figure 6: Hα interference filter photograph of CTB 80 by Van den Bergh (1980) superimposed on 49 cm radio isophotes plus hand-drawn Hα+[NII] filaments by Blair et al. (1984). Moreover, a number of filamentary features of the optical image have been enhanced by hand. Some of them may or may not be artefacts (noise). They all fall inside a 1° circle around some point to the east of the present 'core', perhaps above the letters 'CTB'. The radio morphology may be due to the 10^5 yr-old pulsar.

Chinese records. The age of MSH 15-52 is identified with that of its pulsar; its Sedov age is probably meaningless; its radius is reasonable for an age of $10^{3.2}$ yr (Kundt, 1985b). The ages of the Cygnus Loop and of W50 (Kundt, 1987b) are less certain; estimates are offered by the measured radial velocities of the ejecta.

CTB 80 with its recently detected pulsar of $P/2\dot{P} = 10^5$ yr (Kulkarni et al., 1988) should no longer be recognizable as a SNS and indeed, a shell has been found only recently from IRAS data (Strom, this volume; Fesen et al., 1988). When I tentatively identified an optical Hα+[NII]+[SII] shell on Van den Bergh's (1980) Hα-enhanced photograph during the workshop, the audience sounded sceptical. Yet my interpretation in fig. 6 correlates strongly with the IRAS data. I conclude that

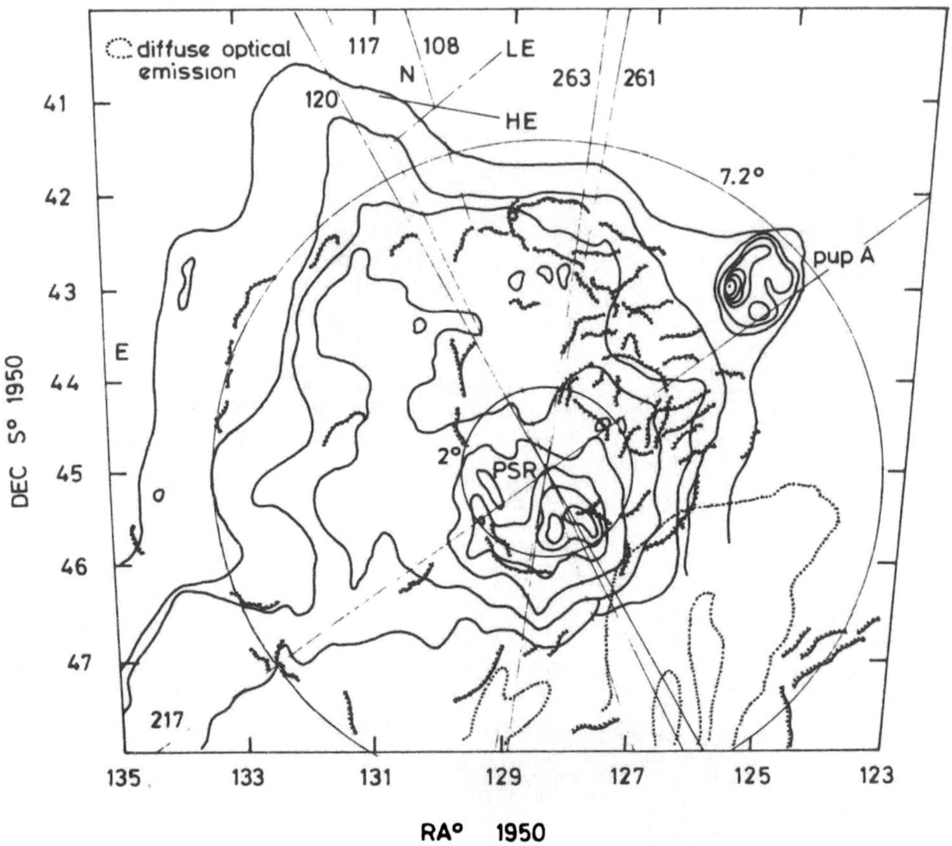

Figure 7: Radio isophotes and optical filaments of the Vela SNS, copied from Smith (1978), who also drew the circles around the Vela PSR. The large circle estimates the extent of the SNS (from which Pup A should be subtracted), the small circle that of the harder source.

1) a 10^5 yr old shell cannot easily be recognized, and that 2) the radio source CTB 80 is due solely to the pulsar.

3. Location of the Stellar Remnant

If SN explosions give birth to a neutron star - as they are likely to do in the majority of cases (cf. Seward, 1985; Srinivasan, this volume; Kundt, 1985a) - then where should we search for it? Note that the beaming factor of pulsars at HWZI may be near unity (Kundt, 1985a) but that more than 50% of all neutron stars are expected to be born binary (Blair & Candy, 1985) and therefore not to act as pulsars because the wind of the companion star screens them.

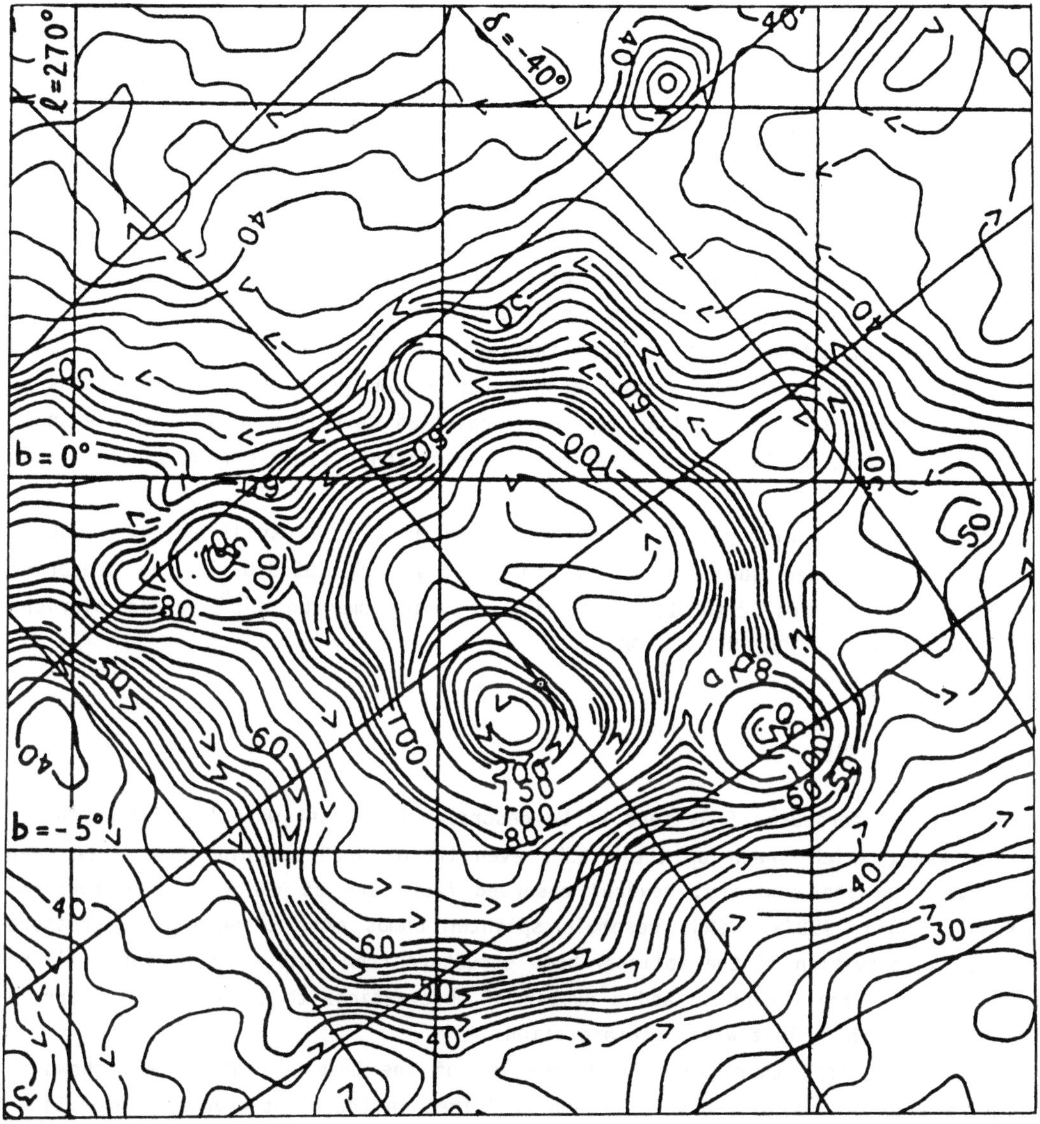

Figure 8: Enlargement from the 408 MHz survey by Haslam et al. (1982) containing the Vela SNS. The Vela PSR is located (and marked) at (ℓ,b) = (263°.6,-2°.8). Unlike fig. 7, this map is oriented w.r.t. the galactic disk; the local spiral arm is centered on b = -2°. This map suggests - like fig. 7 - that the pulsar is located at the center of its shell.

When a SNS (in the warm component) expands at an average rate $v = r/t$ of $10^{3.5}$ Km/s (in order to reach a radius of 30 pc in 10^4 yr) and the neutron star receives a transverse kick velocity of some 10^2 Km/s, then its maximal displacement Δr from the divergence center should be $\Delta r/r = v_\perp/v_{SNS} = 10^{-1.5}/v_{8.5}$, i.e. very small. The stellar remnant should not be searched near the edge of a shell.

A crucial case is the Vela remnant whose PSR looks strongly off-center on standard maps whereas the PSR hardly moves (Bignami & Caraveo, 1988). But where is the outer edge of the SNS? Fig. 7 (from Smith, 1978) demonstrates that the PSR may sit smack at the center when all the outlying optical filaments are properly taken care of. The same conclusion can be drawn from the radio survey by Haslam et al. (1982), fig. 8. Haslam estimates a spiral-arm background of $\gtrsim 40\%$ (by comparison with the antipodal sky map which shows the Cygnus arm) and an arm center offset of 2^{0} towards negative latitudes. After background subtraction, the southwest part of the radio shell is as bright as its opposite half.

A different estimate obtains when the shell is 10^{5} yr old so that $v = r/t$ drops to some $10^{7.5}$ cm/s. Now the displacement can grow comparable with the radius, as is indicated for CTB 80. The pulsar may have travelled west through $\approx 1/5$ of the shell radius, cf. fig. 6, in accord with the proper-motion estimates from scintillation data.

This interpretation of as odd-shaped a SNR as CTB 80 may throw new light on similarly odd-shaped sources such as the flying duck (G5.3-0.1), the bedspring (G357.7-0.1), and Kes 32 (G332.4+0.1), all of which may be $\approx 10^{5}$ yr-old plerions (Weiler & Sramek, 1988; Milne, 1987).

4. Do SNSs Sweep?

In the literature, SNSs tend to be described by 'blast waves' or Sedov-Taylor shock waves which sweep up their CSM, even though such sweeping has never been observed. On the contrary, several observed facts are at variance with sweeping (Kundt, 1985b; 1987a): SNSs resemble splinter bombs more closely than pressure bombs. A derivation of the splinter-bomb structure from simple assumptions will be given below. Here I want to add a number of observations to the list which are in disharmony with a blast-wave interpretation.

To begin with, there are the 'chimneys' of the Crab and of Cas A - shown in figs. 9 and 10 - which I have not seen explained in the blast-wave interpretation. My preference is for an overpressure outlet of the relativistic pair plasma (Kundt, 1983).

Second, I see no way of creating a barrel-shaped mass distribution in roughly all SNSs (Manchester, 1987; Caswell, 1988) other than via ejecta out of mutual pressure contact, i.e. via small-filling-factor ejecta.

Third, essentially all well-mapped SNSs have ejecta way outside the confines of the shell, up to three times in radius. For the Crab, the situation is not crystal clear, but Clark et al. (1983) report radial velocities of up to $5 \cdot 10^{3}$ Km s^{-1}, 2.8 times larger than those of the 'outer edge', in agreement with Murdin & Clark's (1981) photoelectric size determination. For Cas A, Fesen et al. (1987) have found fast-moving flocculi at twice the main-shell radius, and Van den Bergh (1971)

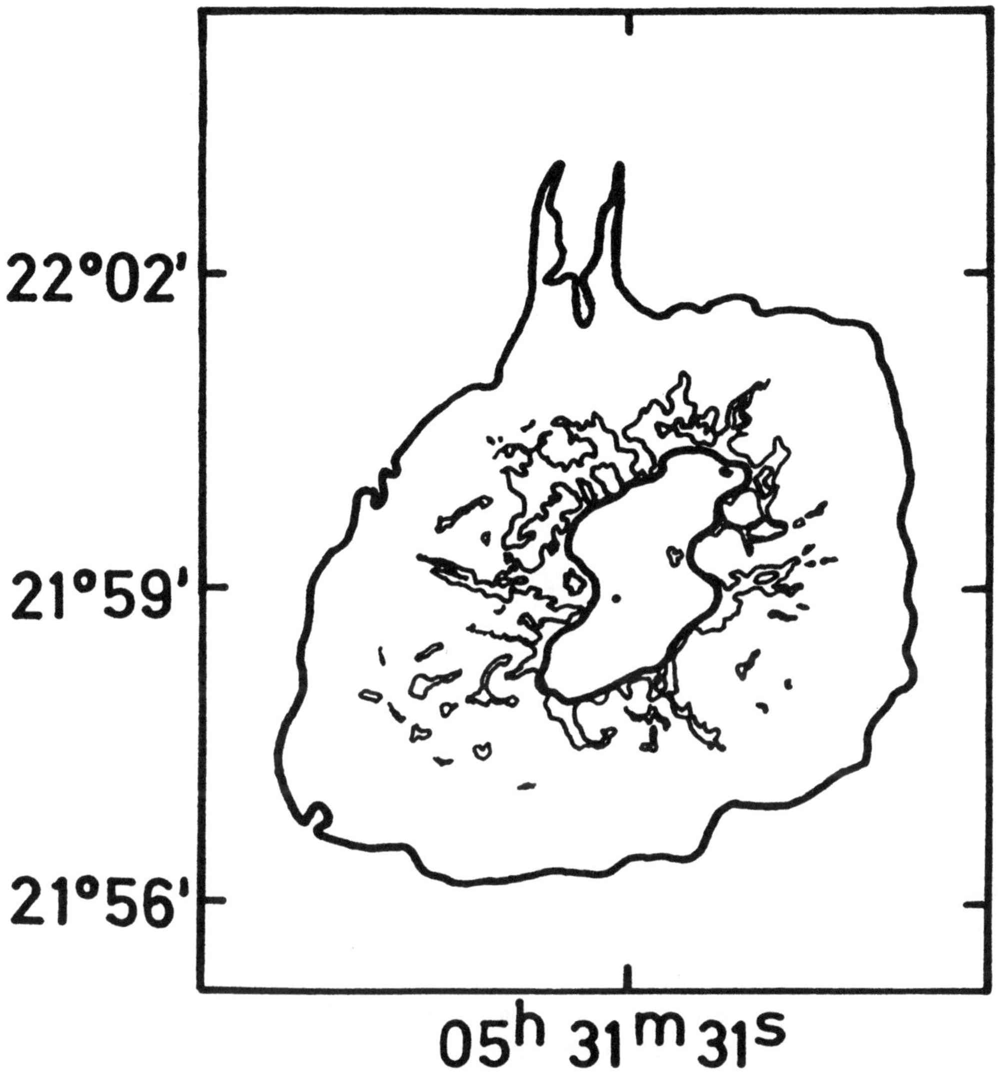

Figure 9: Sketch of the morphology of the Crab nebula which emphasizes the different optical components, copied from Kundt (1983). The outer isophote stems from a deep [OIII] image, the inner one delineates synchrotron and the middle one filamentary radiation. Particularly the 'chimney' may be hard to explain within a blast-wave description.

reports optical emission patches (found by Minkowski) out to 7 arcmin (= 3.5 main-shell radii). Another example (of reported outlying emission) is the Cygnus Loop, see Hester (1987) and references therein.

Fourth, whereas some SNSs appear to have sheetlike condensations (like the Cygnus Loop: Hester, 1987), others (like S147: Kirshner & Arnold, 1979; Fürst & Reich, this volume) appear to be dominated by ropelike condensations (which are no

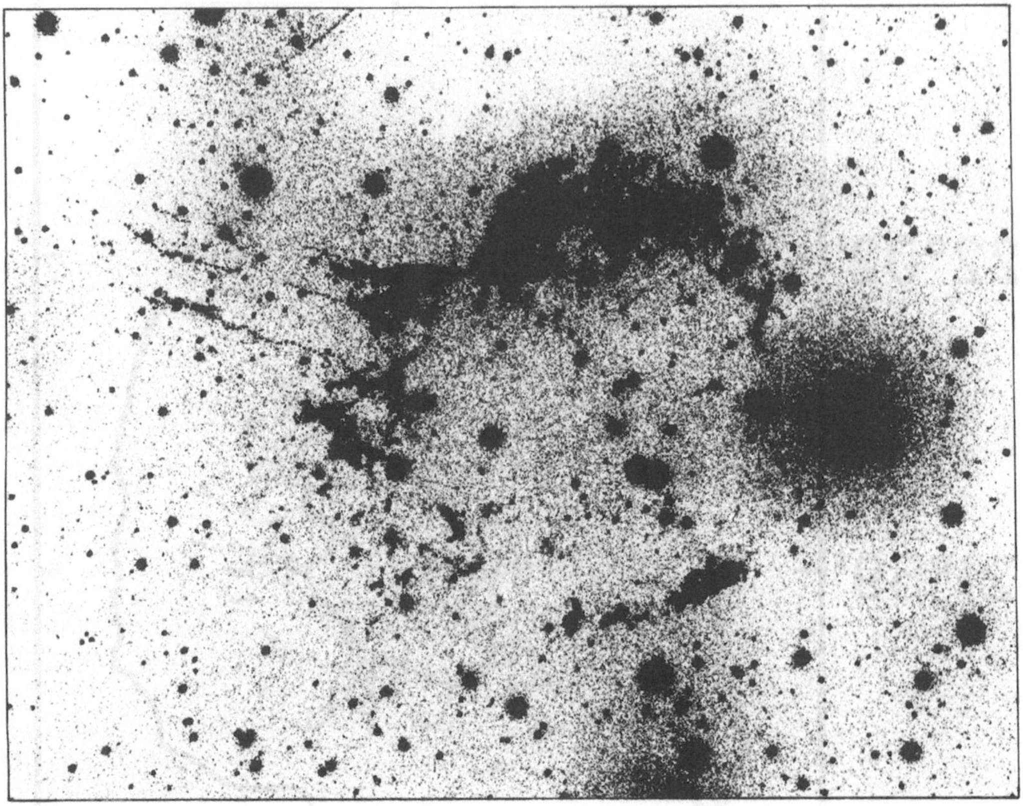

Figure 10: [SII]-centered optical photograph of Cas A taken from Van den Bergh & Kamper (1985). The 'chimney' part - similar to that in fig. 9 - has been enhanced by hand in order to stress the non-wavelike morphology.

natural product of a blast wave). For the Crab, Mac Alpine has found a (radial) ropelike Ni concentration (communicated by Gaskell).

Fifth, I am not aware of any explanation of the 8% post-acceleration of the Crab filaments other than that by Kundt & Krotscheck (1980), via the joint pressure of the multiply reflected 30 Hz waves.

Sixth, a long free-expansion stage has been found statistically by Mills et al. (1984).

If SN ejecta are filamentary, rather than a strong shock, shells like VRO 42.05.01 (= G 166.0+4.3) cannot be explained as formed by the faster propagation (of the blast wave) through a hot sheet (Pineault et al., 1987). It is not even clear to me whether such hot sheets (of significant filling factor) exist anywhere in the Galaxy.

5. Importance of the CSM and Energy Estimates

The predominant temperature of the ISM may be 10^4 K because of its stabilization by hydrogen ionization and because this temperature is indicated by the (large) neutral component entering the heliosphere (Copernican principle). Embedded in this warm component are cold HI regions ($T \lesssim 10^2$ K) and even colder molecular clouds ($T \leq 10$ K), of (total) filling factor $\approx 10\%$. It is not clear to me what fraction of SN explosions take place in either component and whether a hot component ($T \gtrsim 10^5$ K) should be added as a third choice.

It is clear, however, that young SNSs feel neither the warm nor the cold component but the thinning windzone of their progenitor star and its dense, piled-up outer boundary layer. The density of the windzone is expected to be much higher for exploding red giants than for exploding blue giants. This density should be important for the deceleration of the ejecta, both in the splinter interpretation and in the blast wave interpretation: the radiation of a SNS is likely to peak when the ejecta cross the dense boundary layer of the former windzone. Kepler's SNS may be the best example, as discussed by Bandiera in this volume.

The kinetic energy of SNSs cannot be thermalized and radiated: it largely exceeds the observed energy input into the ISM (Heiles, 1987). Bright SNSs radiate $L = 10^{37.5\pm0.5}$ erg s^{-1} so that the integrated radiation output $\int L dt \lesssim 10^{49}$ erg is much smaller than the $10^{51\pm0.5}$ erg of kinetic shell energy. Quite likely, the kinetic energy of a SNS is stored potentially, as an expansion of the galactic disk, and liberated (on average) wherever the ISM contracts, in the form of IR radiation by galactic clouds.

If SN ejecta don't thermalize, the existing estimates of involved masses and kinetic energies are in danger of being underestimates. When a solar mass is ejected at an initial speed v of some 10^4 Km/s, its kinetic energy $E_{kin} = Mv^2/2 = 10^{51}$ erg $(M/M_\odot)\, v_9^2$ exceeds Braun's (1987) $4\cdot10^{50}$ erg. Both young SNSs and SN spectra suggest initial bulk speeds of $v \gtrsim 10^{3.8}$ Km/s and ejected masses above 2 $M_\odot$ (in order to store the radiated light and/or unbind a neutron star binary), resulting in $\gtrsim 10^{51}$ erg of mechanical energy.

Dopita (1988) estimates a mass of $5\cdot10^{\pm0.4}$ $M_\odot$ for the shell of SN1987A.

References

Ashworth, W.B., 1980: J. Hist. Astr. 11, 1

Bartel, N., Ratner, M.I., Rogers, A.E.E. & Shapiro, I.I., Bonometti, R.J., Cohen, N.L., Gorenstein, M.V., Marcaide, J.M. & Preston, R.A., 1988: Astrophys. J. 323, 505

Bignami, G.F. & Caraveo, P., 1988: Astrophys. J. 325, L5

Blaauw, A., 1985, in: Birth and Evolution of Massive Stars and Stellar Groups, eds. Boland & van Woerden, Reidel, p. 211

Blair, D.C. & Candy, B.N., 1985: Mon. Not. R. astron. Soc. 212, 219

Blair, W.P., Kirshner, R.P., Fesen, R.A. & Gull, T.R., 1984: Astrophys. J. 282, 161

Braun, R., 1987: Astron. Astrophys. 171, 233

Caswell, J.L., 1988, in: SNRs and the ISM, IAU 101, eds. Roger & Landecker, Cambridge Univ. Press, p. 269

Clark, D.H., Murdin, P., Wood, R., Gilmozzi, R., Danziger, J. & Furr, A.W., 1983: Mon. Not. R. astron. Soc. 204, 415

Dopita, M.A., 1988: Nature 331, 506

Fesen, R.A., Becker, R.H. & Blair, W.P. 1987: Astrophys. J. 313, 378

Fesen, R.A., Shull, J.M. & Saken, J.M., 1988: Nature 334, 229

Haslam, C.G.T., Salter, C.J., Stoffel, H. & Wilson, W.E., 1982: Astron. Astrophys. Suppl. 47, 1

Heiles, C., 1987: Astrophys. J. 315, 555

Hester, J.J., 1987: Astrophys. J. 314, 187

Kamper, K. & Van den Bergh, S., 1976: Astrophys. J. 32, 351

Kirshner, R.P. & Arnold, C.N. 1979: Astrophys. J. 229, 147

Kirshner, R.P. & Fesen, R.A., 1978: Astrophys. J. 224, L59

Kulkarni, S.R., Clifton, T.C., Backer, D.C., Foster, R.S., Fruchter, A.S. & Taylor, J.H., 1988: Nature 331, 50

Kundt, W., 1980: Ann. N.Y. Acad. Sci. 336, 429

Kundt, W., 1983: Astron. Astrophys. 121, L15

Kundt, W., 1985a: Bull. Astron. Soc. India 13, 12

Kundt, W., 1985b, in: The Crab Nebula and Related SNRs, eds. Kafatos & Henry, Cambridge Univ. Press, p. 151

Kundt, W., 1987a, in: Interstellar Magnetic Fields, eds. Beck & Gräve, Springer, p. 185

Kundt, W., 1987b: Astrophys. Sp. Sci. Lett. 134, 407

Kundt, W., 1988: Comments on Astrophys. 12, 113

Kundt, W. & Krotscheck, E., 1980: Astron. Astrophys. 83, 1

Lyne, A.G. Anderson, B. & Salter, M.J., 1982: Mon. Not. R. astron. Soc. 201, 503

Manchester, R.N., 1987: Astron. Astrophys. 171, 205

Mills, B.Y., Turtle, A.J., Little, A.G. & Durdin, J.M., 1984: Aust. J. Phys. 37, 321

Milne, D.K., 1987: Aust. J. Phys. 40, 771

Murdin, P.G. & Clark, D.H., 1981: Nature 294, 543

Pineault, S., Landecker, T.L. & Routledge, D., 1987: Astrophys. J. 315, 580

Seward, F.D., 1985: Comments on Astrophys. 11, 15

Smith, A., 1978: Mon. Not. R. astron. Soc. 182, 39P

Trimble, V., 1968: Astron. J. 73, 535

Tuffs, R. 1986: Mon. Not. R. astr. Soc. 219, 13

Van den Bergh, S., 1971: Astrophys. J. 165, 259

Van den Bergh, S., 1980: Publ. Astron. Soc. Pacific 92, 768

Van den Bergh, S., 1983: Publ. Astron. Soc. Pacific 95, 388

Van den Bergh, S. & Kamper, K., 1984: Astrophys. J. 280, L51

Van den Bergh, S. & Kamper, K., 1985: Astrophys. J. 293, 537

Weiler, K.W. & Sramek, R.A., 1988: Ann. Rev. Astron. Astrophys., to appear

EVOLUTION OF THE MORPHOLOGY OF SUPERNOVA REMNANTS WITH PULSARS

G. Srinivasan[1,2] and D. Bhattacharya[2]

1. European Southern Observatory
Karl-Schwarzschild-Str. 2
D-8046 Garching bei Munchen
West Germany

2. Raman Research Institute
Bangalore - 560 080, India

1 INTRODUCTION

It is now clear that SNRs come with several distinct morphologies. While the overwhelming majority have the appearance of shells with hollow interiors, a handful are centrally filled like the Crab Nebula, and a few have hybrid morphology. And then there are the "birds", "Struwelpeters" etc.!

For a long time the Crab Nebula was alone in its class, and it was Weiler/1/ who first drew attention to the fact that 3C58 resembled the Crab in many respects. A few years later the existence of a small group of flat spectrum, centrally-filled nebulae was firmly established. It was natural to hypothesize at that stage /2-5/ that these nebulae, too, were powered by a central pulsar, but that the pulsars in them were not beaming towards us.

This raised several interesting questions. For example, if active pulsars produce synchrotron nebulae, and if type II SN are associated with the formation of neutron stars, then should not one expect at least half of the shells to have a pronounced central emission, thus giving them a hybrid morphology? The question could also be turned around and asked as follows: Why is it that the classical plerions like the Crab and 3C58 show no limb brightening? Strangely these questions did not attract much attention in the literature!

The following question was, however, raised by several investigators : why do the shells and the plerions have their distinct morphologies? According to some, the different morphologies should be attributed to different types of supernova explosions /6,7/. It has been suggested in the literature that Type I SN produce shell SNRs with hollow interiors since in most current models no stellar remnant is left behind. Others have argued that type II supernovae will produce plerions since a neutron star is expected to be left behind. Such a distinction is perhaps an oversimplification. While it is reasonable to suppose that those supernovae that do not leave behind pulsars will not produce plerionic nebulae, the presence of a pulsar does not guarantee that the remnant will have a centrally-filled appearance, certainly not at all times. Let me recall that both the Crab Nebula and MSH 15 - 52 harbor active central pulsars and yet the latter is a shell with hardly any central radio emission. We should also mention the suggestion made by Kundt a number of years ago that shells and plerions may be the remnants of the first and second explosions, respectively, in a binary system /8/.

An alternative suggestion was made by Lozinskaya /9/ who advanced the view that plerions and shells were merely different evolutionary stages of all SNRs; in their youth SNRs will have a Crab-like appearance, and as they grow older will turn into classical shells. In this paper we wish to examine this interesting suggestion.

THE APPROACH: - The basic question we wish to address is the following. Consider a type II supernova explosion. A certain amount of mass is ejected at a fairly high velocity, and there is a pulsar at the site of the explosion. What will be the (radio) morphology of the supernova remnant, and how will it evolve with time? To answer this question one will have to isolate and examine the factors that determine the luminosity of the shell and of the pulsar produced nebula.

2 THE EVOLUTION OF THE LUMINOSITY OF THE PLERION

It is now well established that the relativistic particles and the magnetic field responsible for the synchrotron radiation from the Crab Nebula are supplied by the central pulsar. The luminosity of the nebula at any instant is determined by the built up magnetic field in the cavity and the spectrum of the relativistic electrons. Since the walls of the cavity are expanding, one will have to take into account adiabatic losses; radiation losses are not important for radio-emitting electrons. In their classical work Pacini and Salvati /10/ developed a formalism to calculate the evolution of the luminosity of such a pulsar-produced nebula. They showed that most of the magnetic field in the nebula gets built up during the initial characteristic slow-down time scale, τ_0 , of the pulsar; this is $\sim$300 years for the Crab pulsar. During this phase, the luminosity of the nebula initially increases and then slowly decreases with time. After $t = \tau_0$, the pulsar luminosity starts to decrease, and consequently the nebular luminosity begins to decline rather rapidly.

The spectral luminosity of a pulsar-produced nebula depends rather strongly on the initial period of the pulsar, its magnetic field and the expansion velocity. This may be seen in the following expression for the radio luminosity at $t > \tau_0$:

$$L_\nu \propto B_*^{(3-5\gamma)/2} \; P_0^{2(\gamma-2)} \; v^{-3(1+\gamma)/4} \; t^{-2\gamma} \; \nu^{(1-\gamma)/2}$$

Here γ is the slope of the injected particle spectrum ($\sim$1.6 for the Crab nebula). P_0 the initial period, B_* the magnetic field and v the expansion velocity of the nebular boundary.

PULSAR PARAMETERS : - According to conventional wisdom newly born neutron stars should be spinning rapidly with periods milliseconds. However, a careful analysis of the statistics of the observed plerions suggests that the majority of pulsars must be relatively slow rotators at birth /11/. An analysis of the statistics of pulsars also forces one to this conclusion /12-14/. In view of this we have allowed for a range of initial periods in the calculations we shall present below. To be specific we have adopted the magnetic field of the Crab pulsar as the canonical value, but it should be borne in mind that the observed fields of pulsars have a fairly big spread.

THE EXPANSION VELOCITY : - Just as in the discussion of the shell emission, while calculating the secular evolution of the plerion one must take into account the deceleration of the nebular boundary when the mass swept up becomes comparable to the mass ejected. When the nebula decelerates the adiabatic loss rate of the particles, and the magnetic field, decreases. Consequently the luminosity declines less rapidly than in the free expansion phase. Clearly, when this effect sets in depends upon the mass ejected, the initial velocity AND the density of the ambient medium.

3 THE SHELL COMPONENT

Let us now turn our attention to the boundary of the plerion where the swept up mass is accumulating, and the consequent effects. It has been long recognised that in order to explain the observed level of radio emission from the shell SNRs one has to have a mechanism for the acceleration of the electrons and the amplification of the magentic field. Various mechanisms relevant for young remnants are variants of what we shall call "Gull's mechanism". Numerical fluid dynamical models studied by Gull/15/ show that the interaction of the expanding ejecta with the ISM leads to a convective instability in a narrow zone behind the shock. The turbulent energy in this zone is about 1% of the total energy. These turbulent cells can amplify the preexisting magnetic field by twisting and tangling the field lines. Charged particles can be accelerated by the Fermi mechanism in this turbulent region. Calculations show that rough equipartition is achieved between the energy densities in turbulence, magnetic field and relativistic particles.

In the final analysis the luminosity of the shell is determined by

* the energy of the explosion

* the amount of mass ejected, and

* the density of the surrounding medium

THE INTERSTELLAR MEDIUM : - Of the three parameters that determine the luminosity, the one that concerns us most is the density of the ambient medium. Whilst one may be safe in assuming a standard value for the blast energy and the mass ejected, the same cannot be said for the "standard model of the ISM" that pervades the SNR literature !! It has become customary to assume that all supernovae go off in a uniform medium with a particle density of the order of 1 atom/cc. There is mounting evidence that this is not a good assumption.

THE STANDARD MODEL : - The "standard model" of the ISM that emerged from 21 cm observations is the so called "raisin pudding model" in which small cool clouds are in pressure equilibrium with a 'warm' diffuse INTERCLOUD MEDIUM whose density is ~1 atom/cc /16/. It is this intercloud medium that is important in the context of the evolution of SNRs.

THE CORONAL GAS : - While the existence of such a "warm" medium cannot be denied, its filling factor is not at all clear. Recent ultraviolet and x-ray observations /17-20/ have revealed the presence of a medium of much lower density and higher temperature ($n \sim 3.10^{-3}$ atoms/cc and $T \sim 10^{6}$ K). The existence of such a Coronal gas had in fact been predicted by Spitzer way back in 1956. According to McKee and Ostriker this gas must have a filling factor ~70% /21/. Although it has not been possible to determine the filling factor of this component either from observations, there is no question that it is there and one should examine carefully how an SNR will evolve in such a medium. It has been pointed out by several authors /22-24/ that a shell expanding in a low density medium will turn on later, and its peak luminosity will be smaller than if it were expanding in a dense medium. If, in addition, there is a central pulsar then the resultant morphology could be quite different.

STELLAR WIND BUBBLES : - Quite apart from the coronal gas there are reasons to believe that a large fraction of the type II supernovae may go off in low density regions. In recent years it has been well established that stars lose a fraction of their mass by winds before they end their lives. For massive stars, the winds during the main sequence and the blue supergiant phases are of such high velocity that they are expected to excavate a cavity in the interstellar medium. If such a bubble remains stable till the star explodes (that is, if it does not collapse) the ejecta will be initially expanding not in the original interstellar medium, but in the low density bubble.

Clearly, this is not the place to go into the details of the theoretical predictions concerning stellar wind bubbles or their observational status. We shall merely remark that various considerations suggest that the radius of such a bubble can be > 10 pc and the density inside the bubble is expected to be ~ 0.01 atom/cc, and the observational evidence in support of this is mounting /25-29/. It may be worth mentioning that the presently available evidence suggests that the recent supernova in LMC may be an example of a supernova going off in such a bubble.

4 MODEL CALCULATIONS

Let us finally turn to the model calculations. We assume

1. a value for the mass ejected

2. an initial velocity for the expanding ejecta

3. a model for the medium into which the ejecta is expanding

4. a central pulsar with a certain initial period.

We then calculate the evolution of the surface brightness of the plerion produced by the pulsar, and of the shell, assuming them to evolve independently of each other.

To calculate the luminosity of the plerionic component we have used the formalism developed by Pacini and Salvati.

Since a similar simple analytical approach does not exist to estimate the luminosity of the radio emission from the shell, and since detailed fluid dynamical calculations do not seem warranted at this stage, we have adopted the following simple approach.

In order to calculate the luminosity of the shell one has to know the particle distribution N(E,t) and the magnetic field B. Gull has presented /15/ in graphical form the development of the magnetic field and the turbulent energy for certain assumed values for the energy of the blast, the mass ejected and the density of the ambient medium. He has also given how the radius evolves with time in his models. Using this one can express E_{turb}/E_{tot} and $B_s[M_{ej}/n_o E_{tot}]^{1/2}$ as a function of the ratio of the swept up mass to the mass ejected. These particular combinations obtain because of some scaling relations in Gull's model. Thus, from the graphical results for the random energy and the magnetic field as a function of time, one can obtain a relation for these quantities as a function of the mass ratio.

One can thus use Gull's results for OTHER VALUES of the blast energy, the mass ejected and the ambient density. What we have outlined is just a cheap way to calculate the evolution of the radio luminosity of the shell component, without having to do a full fledged calculation for every choice of the parameters. That would not only be a formidable undertaking, but also not worth it at this stage!

5 RESULTS

Let us illustrate the results with the aid of the next few figures.

Figure 1 shows the evolution of the Crab nebula. The mass in the ejecta is 1 $M_{\odot}$, and the initial velocity is assumed to be the presently observed one. We see that the Crab nebula will always be a plerion! The shell component will never become as bright as the central component, let alone brighter. The main reason for this is the very low kinetic energy of expansion ($\sim 10^{49}$ ergs as opposed to the typical value of $\sim 10^{51}$ ergs).

In figure 2 the initial expansion velocity is 10,000 km/s, but the density of the ambient medium is still 1 atom/cc. The plerions produced by three pulsars with initial periods of 3,16 and 100

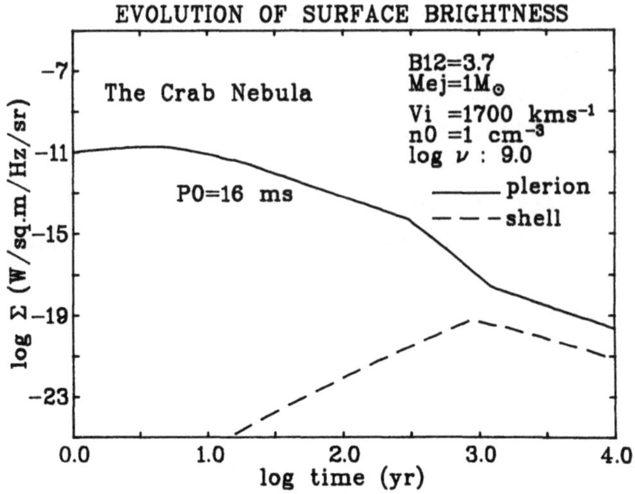

Fig. 1: Evolution of the shell and the plerionic components of the Crab Nebula, assuming an ambient density of 1H atom/cc.

milliseconds are displayed in the same figure. The track labelled "16 ms" corresponds to the nebula that would be produced by the Crab pulsar if placed inside a rapidly expanding cavity. We see that because of the high velocity the shell emission will build up rather quickly and eventually its surface brightness will become comparable to that of the central component. Thus such a remnant will evolve from a plerion to a hybrid SNR.

We also see in this figure how important the initial spin period of the pulsar is in determining the morphology. To nobody's surprise the plerion produced by a 100 ms pulsar will never be very bright. What is interesting is that the plerion produced by a 3 ms pulsar, although initially very bright, will become less luminous than that produced by a 16 ms pulsar. The reason for this is easy to see. A fast pulsar will dump most of its energy very early on, and therefore adiabatic losses are more severe. The main message to be extracted from this figure is the following.

> After the shell emission builds up, the highest ratio of plerion-to-shell surface brightness obtains when the initial slow-down time scale of the central pulsar is roughly equal to the time scale for the deceleration of the ejecta to set

in. For the standard value of the energy of the explosion this happens for a CRAB-LIKE PULSAR. We shall see in the next figure that this conclusion remains valid for expansion into lower density media also.

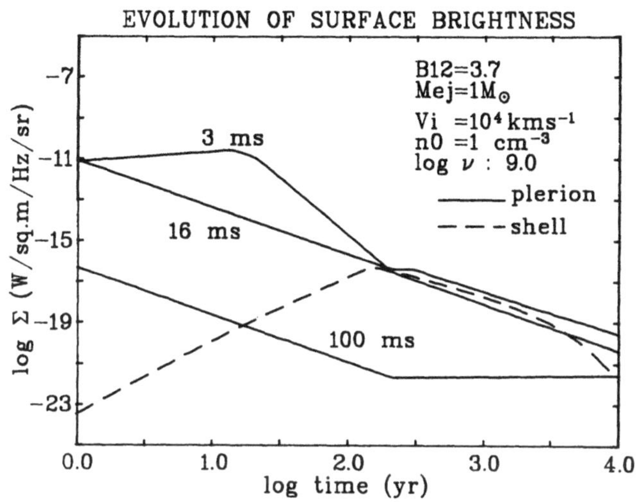

Fig. 2: The evolution of the surface brightness at 1GHz of the plerionic and shell components of a supernova remnant expanding in an uniform medium of density 1H atom/cc. Plerionic components produced by pulsars with initial rotation periods 3, 16 and 100 ms are shown. 10^{12} B12 gauss is the magnetic field assumed for these pulsars. Mej is the ejected mass, Vi the expansion velocity before deceleration, and n0 is the density of the ambient medium.

Figure 3 shows the evolution of the morphology when the density of the ambient medium is 0.01 atoms/cc, appropriate for the interior of a stellar wind bubble; all other parameters have the same value as in the previous figure. We see that the onset of radio emission from the shell is delayed, and the maximum surface brightness reached is also smaller. And yet, the plerion does not dominate at late times. The reason is the following. When the plerion is expanding freely its luminosity begins to decline quite rapidly when the pulsar luminosity begins to decline. When the nebula decelerates due to the mass accumulated the rate of decline of the luminosity is vastly reduced. In the present case, since the density is much smaller than in the previous case the plerion luminosity decreases rapidly for a longer time, so that by the time the shell builds up the plerion is at most as bright as the shell.

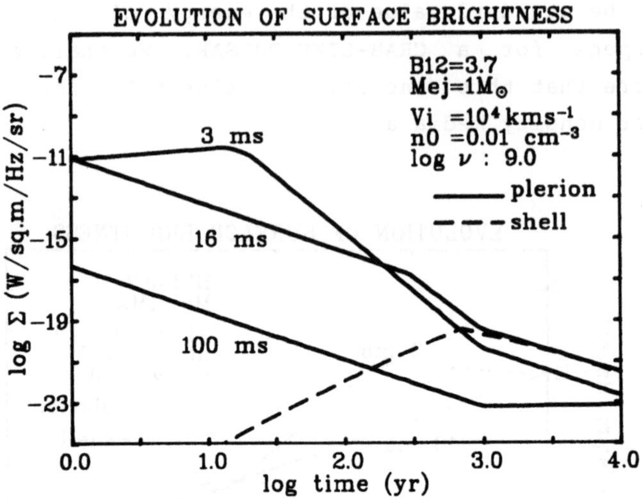

Fig. 3: Same as fig. 2, but for ambient density of 0.01H atom/cc, appropriate for that of a stellar wind bubble.

Since the trends are the same for expansion in the coronal gas, we shall not show the corresponding figures here. We shall, instead, discuss the interesting remnant MSH 15 - 52 which poses some intriguing problems.

6 MSH 15 - 52

The relevant observational facts are as follows /30,31/:

1. It is a remnant with a high-field pulsar in the centre.

2. Although there is no detectable radio plerion surrounding the pulsar, there is a bright X-ray plerion.

3. The spindown age of the pulsar is only ~1600 years, and yet

4. The standard "Sedov age" for the shell is $\sim 10^4$ years.

There are thus many puzzles to be explained!

* What is the true age of the remnant?

* Is the pulsar much older than 1600 years?

* With such an active pulsar why is the radio surface brightness of the plerion so low?

With particular reference to the age discrepancy, Blandford et.al. /32/ have suggested that the NEUTRON STAR may be as old as the remnant ($\sim 10^4$ years), but that it turned on as a pulsar only recently. We wish to suggest an alternative scenario which simultaneously explains all the questions mentioned above.

BASIC ASSUMPTIONS :

(1) This SNR is in the vicinity of a complicated OB association. It is, therefore, likely that its progenitor was a massive star with a strong stellar wind. Such a star would have produced a low density bubble around itself, and we assume that the supernova that produced this remnant went off in such a bubble. In such a situation the remnant could easily have expanded to the present size in 1600 years.

(2) Let us assume that the mass ejected was $\sim 2\ M_\odot$, and that the initial velocity was $\sim 12{,}000$ km/s. These reproduce the observed size of the remnant, if one takes into account deceleration. The surface brightness of the shell is also reasonably well reproduced by this choice of the initial conditions (see Figure 4a).

In order to explain why the radio luminosity of the plerion is so low one has to have an estimate for the INITIAL PERIOD of the pulsar. Since we have fixed the dynamics of the nebular boundary, and we know the magnetic field of the pulsar, we can estimate the initial period by fitting the luminosity of the X-RAY PLERION! As will be seen from Figure 4b, the presently observed X-ray luminosity is best fit by an initial period of 6 milliseconds. Given this estimate, one is in a position to compare the expected surface brightness of the radio plerion with that of the shell. This is shown in figure 4a. We see that the surface brightness of the plerion will be $< 10^{-23}$ W/m^2/Hz/sr,

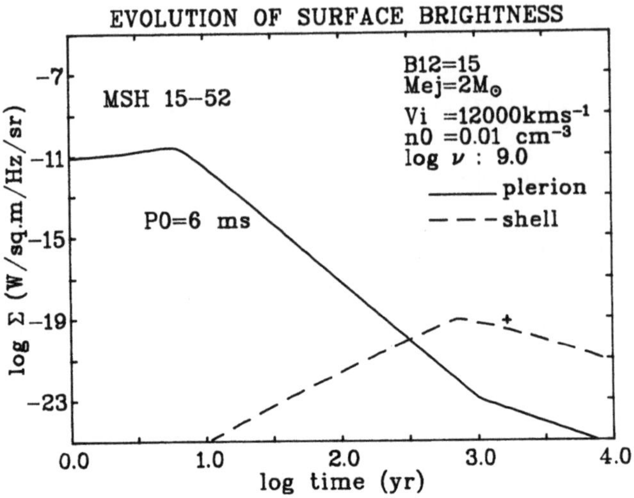

Fig. 4(a): A model evolution of the radio surface brightness of the shell and the plerionic components of the supernova remnant MSH 15-52. It is assumed that the remnant is expanding in a stellar wind bubble. The observed value of Σ is shown by a '+' mark.

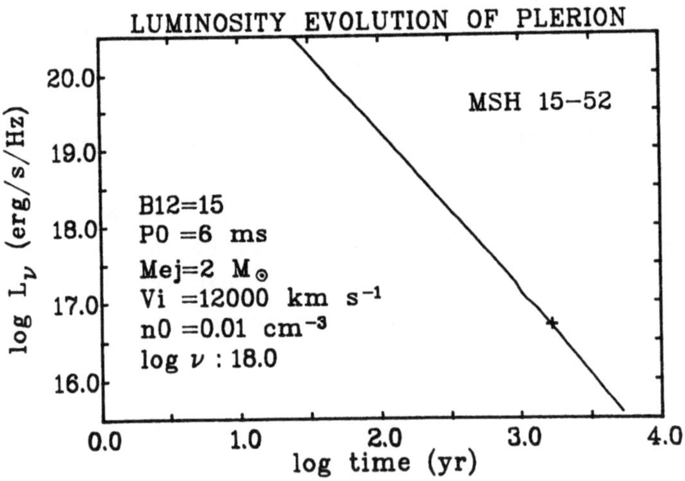

Fig. 4(b): A model evolution of the X-ray luminosity of the plerionic component of the supernova remnant MSH 15-52. The observed X-ray luminosity is indicated by the '+' mark.

which is below the level of detectability. Thus, MSH 15 - 52 could be an example of a remnant that initially was a bright plerion, but evolved to become a shell with a hollow interior.

7 SUMMARY AND CONCLUSIONS

The main question we have addressed in this paper concerns the radio morphology of supernova remnants with central pulsars.

It is useful to list separately the expected morphology for three types of central pulsars :

1. "fast" pulsars : initial period ~ 5 ms

2. "slow" pulsars : initial period > 100 ms

3. "Crab pulsars" : $P_o \sim 20$ ms and $B_* \sim 5.10^{12}$ gauss.

1. STRONG EXPLOSIONS: - Let us first consider strong explosions in which the energy released is $\sim 10^{51}$ ergs. For all the three pulsars mentioned above the remnant will be centrally concentrated early on, but the morphology will change later. By the time the shell builds up, the surface brightness of the plerions produced by the "fast" and "slow" pulsars would have dropped below that of the shell, so these remnants will turn from plerions to shells. MSH 15 - 52 may be an example of such a transformation of morphology.

In the case of a "Crab-like" pulsar, the plerion will still be strong when the shell builds up so that the remnant will now have hybrid morphology - a shell with bright central emission.

2. WEAK EXPLOSIONS : - Let us turn next to remnants produced by "weak" supernovae. Available evidence suggests that only $\sim 10^{49}$ ergs was released in SN 1054. In such cases both the "fast" and "medium-fast" pulsars will have a significant dynamical effect on the nebular boundary, i.e. the remnant will be "pulsar-driven" like the Crab nebula. Like in the case of a strong explosion, remnants harboring "fast" pulsars will be bright plerions in their youth, but transform to shells in their old age. The remnant produced by a "Crab-like" pulsar will always be a plerion. Moreover, the plerion will be bright and long lived.

3. UNIQUENESS OF THE CRAB NEBULA : - From the various cases mentioned above we see that long-lived and bright plerions are expected only in those rare cases when the initial period of the pulsar is ~ 20 ms and its magnetic field $\sim 10^{12.5}$ G. The particular nature of the Crab must be understood in terms of the Crab pulsar having just these characteristics!

4. INITIAL ROTATION OF PULSARS : - From the absence of significant central emission in the overwhelming majority of SNRs we must conclude that the majority of pulsars must be rather slow rotators at birth.

8 REFERENCES

1. Weiler, K.W., 1969, Ph. D. Thesis, California Institute of Technology

2. Radhakrishnan, V., Srinivasan, G., 1978, paper presented at the Asian-South-Pacific Regional Meeting in Astronomy, Wellington, unpublished

3. Radhakrishnan, V., Srinivasan, G., 1980, J. Astrophys. Astr., 1, 25-32

4. Weiler, K.W., Shaver, P.A., 1978, Astr. Astrophys., **70**, 389-397

5. Weiler, K.W., Panagia, N., 1980, Astr. Astrophys., **90**, 269-282

6. Shklovskii, I.S., 1980, Publ. Astr. Soc. Pacific, **92**, 125-126

7. Weiler, K.W., 1983, in IAU Symp. 101: Supernova Remnants and Their X-ray Emission, ed. P. Gorenstein and J. Danziger, D. Reidel, Dordrecht, p.299-320

8. Kundt, W., 1980, Ann. N. Y. Acad. Sci., **336**, 429-441

9. Lozinskaya, T.A., 1980, Sov. Astr., **24**, 407-412

10. Pacini, F., Salvati, M., 1973, Astrophys. J., **186**, 249-265

11. Srinivasan, G., Bhattacharya, D., Dwarakanath, K.S., 1984, J. Astrophys. Astr., **5**, 403-423

12. Vivekanand, M., Narayan, R., 1981, J. Astrophys. Astr., **2**, 315-337

13. Chevalier, R.A., Emmering, R.T., 1986, Astrophys. J., **304**, 140-153

14. Narayan, R., 1987, Astrophys. J., **319**, 162-179

15. Gull, S.F., 1973, Mon. Not. R. astr. Soc., **161**, 47-69

16. Radhakrishnan, V., Murray, J.D., Lockhart, P., Whittle, R.P.J., 1972, Astrophys. J. Suppl. Ser., **24**, 15-47

17. Rogerson, J.B., Spitzer, L., Drake, J.F., Dressler, K., Jenkins, E.B., Morton, D.C., York, D.G., 1973, Astrophys. J., **181**, L97-L102

18. Jenkins, E.B., Meloy, D.A., 1974, Astrophys. J., **193**, L121-L125

19. York, D.G., 1974, Astrophys. J., **193**, L127-L131

20. Cowie, L.L., Songaila, A., 1986, Ann. Rev. Astr. Astrophys., **24**, 499-535

21. McKee, C.F., Ostriker, J.P., 1977, Astrophys. J., **218**, 148-169

22. Lozinskaya, T.A., 1979, Sov. Astr., **23**, 506-507

23. Higdon, J.C., Lingenfelter, R.E., 1980, Astrophys. J., **239**, 867-872

24. Srinivasan, G., Dwarakanath, K.S., 1982, J. Astrophys. Astr., 3, 351-361

25. Weaver, R., McCray, R., Castor, J., Shapiro, P., Moore, R., 1977, Astrophys. J., **218**, 377-395

26. Georgelin, Y.M., Georgelin, Y.P., Laval, A., Monnet, G., Rosado, M., 1983, Astr. Astrophys. Suppl. Ser., **54**, 459-469

27. van der Bij, M.D.P., Arnal, E.M., 1986, Astrophys. Lett., **25**, 119-125

28. Rosado, M., 1986, Astr. Astrophys., **160**, 211-228

29. Laval, A., Boulesteix, J., Georgelin, Y.P., Georgelin, Y.M., Marcelin, M., 1987,Astr. Astrophys., **175**, 199-207

30. Seward, F.D., Harnden, F.R., Murdin, P., Clark, D.H., 1983, Astrophys. J., **267**, 698-710

31. Manchester, R.N., Durdin, J.M., 1983, in IAU Symp. 101: Supernova Remnants and Their X-ray Emission, ed. P. Gorenstein and J. Danziger, D. Reidel, Dordrecht, p.421-427

32. Blandford, R.D., Applegate, J.H., Hernquist, L., 1983, Mon. Not. R. astr. Soc., **204**, 1025-1048

ON THE DISTANCES OF THE REMNANTS OF HISTORICAL TYPE I SUPERNOVAE

Richard G. Strom
Netherlands Foundation for Radio Astronomy
Postbus 2, 7990 AA Dwingeloo

Because the work described in this talk has been recently published (Strom, 1988) only an extended summary will be given here. Distances to supernova remnants are important because so many fundamental parameters (diameter, transverse speed, luminosity, density, etc.) depend upon the distance. Reliable distances to the remnants of historical supernovae are particularly important because we know their ages, something about the initiating event, and are often able to observe changes as they are relatively young. Such objects are an important test bed for modelling spherical shocks.

The distance determination method depends upon the well-known Sedov/Taylor similarity solution for an adiabatic, spherical shock (Sedov, 1959), according to which the shock radius will vary with time as,

$$r \propto (E_o/N_o)^{\frac{1}{5}}\ t^{\frac{2}{5}}$$

where E_o is the initial energy released by the explosion and N_o is the ambient density. By comparing the angular radii of remnants of known age, this equation can be used to determine relative distances. Ambient densities have been derived from X-ray measurements, and the initial energies in the Type I supernovae chosen have been assumed equal. The weak dependence of radius upon E_o and N_o means that the distances so derived are not terribly sensitive to these somewhat uncertain parameters, in particular the equal energy assumption.

The following shell remnants of Type I supernovae were included: RCW86 (SN185), PKS1459-41 (SN1006), Tycho (SN1572) and Kepler (SN1604). In several the measured expansion rate is consistent with that predicted by Sedov/Taylor, justifying the assumption that the shocks are indeed adiabatic. The relative distances have been calibrated using other, independent, determinations. The results are summarized in Figure 1. The method has the advantages of a solid physical basis, and should in particular be more reliable than the Σ-D approach, it can provide distances for the all important young remnants, and should help tie down nearby ones for which the fractional uncertainty in other methods is large. Further details can be found in Strom (1988).

References

Sedov, L., 1959. Similarity and Dimensional Methods in Mechanics, Academic Press, New York.

Strom, R.G., 1988. Mon. Not. Roy. Astron. Soc. 230, 331.

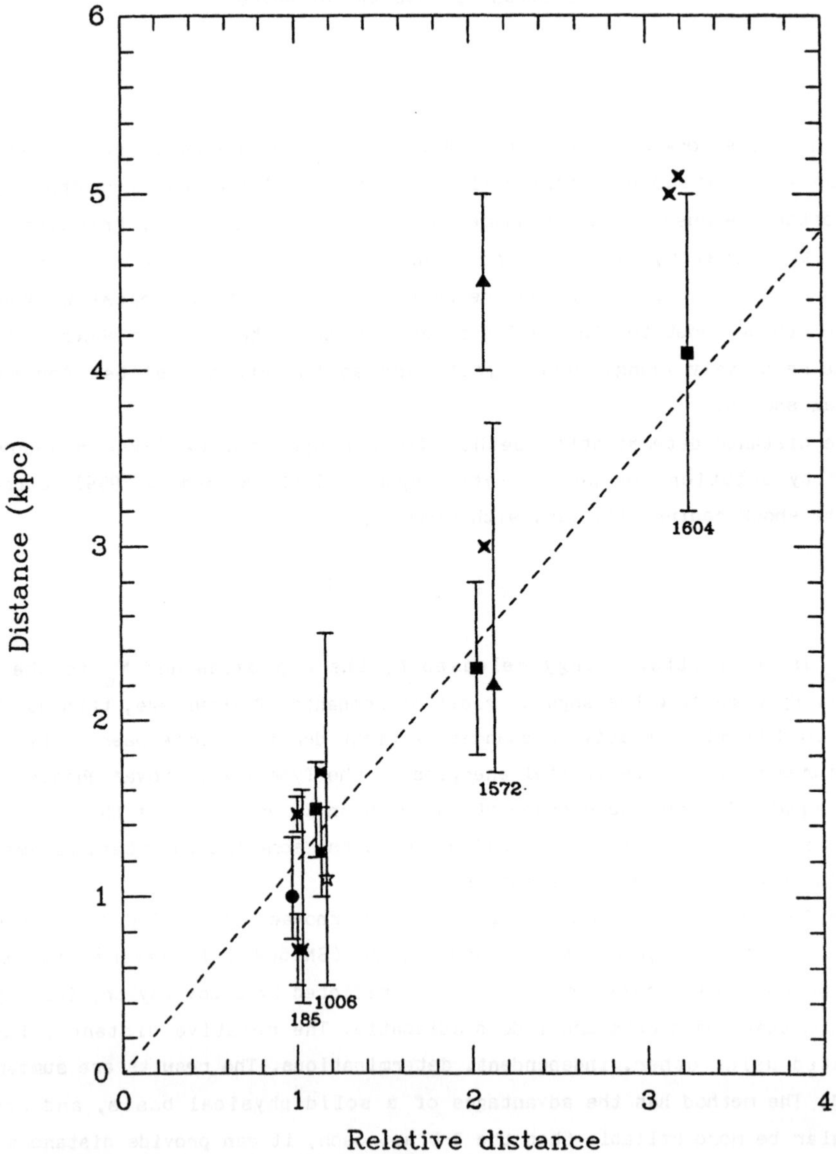

Figure 1. Relative distances of four historical supernova remnants obtained by the method described and compared with various other distance determinations. The points are slightly offset along the abscissa for clarity.

SPECTRAL INDEX VARIATIONS IN SUPERNOVA REMNANTS

E. Fürst, W. Reich

Max-Planck-Institut für Radioastronomnie
Auf dem Hügel 69
D-5300 Bonn 1
Federal Republic of Germany

Summary

Spectral index variations in supernova remnants are reviewed. In case of S147 the total emission is decomposed into small-scale and large-scale diffuse emission. Both components have different radio spectra. Possible consequences on the nature of the optical filaments in S147 are indicated.

1. Introduction

The radio spectrum of supernova remnants (SNRs) is known to be due to synchrotron radiation and hence governed by the energy distribution of relativistic electrons. There are several proposed sources for such electrons: A compact stellar remnant (pulsar, binary neutron star, etc.) feeding the nebular remnant, acceleration of electrons within the supernova shock front, the compression of cosmic ray electrons by the SNR shock wave and magnetic reconnections behind filaments. It is not surprising that differing local conditions result in different radio flux density spectral indices α ($S_\nu \sim \nu^\alpha$). REICH and FÜRST [1] have shown on the basis of an enlarged sample of SNRs that α varies from source to source between ≈ 0.0 and ≈ -0.9. The mean value of α for filled-center SNRs is ≈ -0.1 and about ≈ -0.5 for shell-type SNRs, reflecting the different acceleration mechanisms for the relativistic electrons.

Contrary to the large variance of α for different SNRs, a variation of α with frequency and/or with location inside an individual SNR is known only for a very limited number of objects. In Section 2 we will present a short summary of the observational evidence for spectral index variations. Most emphasis will be devoted to the shell-type source G180.0-1.7 (S147) which is one of the few objects where both, the dependence on frequency and location, have been found.

2. Observational evidence for spectral index variations in SNRs

A variation of the spectral index across SNRs is most obvious for the class of composite SNRs. The filled-center portion, which is powered by the central stellar remnant, is manifested by its flat radio spectrum. The radio spectrum of the shell-type part, interacting with the interstellar medium is relatively steep (G29.7−0.3, G68.9+2.8, G236.9−3.3, G326.3−1.8 [2]; G11.2−0.4 [3]; G18.95−1.1 [4]).

Within the small sample of filled-center SNRs only two are known to show a spectral bend: G27.8+0.6 [5] near 5 GHz with $\alpha_{<5GHz} = -0.3$ and $\alpha_{>5GHz} \approx -1.0$, G74.9+1.2 [3] between 11 GHz and 30 GHz, where $\alpha_{<11GHz} = -0.26$ and $\alpha_{>11GHz} \approx -1.1$. For this type of SNRs a spectral bend is often explained in terms of an evolutionary model proposed by WEILER and PANAGIA [6], relating the bend-frequency to the age of the remnant.

Most of the SNRs fall into the class of shell-type SNRs. But even for this large sample only a few SNRs have been found to show a spectral index variation, among them Cygnus Loop and HB9 as discussed by DE NOYER [7], G33.2−0.6 [8], S147 [9,10], IC443 [11,12]. In case of the Cygnus Loop the small effect of steepening beyond 1 GHz is probably concentrated in the northern part of the remnant [13]. The spectral bend found for HB9 at ≈1 GHz is doubted by DWARAKANATH et al. [14] after recalibrating low frequency data, but the flux density at high frequencies (≈5 GHz) is a factor 2 too low to be consistent with the low frequency spectrum. In case of IC443 the spectrum of the integrated radio flux density is straight within the errors of the individual measurements. A spatial variation of α has been reported by KUNDU and VELUSAMY [11] and GREEN [12]. The radio spectrum of the south-west emission of IC443 is about $\Delta\alpha = 0.2$ steeper than that of the bright rim in the north-east. There are also small areas with $\alpha \approx -0.2$, but this emission is probably affected by hitherto unrecognized HII emission [12]. In no SNR except S147 has it been possible to relate steep and flat spectra to physical entities like optical filaments. S147 will, therefore, be discussed in more detail.

2.1 Observations of the supernova remnant S147

For S147, FÜRST and REICH [10] supported previous observations of a spectral bend at about 1 GHz [9]. The most recent radio spectrum is shown in Figure 1. The flux density value at 5 GHz is calculated from the observations of a part of the remnant [15]. Recently, this map was extended to the whole remnant at a frequency of 4.75 GHz [10]. The integrated flux density at 4.75 GHz is 16±5 Jy after subtracting the small diameter sources listed by FÜRST et al. [16]. The new value supports the detected bend at 1 GHz. In addition, FÜRST and REICH [10] obtained a spectral index map, which shows significant variations over the whole source. Areas of flat radio spectra ($\alpha \approx -0.5$) coincide with optical filaments, while steep radio spectra ($\alpha \approx -1.0$) are found in areas of diffuse emission with no detectable optical counterparts. Instead of calculating spectral indices from flux density measurements, it is often more appropriate to use the method of differential temperature plots (TT-plots). For this purpose we convolved the original 2.695 GHz map to the HPBW (9.'4) of the 1.425 GHz map and scaled both data sets to main beam brightness temperature T_B. The relation between both temperatures is given by: $T_{B1}(x,y) = \text{const.} + (\nu_2/\nu_1)^{-\alpha} \times T_{B2}(x,y)$. The

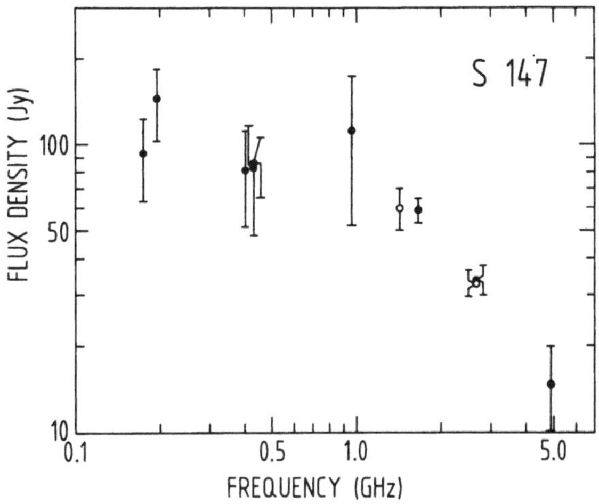

Fig. 1: Spectrum of the integrated radio flux density of S147 after removing point sources in the list published by FÜRST et al. [16]

constant includes the baselevels of both maps which are assumed to be independent of position (x,y) over the source. Plotting T_{B1} versus T_{B2} yields the temperature spectral index less affected than by baselevels comparing flux densities point-wise. We decomposed both maps into small-scale structures and large-scale diffuse emission using the "background filtering" method introduced by SOFUE and REICH [17]. If structures > 25' are considered as diffuse we obtain a separation as shown in Figure 2 for 1.425 GHz

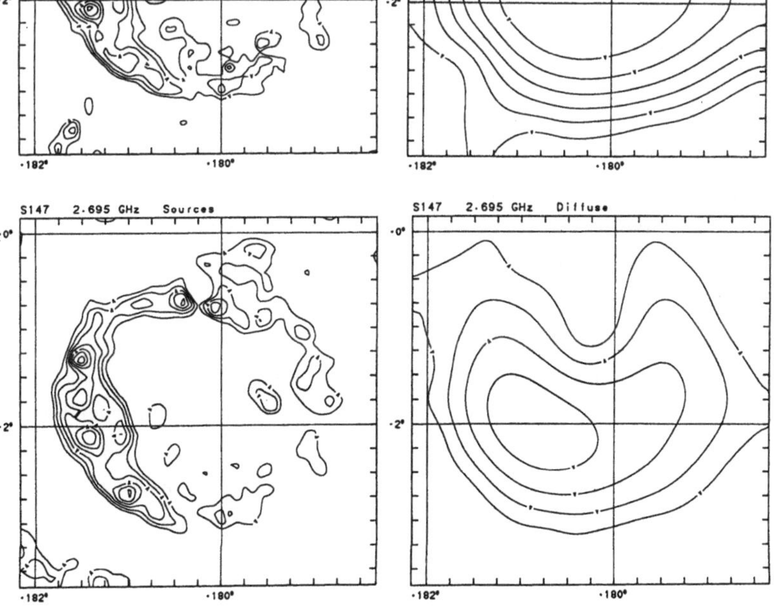

Fig. 2: Contour maps of S147 after decomposition into small-scale structures (< 25', lower left at 2.695 GHz, contour steps 15 mK T_B beginning at 30 mK T_B; upper left at 1.425 GHz, contour steps 70 mK T_B beginning at 140 mK T_B) and diffuse emission (scale size > 25', lower right at 2.695 GHz, contour steps 7.5 mK T_B beginning at 15 mK T_B; upper right at 1.425 GHz, contour steps 35 mK T_B beginning at 70 mK T_B). The half power beam width of all maps is 9!4.

and 2.695 GHz. At 2.695 GHz about 60% of the total flux density is concentrated in the diffuse component. We performed differential temperature plots for both components and obtained: The flux density spectral index for the small-scale structure is $\alpha = -0.4 \pm 0.15$. The small-scale features are radio counterparts of the optically visible filaments. For the diffuse emission we obtained a flux density spectral index of $\alpha = -0.9 \pm 0.15$.

A similar separation at 4.75 GHz yields about 50% of the total flux density in either component. The spectral index derived from the differential temperature plot is about $\alpha = -0.5$ for the small-scale structure [10] and about $\alpha = -1.0$ for the diffuse component at frequencies between 2.695 GHz and 4.75 GHz. The diffuse component cannot be related to visible optical filaments. The observations show a significant difference for the flux density spectral index ($\Delta\alpha \approx 0.5$) for both components. Comparing this result with the integrated radio spectrum in Figure 1, it is obvious that the radio counterparts of the optical filaments do not show a bend in their radio spectra up to 5 GHz [10]. The detected bend is entirely due to the diffuse emission, which comprises > 60% of the total flux at 2.695 GHz.

In case of S147, the detected radio spectral index variations are of significant order and can be attributed to different optical features. The discussion will, therefore, concentrate on this SNR.

3. Discussion

3.1 Proposed models for spectral bends in supernova remnants

Several mechanisms have been proposed to explain variations of the radio spectral index. Synchrotron losses, particularly of high energetic electrons, have been ruled out because unreasonably high magnetic fields are necessary to obtain a spectral bend near 1 GHz during the life-time of a SNR (some 10^4 years). The compression of cosmic ray electrons, originally proposed by van der Laan [18], has been discussed by DE NOYER [7]. Depending on the compression ratio and the energy change of the electrons during the compression, the detected bend in the galactic background emission (at $\approx$ 200 MHz [19]) is shifted towards higher frequencies. A shift to 1 GHz requires a compression ratio of 3–6. In case of S147, a bend is only observed for the diffuse emission which likely has a lower compression ratio than the small-scale structures. Hence, the compression of cosmic ray electrons is in qualitative agreement with the spectral index observations. Another explanation is given by the proposed shock acceleration of electrons in the SNR-shell. The spectral index of the accelerated electrons depends on the compression ratio [20], on the spatial diffusion coefficient and a possible action of the second order Fermi acceleration [21]. If we assume constant expansion velocity and different plasma parameters β (different compression ratios), the net effect of second order Fermi acceleration is a flattening of the radio spectrum with decreasing compression ratio. This is not observed in S147. If the spectral index variation is due to shock acceleration, it is either caused by first order Fermi acceleration at low compression ratios and/or by an appropriate ratio between acceleration time and diffusion time. In the latter case we expect an exponential bend which may be consistent with the data for S147, but is inconsistent with the data for HB9 and the Cygnus Loop. In any case a bend towards a steep spectrum at about 1 GHz requires a relatively low compression.

3.2 The 2-component radio structure

The diffuse emission, as obtained by the method described above, may be alternatively explained in terms of: Relativistic electrons, more or less evenly distributed in a thick shell, overlapping radio emission of faint, optically not detectable sheets or ropes, or overlapping of radio emission of optical sheets seen face-on. The small-scale emission structure may be associated either with optical ropes or with optical sheets seen edge-on.

However, the very different radio spectra of the diffuse and small-scale radio structures are inconsistent with the explanation of both components by sheets. At 2.695 GHz about 40% of the total emission is concentrated in the small-scale radio features, which are excellently correlated with optically visible features. If we explain these optical features as sheets, seen edge-on, we would expect a significant diffuse emission from sheets seen face-on with the same radio spectrum. At 4.75 GHz we then would expect an increasing fraction of diffuse emission, contrary to the observed value of ≈ 50%.

The presented observations of spectral index variation in S147 are best in agreement if the radio emission of the small-scale structures is due to highly compressed ropes (any bend is shifted to very high frequencies), and a diffuse emission caused by either less compressed ropes (a large number overlaps and merges to produce the diffuse structure) or by more or less evenly distributed electrons. The origin of the bend in the electron energy spectrum corresponding to the radio bend at 1 GHz is still unclear. In particular, a more careful study of acceleration and diffusion time scales seems to be necessary.

References

1. Reich, W., Fürst, E.: 1988, this volume
2. Weiler, K.W.: 1983, in *Supernova Remnants and their X-ray Emission*, IAU Symp. 101, eds. J. Danziger and P. Gorenstein, Reidel, Dordrecht, p. 299
3. Morsi, H.W., Reich, W.: 1987, Astron. Astrophys. Suppl. 163, 313
4. Fürst, E., Hummel, E., Reich, W., Sofue, Y., Sieber, W., Reif, K., Dettmar, R.-J.: 1988, Astron. Astrophys., submitted
5. Reich, W., Fürst, E., Sofue, Y.: 1984, Astron. Astrophys. 133, L4

6. Weiler, K.W., Panagia, N.: 1979, Astron. Astrophys. 90, 269

7. de Noyer, L.K.: 1974, Astron. J. 79, 1253

8. Reich, W.: 1982, Astron. Astrophys. 106, 314

9. Kundu, M.R., Angerhofer, P.E., Fürst, E., Hirth, W.: 1980, Astron. Astrophys. 92, 225

10. Fürst, E., Reich, W.: 1986, Astron. Astrophys. 163, 185

11. Kundu, M.R., Velusamy, T.: 1968, Monthly Notices Roy. Astron. Soc. 140, 173

12. Green, D.A.: 1986, Monthly Notices Roy. Astron. Soc. 221, 473

13. Sastry, Ch.V., Dwarakanath, K.S., Shevgaonkar, R.K.: 1981, J. Astrophys. Astr. 2, 339

14. Dwarakanath, K.S., Shevgaonkar, R.K., Sastry, Ch.V.: 1982, J. Astrophys. Astr. 3, 207

15. Sofue, Y., Fürst, E., Hirth, W.: 1980, Publ. Astron. Soc. Japan 32, 1

16. Fürst, E., Reich, W., Beck, R., Hirth, W., Angerhofer, P.: 1982, Astron. Astrophys. 115, 428

17. Sofue, Y., Reich, W.: 1979, Astron. Astrophys. Suppl. 38, 251

18. van der Laan, H.: 1962, Monthly Notices Roy. Astron. Soc. 124, 125

19. Bridle, A.H.: 1967, Monthly Notices Roy. Astron. Soc. 136, 219

20. Bell, A.R.: 1978, Monthly Notices Roy. Astron. Soc. 182, 147 and 143

21. Dröge, W., Lerche, I., Schlickeiser, R.: 1987, Astron. Astrophys. 178, 252

STATISTICAL STUDIES OF SNRs: SELECTION EFFECTS

D A Green
National Research Council, Herzberg Institute of Astrophysics,
Dominion Radio Astrophysical Observatory,
P.O. Box 248, Penticton, B.C. V2A 6K3, Canada.

Summary: The selection effects applicable to the identification of Galactic SNRs are discussed, with particular reference to the limitations that they impose upon statistical studies of SNRs, including the $\Sigma - D$ and $N(< D) - D$ relations.

1 Selection effects

The identification of SNRs is generally made from radio surveys of the Galactic plane, and is basically limited by two selection effects: firstly, the surface brightness of the remnant must be above the sensitivity limit of the observations, and be readily distinguished from the Galactic background emission, and secondly, in general, the angular size of the remnant must be at least several times the beamwidth of the observations.

1.1 The surface brightness selection effect

It is not possible to quote a single surface brightness completeness limit for current catalogues of SNRs, not only because the background emission varies in different regions of the Galactic plane, but also because different regions have been surveyed with different instruments. However, most SNRs so far recognised have been identified from the Molonglo 408-MHz and Parkes 5-GHz survey of most of the Galactic plane, and Clark & Caswell (1976) in their statistical study of SNRs following from the Molonglo-Parkes survey conclude that their lists of SNRs are complete over most of the survey region to a surface brightness which corresponds to $\Sigma_{1\ \mathrm{GHz}} \approx 8 \times 10^{-21}$ W m^{-2}Hz^{-1}sr^{-1}(the surface brightness of a remnant having a flux density of S_ν, in Jy, at a frequency ν, and an area equal to that of a circle of angular diameter θ, in arcminutes, is $\Sigma_\nu \approx 1.5 \times 10^{-19} S_\nu/\theta^2$ W m^{-2}Hz^{-1}sr^{-1}). Many SNRs have been identified since the Molonglo-Parkes survey; mostly these are faint remnants found from a survey with the Effelsberg 100-m telescope at 2.7 GHz (Reich *et al.* 1984; Reich *et al.* 1988). Reich *et al.* (1988) discuss the new remnants identified from the Effelsberg survey, and give a detection limit close to $\Sigma_{1\ \mathrm{GHz}} \approx 2 \times 10^{-22}$ W m^{-2}Hz^{-1}sr^{-1}, although this applies only to regions where the background emission is small ($|b| > 0\overset{\circ}{.}5$).

Faint remnants are more easily detected in regions of the Galactic plane where the background emission is lower, and it may be difficult to disentangle this bias from any intrinsic variation in the properties of SNRs with distance from the Galactic plane. Also, remnants in the region $90° < l < 270°$, which are all outside the solar circle, are easier to detect than those in other regions, and this may bias studies of the distribution of SNRs with Galactocentric radius.

1.2 The Angular size selection effect

Although incompleteness with respect to faint remnants has always been recognised, the incompleteness of catalogues of Galactic SNRs with respect to small angular size remnants has not been so well appreciated.

This is, basically, because it was not perceived as a problem when Clark & Caswell (1976) made their studies, as the distance estimates then available for known young remnants made it unlikely that any young but distant (*i.e.*, small angular size) remnants had been missed, even if they were at the other side of the Galaxy. But the subsequent downwards revision (Green 1984 and references therein) of the distances for several historical SNRs means that this is *not* the case. A lower limit of ≈ 8 arcmin to the angular size of remnants identified from the Molonglo-Parkes survey is suggested, as the resolution of the Molonglo 408-MHz survey is 3 arcmin. This in turn implies that, from the Molonglo-Parkes survey, in at least half the Galaxy, any remnants with diameters up to at least 20 pc could not have been identified. A somewhat larger limit on diameters is applicable to remnants identifiable from the Effelsberg survey, which has a resolution of 4.6 arcmin.

Although it is implausible (see Green 1984 for a fuller discussion) that any high surface brightness remnants like G111.7−2.1 or G184.6−5.8 (Cassiopeia A and the Crab Nebula) would have been missed, any like the young historical remnants G327.6+14.6, G130.7+3.1, G120.1+1.4 and G4.5+6.8 (the remnants of the SN of AD1006, AD1181, AD1572 and AD1604) at the other side of the Galaxy would have flux densities of only a few Janskys at most and diameters of less than a few arcminutes, and would have been overlooked in the available searches. Indeed, high resolution observations of some compact Galactic plane radio sources made in search for young but distant remnants have revealed two such objects (Green & Gull 1984, Reich *et al.* 1985), and others presumably await identification, particularly in the first and fourth quadrants of the Galaxy (see Green 1985; Fich 1986).

1.3 'Filled-centre' and 'composite' remnants

The identification of 'filled-centre' and 'composite' remnants is limited by a surface brightness selection effect at a level similar to that applicable to 'shell' remnants. The angular size selection effect is, however, more of a problem for 'filled-centre' remnants (and for 'composite' remnants with dominant 'filled-centre' components), because even if such an object is resolved, its amorphous structure and its apparent flat radio spectrum means that it is difficult to distinguish it from an H II region. Either significant polarization at radio wavelengths, a low infrared to radio flux ratio (as recently discussed by Fürst, Reich & Sofue 1987) or a good limit on the absence of radio recombination lines are the most likely additional observational evidence that support a 'filled-centre' remnant rather than an H II region identification (although for distant objects in the first and fourth quadrants depolarization limits the use of the first of these methods of discrimination).

2 The $\Sigma - D$ relation

The fact that there are no high surface brightness SNRs with large diameters has often been expressed in terms of a '$\Sigma - D$' relation (usually of the form $\Sigma \propto D^{-n}$, relating surface brightness, Σ, to linear diameter, D). If such a relation can be calibrated with remnants at known distances, then diameters and hence distances can be derived for every SNR from its observed surface brightness. For some time it appeared that the properties SNRs at known distances were well correlated in the $\Sigma - D$ plane (*e.g.*,

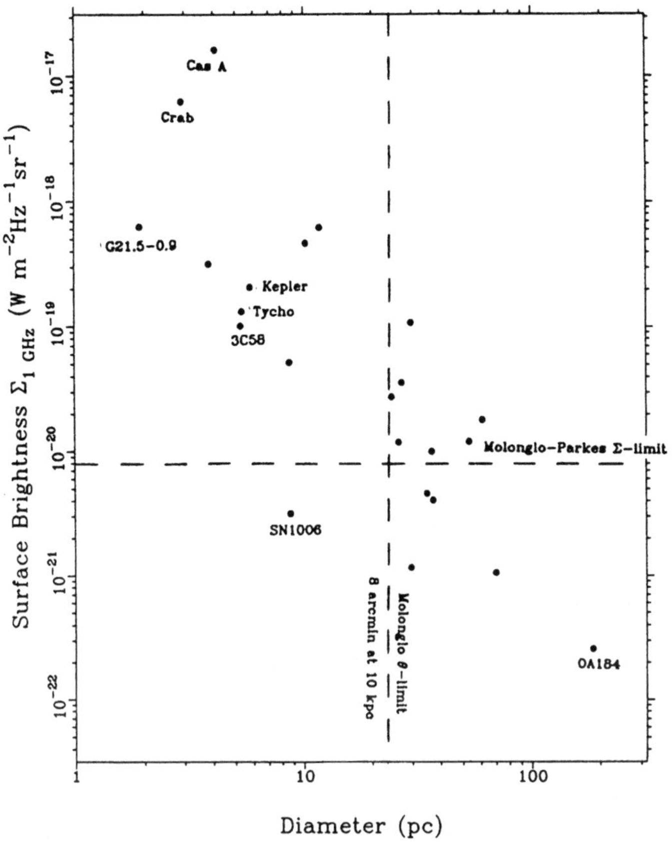

Figure 1: Plot of Galactic SNRs with known distances (mostly from Green 1984) in the $\Sigma - D$ plane. Also marked are the selection limits for SNRs from existing surveys (see text for further details).

Clark & Caswell 1976; Milne 1979), thus allowing useful diameter estimates to be made for other SNRs. However, more recent studies (Green 1984; Berkhuijsen 1986) show that correlation in the $\Sigma - D$ plane is poor, and the full extent of the range of properties of Galactic SNRs in the $\Sigma - D$ plane is not currently known, as the identification of faint and/or small remnants is strongly selected against by current Σ- and θ-selection effects (see Figure 1; also see Berkhuijsen 1986 who includes not only SNRs in the Galaxy and Magellanic clouds, but also faint radio loops in the Galaxy and bright compact radio sources in some nearby galaxies, which have been identified as very old and very young SNRs respectively). It is worth noting that G327.6+14.6 (the remnant of the SN of AD1006, which has a diameter of only $\approx$ 9 pc for a distance of $\approx$ 1 kpc) is fainter than the completeness limit for the Molonglo-Parkes survey, nor would it be sufficiently resolved by the Molonglo observations to be identified as a SNR unless it was closer than $\approx$ 4 kpc — G327.6+14.6 has only been detected because of our knowledge of its approximate position from historical observations and the fact that it is nearby and lies well away from the Galactic plane.

For a given surface-brightness, only an *upper limit* to the diameter of an SNR can be deduced with any confidence (see Green 1984 for further discussions of remnants in the Galaxy; and Berkhuijsen 1986 for a discussion of the 'maximum observable diameter' for remnants both in the Galaxy and external galaxies). This boundary in the $\Sigma - D$ plane is *not* the result of any observational selection effects, and represents some intrinsic limit to the luminosity of radio emission from SNRs, although it does not necessarily represent the evolutionary path of any individual SNR in the $\Sigma - D$ plane.

It may be possible to make improved distance estimates for SNRs from a $\Sigma - D$ relation if a particular class of SNRs could be identified which shows a smaller range of properties than is apparent for all SNRs considered together (although the Σ- and θ-selection effects would likely still pose problems). Huang & Thaddeus (1985) have suggested that 'shell' SNRs associated with giant molecular clouds (presumably the remnants of the 'type II' SN explosions of young, massive stars) are such a class. But, for this particular sub-set of SNRs, distance estimates from a $\Sigma - D$ relation are generally *not* required, as a distance for each remnant will be available from the radial velocity of CO emission from the associated giant molecular cloud. Tuohy *et al.* (1982) suggest that remnants with optical emission dominated by Balmer lines (which they identify as the remnants of 'type I' SN) may form a class with a small range of properties (see also Strom 1988). Unfortunately optical classification for Galactic remnants are limited to nearby objects only, because of obscuration, and also these 'Balmer dominated' remnants appear to be fainter than other remnants with similar diameters, which means that the Σ-selection effect in particular is a serious problem.

3 The N – D relation

The cumulative number-diameter ($N(< D) - D$) relation is potentially a very useful statistical tool for studying the dynamics of SNRs. If all SNR evolve similarly, then the number of remnants, $N(< D)$, with diameter up to a given value, D, is simply the age of the remnant at the limiting diameter, $t(D)$, divided by the mean time between remnant producing SN, τ. For remnants in the free expansion phase, $D \propto t$, so $N(< D) \propto D$; in the Sedov (or adiabatic) phase, $D \propto t^{2/5}$ (with the constant of proportionality depending on the explosion energy of the SN, E_0, and the density of the ISM, n_0), so $N(< D) \propto D^{5/2}$; and $N(< D) \propto D^{>5/2}$ for remnants in the radiative phase. Thus, for complete sample of remnants with known diameters up to a particular value, the dynamical evolution of SNRs can be studied, and if a regime could be found where the Sedov phase was applicable, then constraints on E_0, n_0 and τ could also be derived. However, SN and the media into which they expand are not uniform, and accurate diameters for each remnant are not generally available. Moreover current catalogues of SNRs are not complete to a limiting diameter, as and the spread of properties of remnants in the $\Sigma - D$ plane means that the Σ-limit corresponds to an appreciable range of physical diameters, and θ-selection effect means that small remnants are missing. Consequently there are large uncertainties in any deductions made from $N - D$ studies of SNRs.

4 Other Comments

An important consequence of the Σ-selection effect, combined with the fact that there is a much larger population of old, larger SNRs in the Galaxy awaiting identification than there is of young small SNRs, is that the larger remnants in current catalogues are likely to be biased towards the objects of higher than average brightness (or equivalently energy) for their diameter. Indeed it is apparent that there is a considerably larger range of surface brightnesses of identified Galactic remnants with small diameters than there is for those with large diameters. Since it is unlikely that the evolution of SNRs in a variety of interstellar environments will reduce the spread of their properties, this shows that known remnants are biased towards brighter, more easily identified objects as diameter increases. This effect is probably largely responsible for the apparent increase (*e.g.*, Berkhuijsen 1988, and references therein) in the deduced energy content of SNRs with diameter.

Statistical studies of distance-dependant properties of SNRs can be made directly for remnants in the Magellanic clouds (or other external galaxies), as they are all at essentially the same distance, whereas uncertain distance estimates are required for similar studies of Galactic SNRs. In particular, studies of the luminosity of remnants can readily be made, which avoids the '$1/D^2$' bias that is implicit in $\Sigma - D$ studies (*i.e.*, for a remnant of luminosity L, at a distance d, $L \propto Sd^2$, whereas $\Sigma \propto S/\theta^2$, so $\Sigma \propto L/(\theta d)^2$, or $\Sigma \propto L/D^2$). Incompleteness of current catalogues is, however, also a problem for statistical studies of SNRs in the Magellanic Clouds, and in particular the current Σ-selection effects applicable to the identification of remnants in Magellanic Clouds (in this case from X-ray rather than radio surveys) poses serious problems for $N(<D) - D$ studies (see Green 1984 for further discussion).

References

Berkhuijsen, E.M., 1986. *Astr. Astrophys.*, **166**, 257.

Berkhuijsen, E.M., 1988. *Astr. Astrophys.*, **192**, 299.

Clark, D.H. & Caswell, J.L., 1976. *Mon. Not. R. astr. Soc.*, **174**, 274.

Fich, M., 1986. *Astr. J.*, **92**, 787.

Fürst, E., Reich, W. & Sofue, Y., 1987. *Astr. Astrophys. Suppl.*, **71**, 63.

Green, D.A., 1984. *Mon. Not. R. astr. Soc.*, **209**, 499.

Green, D.A., 1985. *Mon. Not. R. astr. Soc.*, **216**, 691.

Green, D.A. & Gull, S.F., 1984. *Nature*, **312**, 527.

Huang, Y.-L. & Thaddeus, P., 1985. *Astrophys. J.*, **295**, L13.

Milne, D.K., 1979. *Aust. J. Phys.*, **32**, 83.

Reich, W., Fürst, E., Steffen, P., Reif, K. & Haslam, C.G.T., 1984. *Astr. Astrophys.*, **58**, 197.

Reich, W., Fürst, E., Altenhoff, W.J., Reich, P. & Junkes, N., 1985. *Astr. Astrophys.*, **151**, L10.

Reich, W., Fürst, E., Reich, P. & Junkes, N., 1988. In: *Supernova Remnants and the Interstellar Medium*, (IAU Colloquium 101), p.293, eds. Roger, R.S. & Landecker, T.L., (Cambridge University Press, England).

Strom, R.G., 1988. *Mon. Not. R. astr. Soc.*, **230**, 331.

Tuohy, I.R., Dopita, M.A., Mathewson, D.S., Long, K.S. & Helfand, D.J., 1982. *Astrophys. J.*, **261**, 473.

THE GALACTIC DISTRIBUTION OF RADIO SUPERNOVA REMNANTS

Sidney van den Bergh
Dominion Astrophysical Observatory,
National Research Council of Canada

The Galactic distributions of radio supernova remnants, CO emission, 100μm infrared emission and COS B γ-rays all exhibit the same nuclear disc structure. This result shows that the large scale distribution of Galactic radio supernova remnants essentially reflects the distribution of interstellar gas rather than that of the supernova progenitors.

In recent years catalogues of galactic supernova remnants have been published by van den Bergh (1983) and by Green (1984, 1987). For supernova remnants with a 1GHz flux density $F > 12$Jy the number of remnants is well represented by the relation

$$\log n = 2.85 - 0.85 \log F. \quad (1)$$

For $F < 12$Jy departures from this power law relationship probably indicate incompleteness of the sample. In Fig. 1 the Galactic distribution of supernova remnants with $F > 12$Jy in the catalogue of Green (1987) is shown for three ranges of remnant surface brightness Σ expressed in units Jy deg^{-2}. After correcting for the dependence of the surface brightness of supernova remnants on

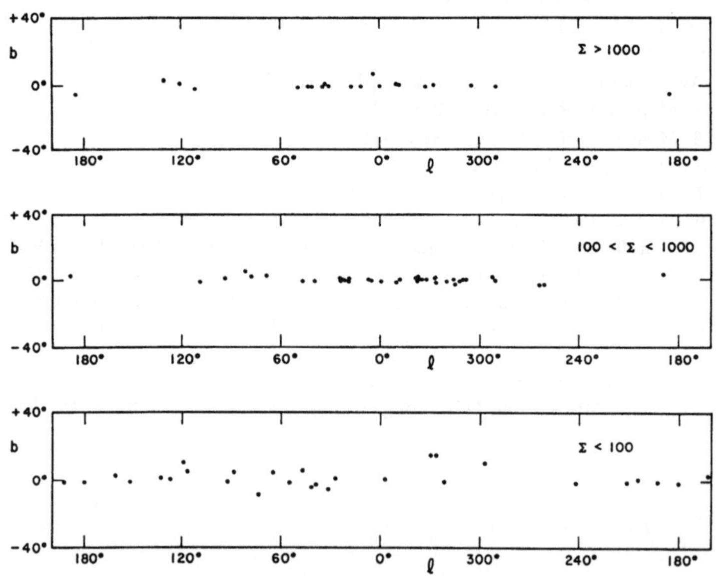

Figure 1. Distribution of supernova remnants of high (upper panel), intermediate (central panel) and low (lower panel) surface brightness in Galactic coordinates. Intrinsically luminous SNR's of high- and intermediate-surface brightness are seen to be concentrated in a thin nuclear disc.

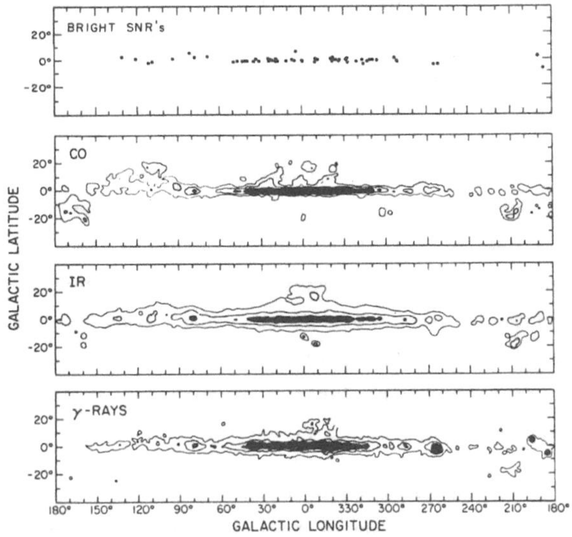

Figure 2. Comparison of the distribution of radio supernova remnants, CO emission, IRAS 100μm infrared emission and COS B γ-rays in the energy range 70MeV-5 GeV. All four diagrams are seen to reveal the same thin nuclear disc of molecular hydrogen gas.

distance from the Galactic plane one finds that $\Sigma \propto D^{-3}$ (Caswell and Lerche 1979). Since the luminosity of a supernova remnant $L = 0.25\pi D^2 \Sigma$ it follows that $L \propto \Sigma^{1/3}$ i.e. the most luminous supernova remnants have the highest surface brightnesses.

Inspection of Fig. 1 shows that the 26 SNR's with $\Sigma < 100$Jy deg^{-2} have a more or less uniform distribution along the Galactic plane with a possibly significant excess of objects in the northern Milky Way. This effect might either be due to a slight incompleteness for low-surface brightness remnants in the southern Milky Way or to a genuine excess of nearby SNR's with $30° < \ell < 130°$ (Aquila to Cassiopeia). Inspection of the upper two panels of Figure 1 shows no obvious difference between the distributions of bright ($\Sigma > 1000$ Jy deg^{-2}), and intermediate brightness ($100 < \Sigma < 1000$Jy deg^{-2}) supernovae. The combined distribution of all radio supernova remnants with $\Sigma > 100$Jy deg^{-2} has therefore been plotted in the upper panel of Fig. 2. The most striking feature of this figure is a pronounced nuclear disk that is not apparent for SNR's of lower-surface brightness. In the figure this nuclear disk is outlined by the remnants of 37 supernovae with $305° < \ell < 55°$ which have $|b| \lesssim 1.0°$. For an assumed distance of 8kpc to the Galactic nucleus a disk extending to 55° from the centre of the Galaxy has a radius of 6.5kpc. (At a distance of 8kpc $|b| <$ 1.0° corresponds to $|z| < 140$pc.

Within the Galactic nuclear disk the distribution of supernova remnants does not depend strongly on Galactic longitude. This indicates that the outer part of the nuclear disk is more densely populated with supernova remnants than is its inner region i.e. the "nuclear disc" is actually an annulus.

It is particularly noteworthy that there is no indication of a population of supernova remnants associated with the nuclear bulge of the Galaxy. Presumably this implies that the remnants of SN Ia, which occur in the nuclear bulge, are exceedingly faint because their precursors exploded in a region where the interstellar gas density is very low. By the same token the strong concentration of all classes of Galactic supernova remnants to the Galactic plane attests to the fact that high-latitude supernovae at large $|z|$ expand themselves out of existence on a short timescale.

It is concluded that the location of Galactic supernova remnants probably tells us more about the distribution of interstellar gas than it does about the distribution of supernova progenitors. This conclusion is strengthened (see Fig. 2) by intercomparison of the distributions of Galactic radio supernova remnants, CO (Dame et al. 1987), IRAS 100μm radiation (from Boulanger quoted by Dame et al. 1987) and γ-rays (Mayer-Hasselwander et al. 1982). All four panels of Fig. 2 show the same prominent thin nuclear disc structure which is, presumably, due to a dense flat ring of molecular hydrogen gas extending from about 3.5 to 6.5kpc from the Galactic centre.

REFERENCES

Caswell, J.L. and Lerche, I. 1979, Mon. Not. R. astr. Soc. 187, 201.
Dame, T.M., Ungerechts, H., Cohen, R.S., de Geus, E.J., Grenier, I.A., May, J., Murphy, D.C., Nyman, L.-A. and Thaddeus, P. 1987, Astrophys.J. 322, 706.
Green, D.A. 1984, Mon. Not. R. astr. Soc. 209, 449.
Green, D.A. 1987, Revised and updated version of Green 1984 (unpublished).
Mayer-Hasselwander, H.A. et al. 1982, Astron. Astrophys. 105, 164.
van den Bergh, S. 1983, in Supernova Remnants and their X-Ray Emission, eds. J. Danziger and P. Gorenstein, Reidel, Dordrecht, p.597.

DISCUSSION

Fürst: We have found, from our 11-cm survey, that we are just approaching a sensitivity of 100 Jy per square degree in the inner part of the Galaxy. This implies that earlier surveys must have missed almost all SNRs fainter than this value. Close to the centre of the Galaxy even the sample of remnants with $100 < \Sigma < 1000$ Jy per square degree will be somewhat incomplete.

van den Bergh: So the distribution of SNRs may be even more concentrated towards the Galactic centre than I indicated!

Gaskell: How many supernovae of type II are known to have occurred in irregular galaxies?

van den Bergh: To date SN1987A is one of the few known examples. It would certainly be very worthwhile to undertake a systematic search for such supernovae of below-average luminosity in metal-poor dwarf irregular galaxies.

Kundt: Have you tried to correct the frequency of SNII for the below-average luminosity of objects such as SN1987A?

van den Bergh: We are in the process of doing this.

Kundt: Is there observational evidence that supernovae of type Ia have occurred in the bulge of our Galaxy?

van den Bergh: No. In elliptical galaxies, which are believed to contain a stellar population similar to that in the nuclear bulge of the Galaxy, one SNIa is observed for every ~ 4000 novae. If the nova rate in the Galactic nuclear bulge is ~ 4 per year one might expect a frequency of one SNIa per millenium in the Galactic nuclear bulge.

Kundt: Wouldn't one expect a larger age for supernova remnants in the halo because of less braking, even though they are less luminous? The "lower" halves of these SNRs should become visible when they hit the Galactic disk.

STATISTICAL PROPERTIES OF RECENTLY NEW IDENTIFIED SUPERNOVA REMNANTS

W. Reich, E. Fürst
Max-Planck-Institut für Radioastronomie
Auf dem Hügel 69
D-5300 Bonn 1
Federal Republic of Germany

Abstract

Several new supernova remnants (SNRs) have been recently identified from sensitive radio continuum surveys of the Galactic plane. The limit in surface brightness is about $\Sigma_{1GHz} = 2 \cdot 10^{-22}$ [W/m^2 Hz sr]. In comparison with previously known SNRs, which are in their majority much brighter, we reexamine the distribution of SNRs in the first Galactic quadrant and some statistical properties.

1. New Supernova Remnants

From deep radio continuum surveys with the Effelsberg 100-m telescope of the first Galactic quadrant at 21 cm and 11 cm wavelength and subsequent observations at shorter wavelengths we have identified so far 32 new SNRs. The identification methods, the list of sources and their parameters as well as additional references have been already published [1]. As an example the new shell-type SNR G8.7-5.0 is shown in Figure 1. More objects are presented by JUNKES et al. [2]. G8.7-5.0 is located at

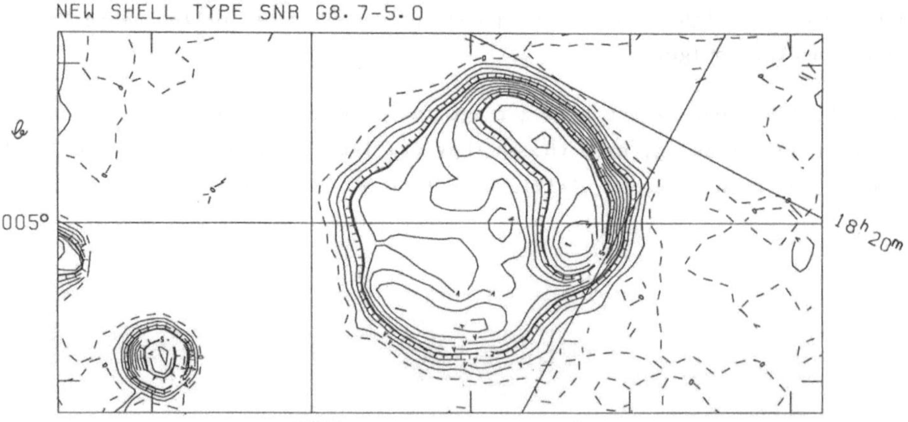

Fig. 1: Contour map at 11 cm wavelength. The contours are 50 mK T_B or 20 mJy/beam area apart up to 500 mK T_B and 100 mK T_B or 40 mJy/beam area above. Contour labels are in K T_B. The angular resolution is 4:27.

the boundary of the 11 cm Effelsberg survey, which covers the Galactic plane for $|b| \leq 5°$. The surface brightness of G8.7−5.0 is $\Sigma_{1GHz} = 10^{-21}$ [W/m² Hz sr], which is about a factor of 5 above our detection limit. The detection limit is strongly dependent on confusion. It is particularly high for $|b| \leq 0.^\circ5$, but also for some regions at higher latitudes, where numerous emission sources overlap. In general for $|b| \geq 0.^\circ5$ the detection limit is $\Sigma_{1GHz} = 2 \cdot 10^{-22}$ [W/m² Hz sr]. Most of our new objects have diameters ≥ 15' (see [1]) and are at least partly resolved with our 4.'3 beam at 11 cm wavelength. More compact objects must be looked at with higher angular resolution to rule out an extragalactic origin.

2. SNR Statistics

In Table 1 we list the types of SNRs located in the surveyed area: $357.^\circ4 \leq \ell \leq 76°$, $|b| \leq 5°$. We distinguish between previously found − "known" − objects and "new" sources which we identified on the basis of our surveys. The fraction of shell-type SNRs (2/3) remains unchanged when the new data are included.

Table 1: Types of SNRs (see text)

	"known"		"new"		Sum
shells	31		22		53
plerions	3		5		8
combined	4		1		5
unclear	9		5		14
	47	+	33	=	80

The distributions of sources with longitude and latitude are shown in Figures 2 and 3. The longitude distribution of sources remains nearly unchanged with the new data included. The latitude distribution seems to be wider now. However, the high

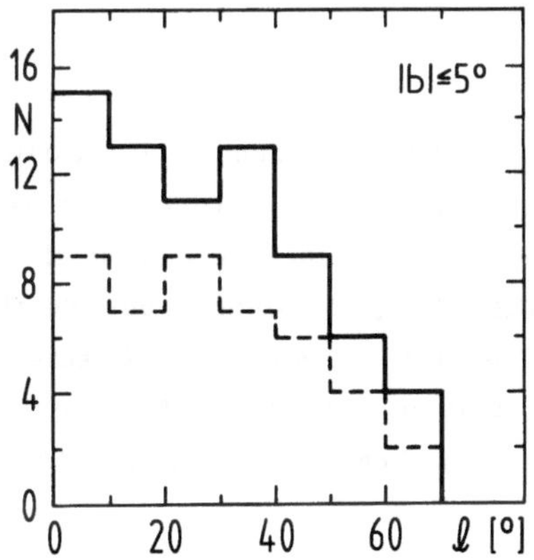

Fig. 2: Number of SNRs as a function of Galactic longitude. The dashed line shows the "known" objects, while the full line shows the number of objects including sources identified on the basis of the Effelsberg Galactic plane surveys.

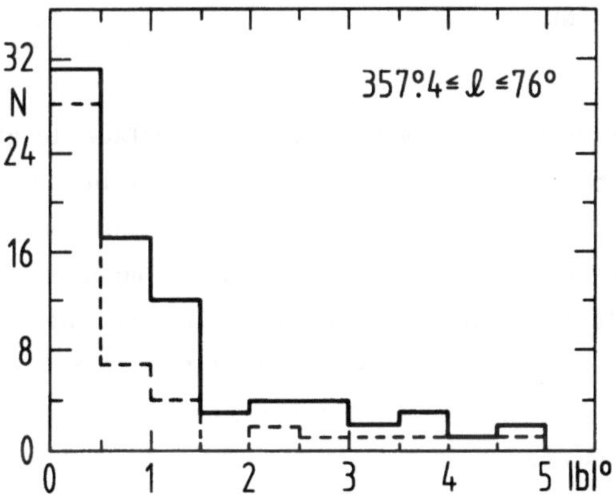

Fig. 3: Same as Figure 2 for Galactic latitude

confusion for $|b| \leq 0.^{\circ}5$ makes it very likely that a number of sources is missing in this interval. For most objects no distances are available. It is, therefore, impossible to convert the latitude distribution into a z-distribution. The number of sources as a function of surface brightness Σ_{1GHz} [W/m² Hz sr] = $1.505 \cdot 10^{-19}$ S_{1GHz} [Jy]/θ^2 [arcmin] is listed in Table 2. One object, G70.68+1.20, has a brightness exceeding Σ_{1GHz} = 10^{-19} [W/m² Hz sr]. G70.68+1.20 is one of the few compact sources in our list [1] which we identify as a SNR [3,4], but this is still subject of controversy.

Table 2: Distribution of Surface Brightness ($|b| \leq 5^{\circ}$)

-log (Σ)	>19	19-20	20-21	21-21.7
"known"	8	20	15	3
"new"	1	2	16	13
all SNRs	9	22	31	16

In total only three new objects with $\Sigma_{1GHz} > 10^{-20}$ [W/m² Hz sr] have been found, while 29 objects have a lower surface brightness. Excluding the latitude range $|b| \leq 0.^{\circ}5$, because of incompleteness due to confusion, the source distribution is listed in Table 3. The number of new objects with $\Sigma_{1GHz} < 10^{-20}$ [W/m² Hz sr] is 26, to be compared with 11 objects known previously. We believe that this sample is fairly

Table 3: Distribution of Surface Brightness ($0.^{\circ}5 \leq |b| \leq 5^{\circ}$)

-log (Σ)	>19	19-20	20-21	21-21.7
"known"	1	5	8	3
"new"	1	1	13	13
all SNRs	2	6	21	16

complete for a wide range of surface brightness and should be subject for further statistical studies like the Σ-D relation and its z-dependence. However, for this purpose distances are needed. The use of HI surveys to search for holes or shells at the positions of SNRs seems to be much more promising for objects slightly outside the Galactic plane with its very complex line emission structures. Recently, we detected a hole in the HI emission associated with G18.95-1.1 providing a distance of 2 kpc [5]. A systematic investigation is in progress using HI data from the Effelsberg 100-m telescope.

The Σ-D relation has been often used to find diameters D or distances d of SNRs. This method is quite uncertain, and we use it here as a first approximation. According to MILNE [6] we use $D\,[\mathrm{pc}] = 4.12 \cdot 10^{-4} \cdot \Sigma_{1GHz}{}^{-0.25}\, \exp(-|z|/214)$ to calculate the diameters for shell-type SNRs. We list the resulting diameters for objects with $|b| \geq 0.^{\circ}5$ in Table 4 and show the cumulative counts N(<D) in Figure 4. The slope of

Table 4: Distribution of SNR Diameters (0°.5 ≤ |b| ≤ 5°)

D (pc)	>10	10-20	20-30	30-40	40-50	50-60	60-70	70-80
"known"	-	2	4	7	1	-	-	1
"new"	-	-	8	10	1	2	-	-
all SNRs	-	2	12	17	2	2	-	1

5/2 for the fitted line is expected for the case of adiabatic expansion and holds for diameters of D ~ 40 pc for the "known" objects and is not altered when the new sources are included. The diagrams show only an offset due to the different absolute numbers. This result is very similar to that obtained by MILNE [6] in 1979, who found agreement with adiabatic expansion up to D ~ 30 pc. Although our source list is increasingly incomplete for large diameter objects at large distances, the drastic deviation of the source counts from the expectation for adiabatic expansion for D > 40 pc seems real. The number of objects with D ≤ 60 pc is more than a factor 2 below the expectation.

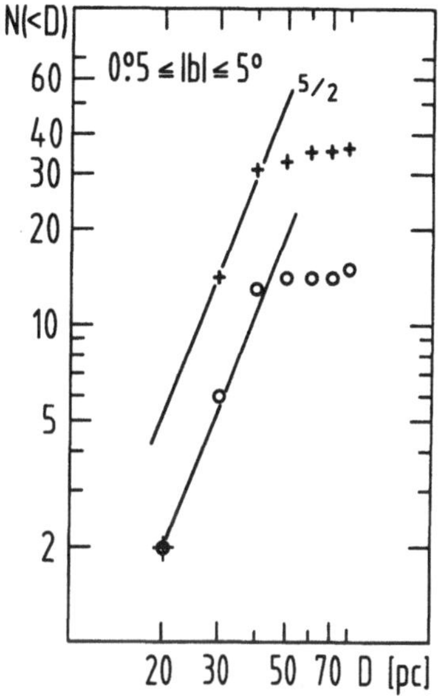

Fig. 4: Cumulative counts N(>D) of shell-type SNRs for 0°.5 ≤ |b| ≤ 5°. An increase of N(>D) ~ $D^{5/2}$ is expected for adiabatic expansion which holds up to D ~ 40 pc.

Using the distances derived from Milne's formula we investigated the d^2- dependence. We find 5 objects up to 2 kpc, 19 up to 4 kpc and 21 up to 5 kpc for our longitude range. Up to 4 kpc these results agree with an even SNR distribution and we expect 92 sources within a complete 4 kpc circle. Assuming the same density of SNRs for the whole Galaxy, we estimate the total number of SNRs to ≈ 800 for a radius of the Galaxy of R = 12 kpc and ≈ 1300 for R = 15 kpc. Explaining the 5/2 power law of Figure 4 as due to adiabatic expansion up to ≈ 40 pc, the age of these remnants is 10000±5000 years. The corresponding birth rate of shell-type SNRs is one in 4 to 19 years.

3. Distribution of Spectral Indices

We show the distribution of spectral indices of the SNRs in Figure 5. The previously well-known peak of the distribution between α = 0.4-0.5 (S ~ $\nu^{-\alpha}$) is not altered, when the new sources are included but the distribution is broadened. Since the fraction of shell-type SNRs was found to be unchanged adding the new data, the larger spread indicates a dependence of α on the surface brightness.

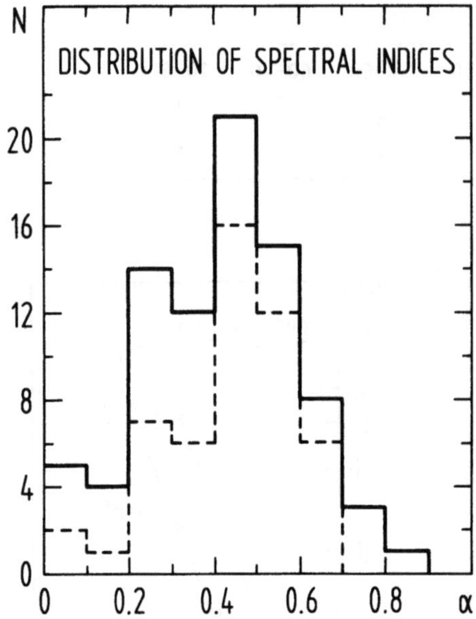

Fig. 5: The number of SNRs N having spectral indices α ($S \sim \nu^{-\alpha}$) are shown for "known" objects (dashed) and for all sources (full line)

4. Conclusion

Our deep radio continuum surveys have significantly increased the number of low surface brightness SNRs in the first Galactic quadrant. The spatial distribution in longitude is however unchanged. The apparent wider latitude distribution is certainly affected by incomplete identification in the Galactic plane. Missing distances of most SNRs is the basic problem in any further statistics. However, using Milne's Σ-D relation as a first approach, the end of the adiabatic phase of shell-type SNRs at 40 pc diameter is indicated. Since most of the new objects are found somewhat outside the Galactic plane, a distance determination on the basis of HI observations seems promising.

References

1. Reich, W., Fürst, E., Reich, P., Junkes, N.: 1986, in *Supernova Remnants and the Interstellar Medium*, IAU Coll. 101, eds. R.S. Roger and T.L. Landecker, Cambridge University Press, p. 293
2. Junkes, N., Fürst, E., Reich, W.: 1988, this volume
3. Reich, W., Fürst, E., Altenhoff, W.J., Reich, P., Junkes, N.: 1985, Astron. Astrophys. 151, L10
4. Reich, W., Junkes, N., Fürst, E.: 1988, this volume
5. Fürst, E., Hummel, E., Reich, W., Sofue, Y., Sieber, W., Reif, K., Dettmar, R.J.: 1988, Astron. Astrophys., submitted
6. Milne, D.: 1979, Australian J. Phys. 32, 83

X-Ray Observations and Non–Equilibrium Ionisation of Supernova Remnants

W. Brinkmann
Max-Planck-Institut für Physik und Astrophysik,
Institut für Extraterrestrische Physik,
8046 Garching, F.R.G.

Abstract:

X-ray observations of young supernova remnants (SNR) provide the most direct tool to study their evolution, their chemical composition, and their interaction with the interstellar medium. We will show that the remnants of type I and type II supernova explosions differ considerably in their physical properties and raise different questions. After a brief review of some X–ray observations of SNR and their modelling we will discuss the role played by non–equilibrium ionisation effects in the interpretation of X–ray spectra.

1. Introduction

In a recent review by van den Bergh [1] 135 objects have been listed which are considered to be galactic supernova remnants (SNR) according to their appearance and their non-thermal radio spectra. Only about 40 of these objects are X-ray emitters. A large compilation of X–ray data and their interpretation can be found in the proceedings of the IAU Symposium 101, held in Venice [2] at the end of the Einstein mission. It was in fact Einstein with its imaging and spectroscopic capability which marked a major step forwards in SN research. Later missions like EXOSAT and Tenma mainly extended the observed spectral range up to $\gtrsim$ 10 keV, allowing a better definition of the X–ray continuum of many of the remnants. They further covered the vital energy range of $\sim$ 6.5 keV where highly excited Fe–lines are to be found.

Whereas radio observations are the most sensitive method to reveal the enormous wealth of complex morphologies of SNR, X–ray observations provide a direct way to obtain information about the explosion itself, the physical state and the chemical composition of the star involved and its circumstellar medium. Especially the well studied historical SNR with known ages play an important role in that context. First we will discuss the state of the art of X–ray observations of SNR, their limitations and prospects using as illustrative examples two well studied remnants, the Crab nebula and Tycho. In the second part we will discuss in some detail one of the most important questions in X-ray spectroscopy of SNR – the role played by non–equilibrium ionisation effects of the hot radiating plasma in the interpretation of the data.

2. Observations

Weiler [3] has estimated that about 80 % of the radio SNR are shell like, 5 % are centrally filled (or Crab - like) and the rest are a combination of both. A similar picture emerges in the X–ray range [4], and the X–ray spectra of these remnants show clearly that we are observing two completely different classes of objects.

The Crab - like or plerionic remnants are fuelled by the interaction with the central rotating, highly magnetized neutron star. The structure, dynamics and the emitted spectrum are predominantly determined by the energy input from the pulsar. As a matter of fact, the detection of the pulsar in the Crab nebula in 1969 solved two long lasting problems: on one hand the question of the origin of the continuous energy supply of $\geq 10^{38}$ erg/sec, required to power the nebular expansion and radiation. On the other hand it confirmed the idea that pulsars are rapidly rotating neutron stars: just this amount of energy will be lost by electromagnetic radiation from a "typical" neutron star with the measured angular velocity of the Crab pulsar. However, how this energy is actually transferred to the radiating particles is still an unsolved problem.

Crab–like remnants show non thermal radiation from the radio to the X-ray range with a centrally - peaked spatial distribution. The best examples of this class of objects are Vela X, 3C58, CTB80, MSH15-52, 0540-693 in the LMC, and the Crab nebula itself. Unfortunately, there are some serious deficiencies in our X–ray observations of these objects. In most cases we lack high quality spectra and high spatial resolution. Therefore we will concentrate in the following mainly on observations of the Crab nebula which is expected to represent the typical object of this class.

The generally accepted explanation is that we are seeing Synchrotron radiation from relativistic electrons, injected by the central pulsar into the surrounding nebular field. The shrinking size of the Crab nebula with increasing photon energy is easily explained as caused by the decreasing lifetime of the higher energetic radiating electrons. This lifetime argument, on the other hand, demands continuous injection of relativistic particles into the nebula. The lifetime of plerionic remnants is therefore determined by the slowing down time scale $\tau = P / 2\dot{P}$ of the pulsar. The object we see now with its low expansion velocity of $\simeq 1000$ km/sec is very likely a pulsar driven secondary phenomenon. So far, all attempts to see traces of the original blast wave of the explosion have failed.

As mentioned above, the enormous energy requirement to balance the losses of the emitted radiation and driving the expansion of the nebula of a few times 10^{38} erg/sec is just matched by the rotational energy loss of the rapidly spinning neutron star. Whereas this energy can be supplied, in principle, by the strong low frequency (30 Hz) electromagnetic wave, generated by a rotating magnetic vacuum dipole, the Synchrotron nature of the radiation requires the continuous injection of relativistic electrons as well, and it is not clear at all how the energy is shared between these two components.

The energy is certainly not injected isotropically into the nebula as the central region of the Crab nebula shows distinct features. First of all we have an emission "hole" around the pulsar, with reduced emission at all frequencies. Secondly, on both sides of the pulsar along the major axis of the nebular we see the so called whisps, a time varying phenomenon which indicates a magneto - hydrodynamic interaction of the pulsar with its surroundings. The question, whether the temporal changes observed [5] represent real material flow motion or only a varying exitation of radiation is so far unsolved.

A rough geometrical picture of the pulsar - nebula interaction proposed by Aschenbach and Brinkmann [6] seems to be confirmed by recent X-ray observations [7]. The observation showed that even at high energies ($E_\nu \gtrsim 50$ keV) the X-ray Crab has a finite extension of $\gtrsim 60''$, in contrast to the simple model of a point like injector [8].

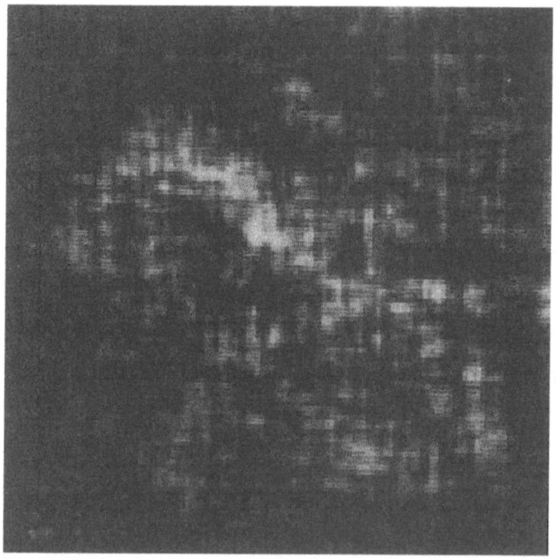

Figure 1: The X-ray emission of the Crab nebula from processed off - pulse Einstein HRI data. For details see [7].

Figure 1 shows the picture obtained by processing the Einstein HRI off-pulse Crab data (see [7]). A ring like structure, including the central hole is clearly visible. The dark parts of the picture contain signals below 10% of the maximum intensity. The overall configuration as well as the ratios of the axes agree closely with the predictions of the above mentioned model, in which electrons are accelerated and then transported outwards, radiationlessly, in the strong electromagnetic wave of the pulsar. When the ram pressure of this relativistic wind equals the thermal pressure of the surrounding nebula a shock forms, producing an amplification of the magnetic field and a randomization of the directed plasma motion [9]. Taking for the pulsar the field configuration of an oblique rotator, the outflow of the electrons will be concentrated near the rotational equatorial plane, producing the toroid-like structure of the emission region. The height of this toroid is determined by comparing the lifetime of the radiating electrons with their effective radial propagation time. The brightness asymmetry in the optical picture near the pulsar and the X-ray offset is explained by an enhancement of the magnetic field strength in the nothwestern parts of the nebula, caused by a compressional effect of the pulsar running into the surrounding medium.

Although in the original model the geometrical configuration of the emitting region is at least qualitatively correct, a detailed comparison of the data with numerical simulations showed that the underlying particle injection and transportation mechanism cannot be as simple as anticipated. It is more probable that a more complex mechanism including particle re-acceleration is involved [10], or that the energy injection mechanism is quite different [11].

The study of the central regions of the Crab nebula shows that the interaction between a pulsar and its surrounding medium is physically and geometrically very complex. None of the theoretical models seem to be able to make definite predictions about the size and shape of other Synchrotron nebulae. Both, the pure vacuum model as well as any simple wind model have various conceptual difficulties, especially as a complete understanding of the most basic ingredient, the pulsar magnetosphere, is still missing.

In the case of the Crab the total energy requirements of pulsar and the surrounding nebula are balanced by the rotational energy loss of the slowing down, spinning neutron star. Most other plerionic X–ray remnants show, however, a total mismatch between the pulsar's rotational energy loss $\dot{E}$ and the "visible" X–ray luminosiy of the system, which is generally below 10% of $\dot{E}$ [12]. Even more extreme is the peculiar remnant MSH15–52 and its central pulsar PSR 1509-58. For the pulsar, a rotational energy loss of the order of 2×10^{37} erg/sec can be estimated, the X–ray luminosity is only 2.8×10^{35} erg/sec [13]. However, in the EXOSAT X–ray band the pulsar has a very hard power law spectrum with a photon index $\alpha \sim 1.1$ which will dominate the system's energy output from the hard X–ray range onwards and will eventually require all the available energy – if not severely cut off below ~ 1 GeV. It is perhaps worth mentioning that there have been claims of a detection of TeV γ–rays from PSR 1509-58 [14]. Although the estimated VHE luminosity of the pulsar amounts to only $\lesssim 3 \times 10^{34}$ erg/sec, some of the "invisible" energy output might be found in these extreme energy bands.

Concluding, it is perhaps fair to say that the coupling between a pulsar and its remnant is still far from being understood. From an observational point of view, X-ray pictures with high spectral and spatial resolution are required, but the chances to achieve this goal in the near future appear, unfortunately, quite low.

In the shell type SNR the emitted radiation is associated with the shock wave driven by the initial explosion into the interstellar medium. The initial kinetic energy of the supernova ejecta is of the order of 100 keV per nucleon. Therefore, when the material is thermalized efficiently through collisions with the ambient medium, the temperatures reached in the shocks give rise to thermal emission mainly in the X-ray range. These remnants are thought to originate from type I supernova explosions where the whole star is disrupted in the explosion. The emitted spectrum thus reflects the physical conditions and the chemical composition of the progenitor star and the ambient interstellar medium.

Best examples for this kind of scenario are the remnants SN 1006, Tycho, and Kepler. For the later two objects the obtained optical light curves clearly indicate type I explosions. The X–ray shapes of the remnants are nearly perfectly spherically symmetric, indicating that the explosion went off in a largely homogeneous medium and no signs for the existence of a central pulsar have been found.

The remnant of the type I explosion observed by Tycho de Brahe in 1572 represents perhaps the most beautiful example of this class. A detailed analysis of the Einstein data [15] confirms roughly the "standard" picture, although the mass derived for the progenitor star is larger than the $\simeq 1.4 M_{\odot}$ expected from the deflagration of a white dwarf [16].

The analysis of various experiments with higher spectral resolution further showed (Becker et al [17], Pravdo et al [18], Hamilton et al [19], and Tsunemi et al [20]) that the chemical abundance cannot be solar, but must be enriched with iron peak elements. Using the data from Nomoto's W7 model [21], and mixing the outermost zones of that model, Itoh [22] was able to reproduce the spectral behaviour observed by Tenma [20] by following numerically the temporal evolution of the hydrodynamical and non - equilibrium ionisation structure up to Tycho's present age.

However, several problems of theoretical as well as observational nature remain open. Later we will discuss some important aspects of deconvolving the observed spectra, here we mention the second class of problems, related to our temporal observational possibilities, where we are faced with the trade off between resolution and counting statistics. Roughly spoken, current detectors have either high spectral resolution but very low spatial resolving power (like the GSPC on EXOSAT or Tenma) or vice versa (like the Einstein HRI). Both types of systems give only limited information about the physical parameters of the SNR and instruments with higher spectral and spatial resolution are required to get unique answers. This is demonstrated in Figure 2 where two line of sight spectra of a non–equilibrium ionization model of Tycho are shown [23].

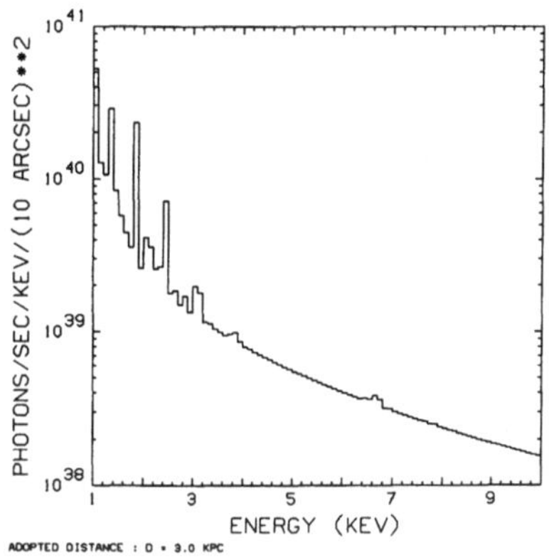

Figure 2a: Line-of-sight photon spectrum, centre of remnant

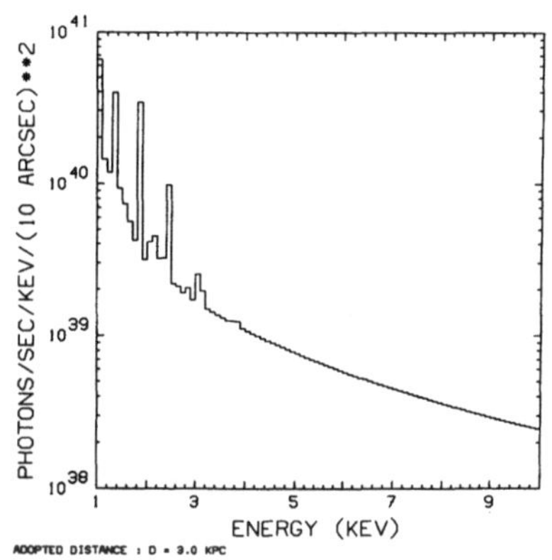

Figure 2b: Line-of-sight photon spectrum, shock region

Plotted are two line-of-sight photon spectra as seen by an instrument with a spatial resolution of 10×10 arc secs at two different positions on the remnant: one (Fig. 2a) near the centre, the other at the position of the shock. Note that although Figure 2b shows the spectrum at the shock, the hottest region of the remnant, no iron line is seen at ~ 6.7 keV, due to the insufficient time to ionize iron to the required stage.

It is evident from Figure 2 that the 'observed' spectra differ considerably, although they are taken from the same model. Trying to fit these spectra, as it would be done during the reduction of observational data, one would conclude that more component thermal bremsstrahlungs fits plus additional lines or ionisational equilibrium models are required with temperatures in the range between 3 keV and $\lesssim$ 15 keV. Although these temperatures are actually in the range encountered in the hydrodynamical calculations, none of them is in any sense representative for this model.

As the equivalent widths of individual lines are dependent on the underlying continuum, the deduced chemical abundance values are clearly model dependent as well. This means, the determination of line strengths without the knowledge of a well defined continuum is highly uncertain. Even the most detailed results of non–solar X–ray spectroscopy so far, the Einstein FPCS measurements of Puppis-A [24] which could be used to perform detailed plasma diagnostics by comparing different line ratios of the same element show this problem: The Einstein energy band extended up to about 4.5 keV and therefore neither the Fe–K line emission nor the high energy continuum could be covered simultaneously. Secondly, the temperature in the remnant is known to vary on spatial scales ≤ 1 arcmin [25] so that plasma parameters determinend from the FPCS with its 3 arcmin by 30 arcmin aperture can only be taken as averages. To pin down the physical parameters of the remnant, high spectral as well as spatial resolution is necessary. As the radial variations of the equivalent widths of the lines are equivalent to their temporal evolution, i.e. their ionisation time $n_e t$, their measurement together with a well defined continuum flux give strong model constraints on the temperature - density - structure of the remnant. A spectral mapping of Tycho with the high spatial resolution assumed in Figure 2 across the remnant would allow a determination of the ionisation history of the matter and, vice versa, give the physical information required to model the object reliably.

3. Non–equilibrium Ionisation

For the modelling and interpretation of the observed spectra several simplifying assumptions have been made in the past: 1. the explosion proceeds spherically symmetric (1–D) into a homogeneous circumstellar medium; 2. the shock parameters are given by a hydrodynamic Sedov–Taylor similarity solution; 3. the plasma is everywhere in thermodynamic equilibrium and 4. the material is always in local ionisation equilibrium at the corresponding temperature.
As result, in general, quite high X–ray emitting masses for the SNR have been estimated, the spectra were described by the superposition of two thermal components (one at low temparatures of about 0.2 – 0.5 keV, the other at a few keV) and a overabundance of oxygen–burning elements as well as an underabundance of iron was found.

The validity of each of the above listed assumptions has been questioned very early on, mainly from theoretical arguments. Especially the assumptions of LTE and local ionisation equilibrium seem not to hold, a view supported by experimental findings. It is well known that the ionisation state of the plasma in the outgoing shock front is not at its equilibrium value at the corresponding temperature, because the time scale to ionize the heavy elements is of the order of $10^4/n_e$ years [26], larger than the age of the remnant.

It seems that the only viable way to obtain reliable estimates of the SNR parameters is to follow simultaneously the hydrodynamical and ionisational evolution of the expanding remnant numerically, starting with the exploding star as initial condition. Given an evolutionary model of a supernova explosion (with well determined chemical composition, density and velocity profiles) the only free parameters for the calculations are the physical and chemical structure of the circumstellar medium. Evolving the hydro – ionisation – structure up to the present age of an experimentally well studied (historical) SNR should allow a determination of the state of the ambient medium — if the initial model is chosen correctly at all!

However in practice several problems have to be overcome (for details see [27]). First of all, due to computational limitations, we are presently only able to run these complex programs for one–dimensional problems, i.e., for nearly spherical remnants. Secondly, as the CPU time spent is directly proportional to the age of the SNR, only relatively young objects can be modelled, which leaves us with few remnants of type I (SN1006, Kepler, Tycho). The second class of problems relates directly to the underlying model. Type I supernovae are thought to be caused by exploding white dwarfs where a deflagration front, starting at the centre of the white dwarf disrupts the whole star [21]. The main uncertainty in these models is, however, the velocity of the deflagration front which, at present, is not calculated from first physical principles but is simulated by a time dependent mixing length theory of convection with a free paramater α, defined as the ratio of the mixing length to the pressure scale height (in particular, the above mentioned W7 model is characterized by $\alpha = 0.7$).

Secondly, the required mixing of the elemental abundances of the outer zones of the model in order to reproduce the observed X-ray spectrum (as well as to fit the observed optical spectra of SN 1981B [28]), indicate that some kind of mixing of matter during the stellar disruption has occurred. This effect is expected because of the decaying nature of the deflagration wave and the hydrodynamics has to be simulated by an at least 2-dimensional hydrocode. In [27] it is shown that the Fe–line region in the observed spectrum of Tycho cannot be fitted at all, if not a substantial mixing of the initial model is done: the calculated equivalent width of the Fe–K line is far too low, despite the fact that the W7 model contains $\sim 0.6\ M_\odot$ of iron ! Satisfactory agreement with the total observed spectrum can, however, be achieved if burned material is mixed into the outer zones of the model containing unburned C and O. It is perhaps worth mentioning that in this case less than $0.1\ M_\odot$ of iron are participating in the X–ray emission, the rest is cold, invisible, demonstrating that there is no "missing iron problem" for type I SNR. In figure 3 we show a simulated photon spectrum, integrated over the whole remnant, which is (when folded through the EXOSAT GSPC detector) in excellent agreement with the observed spectrum of Tycho.

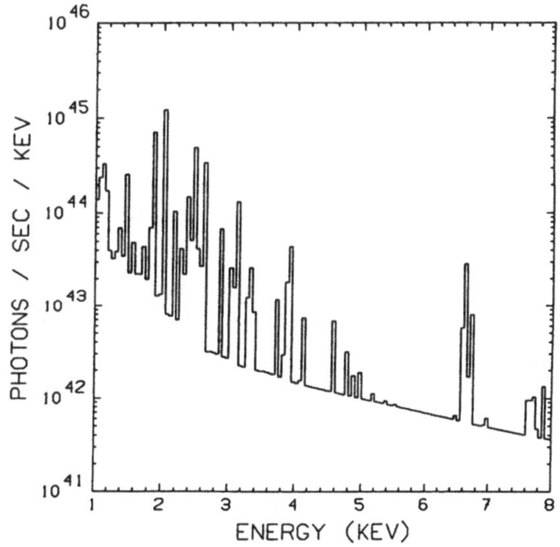

Figure 3 : Simulated photon spectrum, Nomoto's W7 model (see text)

However, there are certain discrepancies between the measured and the calculated spectrum which raise questions related to technicalities (hydro–code) and basic physical uncertainties [27].

Non equilibrium ionization calculations require exact reaction rates for all elements, which in many cases exist only empirically (see [29] and references therein). The biggest uncertainty, however, seems to be the state of the electron gas in the shock front itself: The timescale to equilibrate ion and electron temperatures is of the order of $m_i/m_e\ T_i^{3/2}/n_o$, that is 3×10^5 years for $T_i = 10^8$ and a particle density of $n_o = 10\ cm^{-3}$. McKee [30] suggested that plasma instabilities would cause a much more rapid equilibration, but no specific processes have been worked out. It should be noted that, even if these instabilities are sufficiently fast, they will very likely not produce Maxwellian distribution functions - as always assumed in the determination of the ionization rates.

The shock fronts entering the calculations are assumed to be hydrodynamical, i.e. the flow variables ρ and v vary discontinuously. The jumps are determined from the usual Rankine - Hugeniot conditions, relating the flow variables on either side of the discontinuity, possibly augmented by some numerical viscosity introduced by the hydro codes. It is, however, not clear at all how the structure of a high Mach number ($M \gtrsim 100$) nearly collisionless, ionizing shock is. Particle simulations are at present not able to reach these high Mach numbers at all.

The flat, power law type radio spectrum of most SNR demonstrates the additional presence of relativistic electrons and magnetic fields in the shock regions. Finally, inhomogeneities in the ejecta and in the ambient medium, as well as hydrodynamical instabilities in the outflow would require in many cases two - or three dimensional numerical modelling.

These difficulties indicate that, even given a perfectly well measured X-ray spectrum, the modelling of the supernova remnant expansion incorporates various theoretical uncertainties. It seems as if the gross featureslef the supernova evolution can be represented adequately, but details, requiring an accurate parameter determination which, in turn, can be used to discriminate between different models and possibilities are so far hidden in the 'noise' of imperfect theoretical knowledge.

4. Conclusions

We have shown that X-ray spectra with high spectral and spatial resolution are the best way to obtain better information about both types of supernova remnants.

For SNR of type I explosions it was demonstrated that X-ray spectra integrated over the whole remnant do not reveal the physical parameters of a supernova remnant with the required accuracy. Radially resolved spectra, together with a detailed hydrodynamic simulation and a simultaneous calculation of the ionisation structure seems to be the only way to obtain "physically reliable" parameters for these objects.

Synchrotron nebulae whose best and most extensively studied prototype is the Crab nebula present still a major theoretical challenge, closely linked to our general understanding of the pulsar phenomenon itself.

References

1. S. van den Bergh, 1983, In: Supernova remnants and their X-Ray Emission, IAU symposium no. **101**, eds. J. Danziger and P. Gorenstein, Reidel Publ., Dordrecht, p. 597
2. J. Danziger and P. Gorenstein, Eds., Supernova remnants and their X-Ray Emission, IAU symposium no. **101**, Reidel Publ., Dordrecht, 1983
3. K. Weiler, 1983, In: Supernova remnants and their X-Ray Emission, IAU symposium no. **101**, eds. J. Danziger and P. Gorenstein, Reidel Publ., Dordrecht, p. 299
4. F. Seward, 1988, In: Supernova Remnants and the Interstellar Medium, IAU colloquium no. **101**, eds. R.S. Roger and T.L. Landecker, Cambridge Univ. Press, p.115
5. J.D. Scargle, 1969, Ap.J. 156, p. 401
6. B. Aschenbach and W. Brinkmann, 1975, Astron. Astr. 41, p. 147

7. W. Brinkmann, B. Aschenbach and A. Langmeier, 1985, Nature 313, p. 662

8. K. Makishima, Y. Ogawara, M. Matsuoka, M. Oda, S. Miyamoto, R.M. Pelling, L.E. Peterson, and W.S. Penciesas, 1981, Space Sci. Rev. 30, p. 259

9. M.J. Rees and J.E. Gunn, 1974, Mon. Not. R. astr. Soc. 167, p. 1

10. C.F. Kennel and F.V. Coroniti, 1984, Ap.J. 283, p. 694

11. F.C. Michel, 1985, In: The Crab Nebula, ed. M. Kafatos, Cambridge Univ. Press, p. 55

12. F.D. Seward, 1985, Comments Astrophys. 11, 15

13. E. Trussoni, W. Brinkmann, H. Ögelman, and G. Hasinger, 1988, submitted to Astron. Astr.

14. O.C. De Jager, B.C. Raubenheimer, A.R. North, H.I. De Jager, and G. van Urk, 1988, submitted to ApJ.

15. P. Gorenstein, F. Seward, and W. Tucker, 1983, In: Supernova remnants and their X-Ray Emission, IAU symposium no. **101**, eds. J. Danziger and P. Gorenstein, Reidel Publ., Dordrecht, p. 1

16. K. Nomoto, 1980, In: Type I Supernovae, ed. by J.C. Wheeler, Austin, Univ. Texas, p. 164

17. R.H. Becker, S.S. Holt, B.W. Smith, N.E. White, E.A. Boldt, R.F. Mushotzky, and P.J. Serlemitsos, 1980, Ap.J. 240, L5

18. S.H. Pravdo, B.W. Smith, P.A. Charles, and I.R. Tuohy, 1980, Ap.J. 235, L9

19. A.J.S. Hamilton, C.L. Sarazin, and A.E. Szymkowiak, 1986, Ap.J. 300, p. 713

20. H. Tsunemi, K. Yamashita, K. Masai, S. Hayakawa, and K. Koyama, 1986, Ap.J. 306, p. 248

21. K. Nomoto, F.K. Thielemann, and K. Yokoi, 1984, Ap.J. 286, p. 644

22. H. Itoh, K. Masai, and K. Nomoto, 1988, In: Supernova Remnants and the Interstellar Medium, IAU colloquium no. **101**, eds. R.S. Roger and T.L. Landecker, Cambridge Univ. Press, p.149

23. W. Brinkmann and H.H. Fink, 1988, In: Supernova Remnants and the Interstellar Medium, IAU colloquium no. **101**, eds. R.S. Roger and T.L. Landecker, Cambridge Univ. Press, p.137

24. P.F. Winkler, C.R. Canizares, and B.C. Bromley, 1983, In: Supernova remnants and their X-Ray Emission, IAU symposium no. **101**, eds. J. Danziger and P. Gorenstein, Reidel Publ., Dordrecht, p. 245

25. B. Aschenbach, 1985, Space Sci. Rev. 40, p.447

26. C.R. Canizares, 1984, In:X-ray Astronomy '84 , ed. by Y. Tanaka, Bologna

27. W. Brinkmann, H.H. Fink, and A. Smith, 1988, submitted to Astron Astr.

28. D. Branch, 1984, In: Workshop on Challenges and New Developments in Nucleosynthesis, Chicaco, Univ. of Chicago Press, in press

29. R. Mewe, 1987, Contributions presented at the 33rd Scottish Universities Summer School in Physics at St. Andrews, SRON preprint

30. C.F. McKee, 1974, Ap.J. 188, p. 335

DISCUSSION

Green: Just a comment about the fact that you didn't see any X–ray emission from W50. You may be able to explain that by saying W50 is not a supernova remnant.

Kundt: For heaven's sake..

Claas: What is the main reason for using Nomoto's model ?

Brinkmann: It has been successfully applied to other type I SNR and ist so far the best model for that kind of event.

Shock-Cloud Interactions in Supernova Remnants.

S.A.E.G.Falle & J.R.Giddings,
Department of Applied Mathematical Studies,
The University, Leeds LS2 9JT, U.K.

1. Introduction.

It is now generally accepted (McKee 1987) that the interstellar medium is inhomogeneous on a range of scales from kiloparsecs down to fractions of a parsec. These inhomogeneities will affect the dynamics of a supernova remnant once it has swept up a significant fraction of the ejected mass. If one assumes that magnetic fields do not influence the dynamics, then the only kinds of inhomogeneity that matters are density inhomogeneities, at least as long as the shocks are strong.

For most purposes we can divide these density variations into three kinds, those with scales of the same order as the size of the remnant, those somewhat smaller than the remnant and those too small to be observed. In this article we will not discuss the large scale inhomogeneities since they have already been extensively studied (e.g. Chevalier & Gardner 1974; Falle, Garlick & Pidsley 1984; Tenorio-Tagle, Rozycka & Yorke 1985; MacLow et. al. 1987). We will also ignore clouds so small that neither they nor the shocks resulting from their interaction with the blast wave can be seen. Instead we will concentrate on clouds of intermediate size where there is some chance that the shocks can be directly observed.

If we want to see such interactions occurring, then we should look at older remnants because the shocks are likely to be radiative and so can more easily be observed. Clearly the Cygnus Loop is an ideal candidate since it is nearby and there exists an enormous amount of high quality data in all conceivable wavebands. In fact some of the filamentary structures described by Braun (these proceedings) may be evidence for the kind of process we will be considering.

2. Radiative Shocks.

The filaments in old remnants such as the Cygnus Loop, IC443, S147 etc are usually interpreted as radiative shocks, an idea first explored in detail by Cox (1972) and subsequently by many others. There now exist very detailed calculations of the optical and UV spectrum emitted by plane steady shocks under various circumstances (e.g. Shull & McKee 1979). The success of these models is such that it seems very likely that the observed filaments do indeed correspond to the cooling region behind radiative shocks.

However, there are a number of discrepancies (e.g. Fesen, Blair & Kirshner 1982; Braun these proceedings) that suggest that steady shocks are not the whole story. This is not really surprising since not only are there circumstances, such as the presence of clouds, when the shock speed changes significantly during a cooling time, but there is also an instability which sets in above a shock speed of about 130 km s^{-1} (Innes, Giddings & Falle 1987). Clearly we will have to do better than steady shock models if we want to understand what is really going on.

In the Cygnus Loop, and to a lesser extent in other remnants, the observations now suggest that much of the filamentary structure is due to shocks interacting with clouds which are considerably smaller than the remnant, but large enough to be resolved (Cox 1987). This is in contrast to previous ideas which involved large numbers of very small (< 0.01 pc) invisible clouds (McKee & Cowie 1975; McKee, Cowie & Ostriker 1978). There are a number of objections to the small cloud model, but perhaps the most serious is the existence of continuous changes in spectrum along several filaments in the Cygnus Loop. This is exactly what one would expect if the shocks suffer diffraction and reflection due to density variations on scales of a parsec or so.

Clearly if we want to model this kind of thing in any detail, we need to carry out calculations of the interaction of radiative shocks with clouds of various shapes and densities. There have been a number of such calculations (Chevalier & Theys 1975; Tenorio-Tagle & Rozycka 1984), but they are not particularly reliable largely because the grids are not fine enough to resolve the cooling region. For typical parameters in the Cygnus Loop the cooling length is about a hundredth of the cloud size and no two dimensional calculation to date has come anywhere near the numerical resolution required to simulate this.

Nevertheless, it is now becoming possible to attempt such calculations. What is needed is adaptive grid algorithms which achieve high resolution in the cooling region without degrading the accuracy elsewhere. If implemented on parallel machines based on transputers, these algorithms should be able to simulate two and three dimensional radiative flows reliably. At present we are working along these lines and figure 1 shows some preliminary results for a steady adiabatic flow past a blunt body. This is admittedly a simple problem, but the efficiency is such that these calculations can be performed 50 to 100 times faster than by conventional methods and the memory requirements are reduced by about a factor of 10.

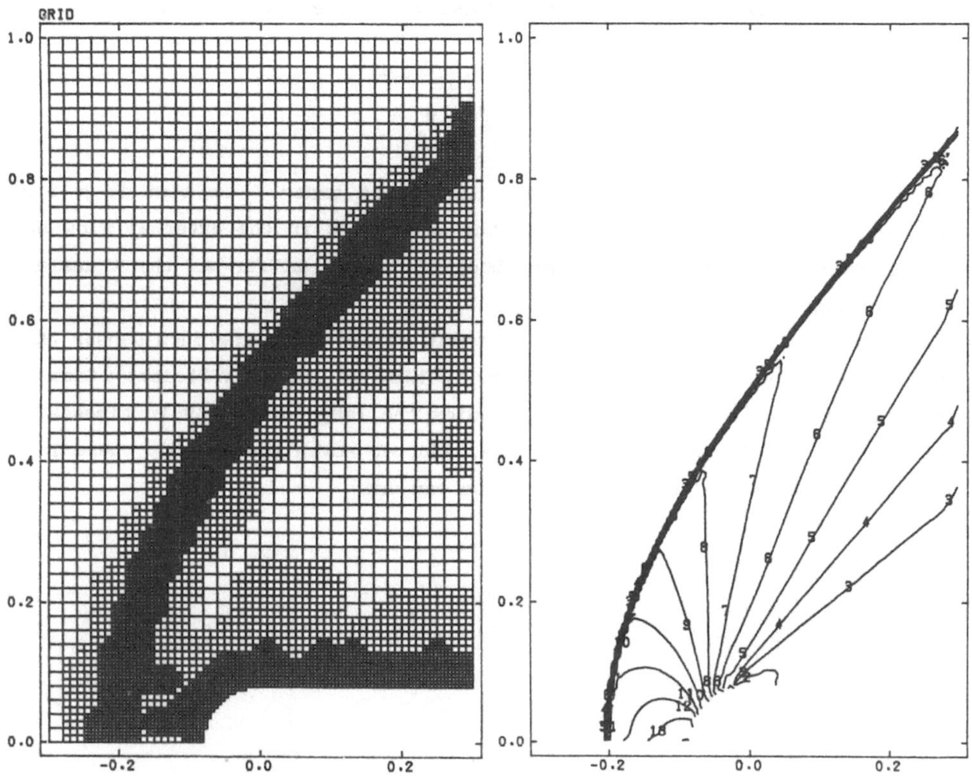

Grid

Pressure Contours

Figure 1. Adaptive 4-level multigrid applied to steady flow past a two dimensional round nosed body. The free stream Mach number is 2.0. The two finest grids are 60x100 and 120x200 but these cover only part of the domain. The integrator is a second order Eulerian Godunov scheme.

These techniques are still at an experimental stage and as yet we have not applied them to time dependent multidimensional problems. However, the principles upon which the algorithms are based are firmly established and we expect great things of them in the near future.

3. Shock Interactions with Rigid Objects.

Let us consider a plane shock with speed V_e in gas with density ρ_e

incident on a stationary cloud with density ρ_c. The speed V_c of the shocks driven into the cloud will be roughly

$$V_c \propto \left[\frac{\rho_e}{\rho_c}\right]^{1/2} V_e .$$

The material speeds in the cloud will also be of order V_c so that if $\rho_c \gg \rho_e$ the cloud deforms slowly compared to the velocities in the external flow and we can regard the cloud as a rigid body, at least as far as the transient stage is concerned.

Because of this it makes sense to treat clouds which are much denser than the ambient medium as rigid bodies. This is more efficient from a computational point of view since it avoids the large disparity in timescales between the interior and exterior flows. Another advantage is that there is a wealth of experimental data upon which we can draw.

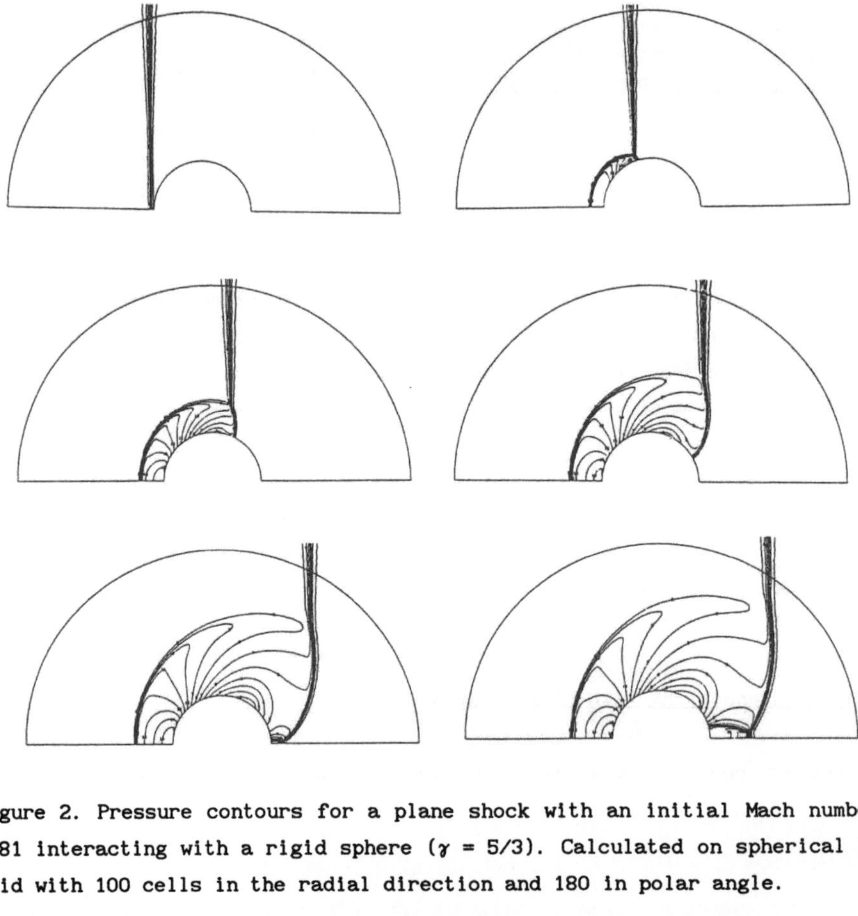

Figure 2. Pressure contours for a plane shock with an initial Mach number of 2.81 interacting with a rigid sphere (γ = 5/3). Calculated on spherical polar grid with 100 cells in the radial direction and 180 in polar angle.

3.1.1 Rigid Sphere.

An obvious case to look at is that of a plane shock hitting a rigid sphere. Figures 2 and 3 show the results obtained with a second order Eulerian Godunov scheme on a high resolution spherical polar grid. The advantages of such a grid are that not only can the sphere be represented exactly, but also that the highest resolution is near the surface of the sphere where most of the action takes place. There is, however, some degradation of the solution due to the non-uniformity of the grid.

Figure 2 shows the sequence of events. Initially there is a regular reflection at the surface which evolves into a Mach reflection at $\theta \simeq 135^0$ (here θ = 0 corresponds to the direction of motion of the incident shock). The reflected shock associated with this eventually becomes a stationary bow shock in this case since the flow behind the incident shock is supersonic in the frame of the obstacle.

Further round the sphere the incident shock is diffracted into a funnel shape similar to that described by Braun (these proceedings) before reflecting off the symmetry axis. Because of the axial symmetry this is a Mach reflection from the very beginning. The Mach disc which moves down the axis initially has zero size, but it grows enough to become significant before the triple point associated with it disappears and it merges with the incident shock.

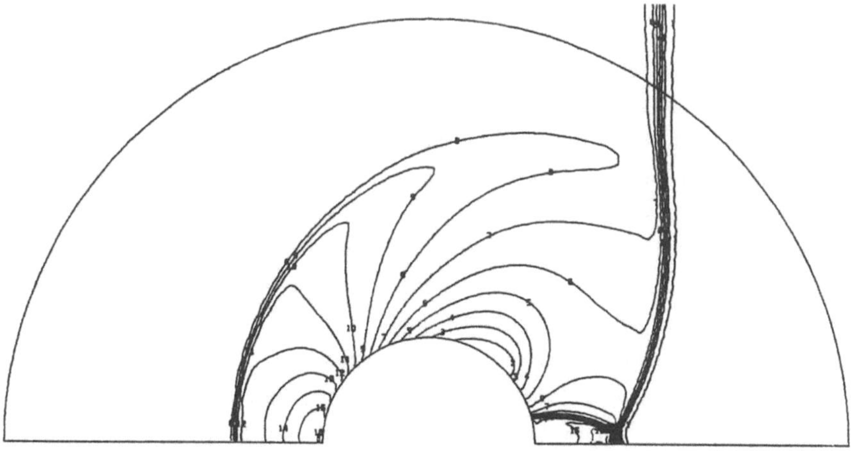

Figure 3. Detail of pressure contours at the latest time calculated.

The solution at the latest time is shown in greater detail in figure 3. This should be compared with the experimental results obtained by Bryson & Gross (1961) for shock interaction with a rigid sphere in air. Most of the features of the experiment are well reproduced by the simulation, the exception being the vortex formed when the slip line produced by the first Mach reflection rolls up. It would, however, be too much to expect to get this sort of thing right at this resolution.

We must now consider how much of this is relevant to supernova remnants. Clearly the flow is essentially the same as the one that arises when a non-radiative shock encounters a very dense spherical cloud and so our results can be used to interpret observations of such shocks (Raymond, Davis, Gull & Parker 1980). The first thing to note is that large portions of both the reflected and diffracted shocks are oblique. This means that the usual practice of interpreting the observations in terms of normal shocks can be very misleading. Fortunately, for non-radiative shocks it is possible to calculate the flow as we have done here and then use the appropriate physics to deduce the spectrum and velocity dispersion. In this way we ought to be able to get a good idea of the nature of some of the filaments in, for example, the Cygnus Loop.

If the shocks are radiative, then life becomes much more difficult. As we pointed out earlier, we cannot as yet compute such flows reliably and so it is probably better to try and guess how cooling modifies the flow. We have little chance of doing this if the cooling length is of the same order as the size of the obstacle or if the shock speeds are in the range in which the radiative instability is important, but we can do something if the cooling length is very short and the shock is stable. The shocks are then isothermal and to some extent one can get an idea of the trends by looking at lower values of the ratio of specific heats γ.

Reducing γ increases the shock compression so that the bow shock is closer to the surface of the sphere. Its shape is therefore similar to that of the obstacle and, since this is true generally, we should be able to deduce something about the shape of a dense cloud if we can identify the bow shock. Another consequence of the greater compression is that the transition to Mach reflection is less likely. For a sphere a Mach shock will always occur both on the surface and on the axis downstream, but as the shock compression tends to infinity the point at which Mach reflection starts approaches $\theta = 90^0$ and the size of the Mach disc on axis tends to zero.

It should be possible to detect the presence of these Mach shocks in supernova remnants provided the cloud is not too small. What we should look for is filaments which split into two and in which the spectrum indicates the existence of three distinct shock velocities. There should also be sharp discontinuities in velocity and temperature which are not associated with pressure jumps since Mach shocks are always accompanied by slip lines across which there is a jump in tangential velocity and temperature but the pressure is continuous.

Finally, although we have assumed that the cloud is rigid, we can say something about what happens to it if we look at the pressure distribution on

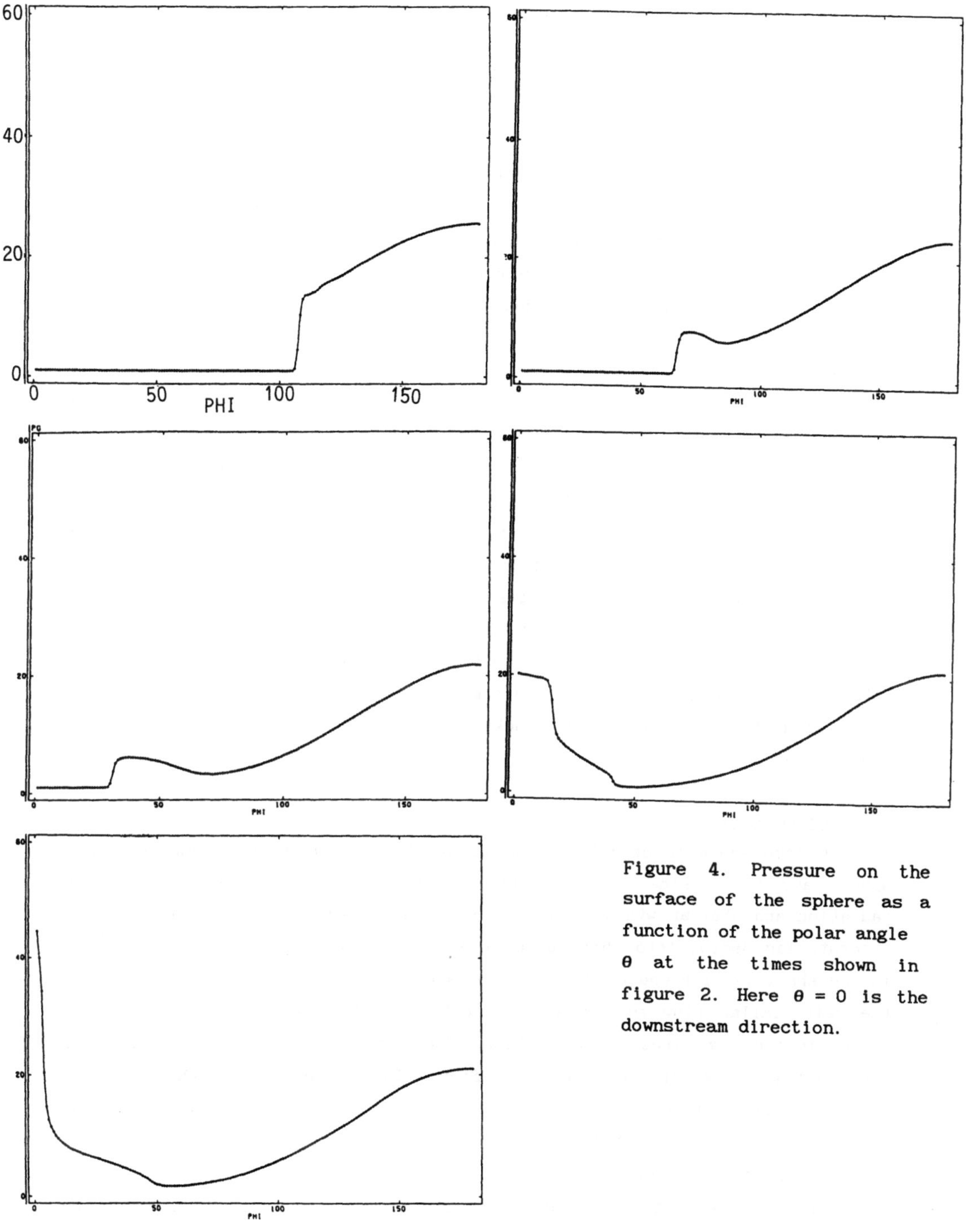

Figure 4. Pressure on the surface of the sphere as a function of the polar angle θ at the times shown in figure 2. Here θ = 0 is the downstream direction.

its surface. Figure 4 shows the surface pressure at the same times as those in figure 2 (with the exception of the earliest time). The most obvious feature is the sharp rise in pressure associated with the incident shock and its reflection on the axis downstream. The lowest pressure occurs at $\theta \simeq 90$ i.e. at the side of the cloud. Because of this the shocks in the cloud will be strongest at the front and back and weakest at the side. If these shocks radiate strongly enough to be seen, they will give rise to filaments oriented perpendicular to the direction of motion of the incident shock. This is of course precisely what is generally observed in the Cygnus Loop but not in S147. This suggests that in the Cygnus Loop we are mostly seeing the shocks in the clouds and in S147 those in the exterior flow.

We would also like to know the eventual fate of the cloud. Initially it is compressed in all directions, but mostly along the symmetry axis so that it becomes disc shaped. Since we are assuming that the cloud does not change its shape significantly during the transient stage of the external flow, we really ought to look at the steady state pressure distribution if we want to know how the cloud deforms. This stage has not been reached in these calculations, but the trend is already clear. The minimum pressures are at the sides and so the cloud will expand sideways. This itself accentuates the pressure difference between the axis and the sides and since these are of the order of the post shock pressure in the cloud, the cloud disrupts on its sound crossing time. This confirms the conclusion that Nittmann, Falle & Gaskell (1982) drew from their simulations of adiabatic shocks hitting clouds with moderate density contrasts.

3.1.2 Knife Edge.

In some sense an object with a sharp edge is the opposite extreme to a sphere and is of some relevance to supernova remnants since the ionising radiation and stellar winds from the supernova progenitor should compress the interstellar medium into thin dense sheets. Again we assume that the density in the sheet is high enough for it to behave like a rigid body. Figure 5 shows the self-similar flow set up when a weak shock hits a knife edge which is perpendicular its direction of motion. These results are also in excellent agreement with experimental results (figure 6) as far as the shape of the reflected and diffracted shocks are concerned. In this case the simulation does contain a vortex, but its behaviour does not really resemble that of the real vortex very closely.

Vortices are a common feature of shock interactions and it would be interesting to see whether there is any evidence for their presence in supernova remnants. In a compressible vortex both the pressure and density can be very low in the centre and so we would expect them to be faint in the

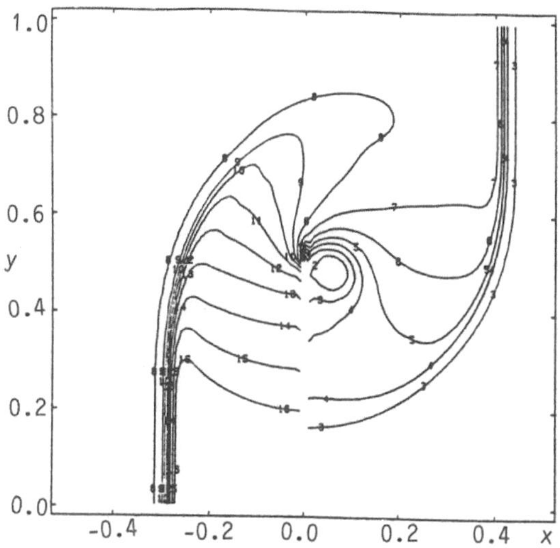

Pressure Contours

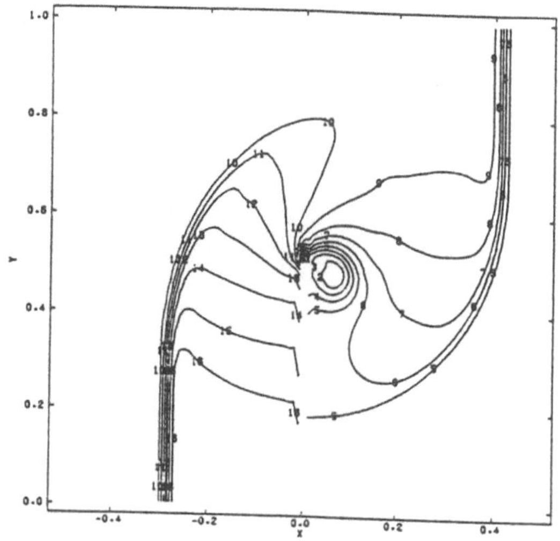

Density Contours

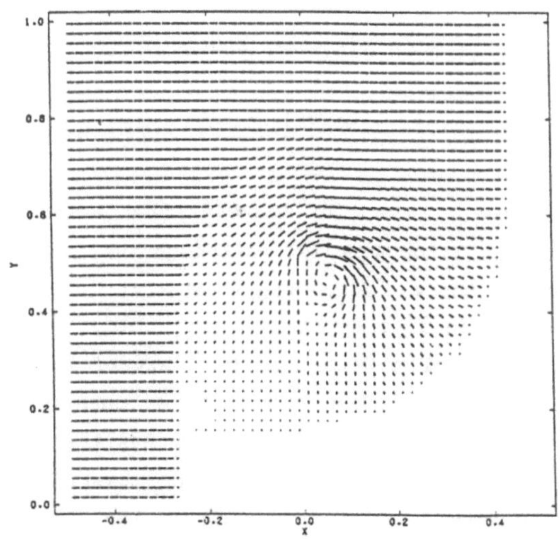

Velocity Vectors

Figure 5. Plane shock with an incident Mach number of 1.38 incident on a knife edge calculated on a 100x100 grid.

optical, UV and X-ray. On the other hand, if the magnetic field is passively convected, it will be stretched and amplified by the differential rotation and this effect is greatest in the eye of the vortex. It follows that the characteristic feature of a vortex is that it is faint in the optical, UV and X-ray, but visible in radio. Mufson et. al. (1986) have obtained a high resolution map of IC443 with the VLA and although the optical and radio filaments generally correlate very well, there are places where they do not.

Obviously one cannot deduce the presence of vortices from this alone, but it should be possible to construct a reasonable model of the radio emission from a vortex which could then be checked against some of these observations.

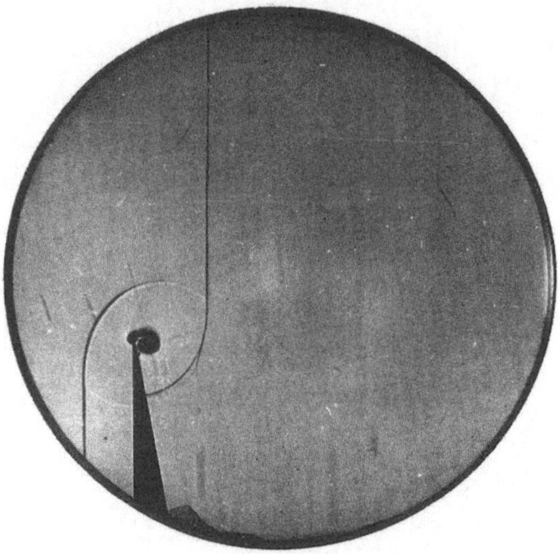

Figure 6. Schlieren photograph of a shock with a Mach number of 1.38 incident on a knife edge (Evans 1977).

4. Conclusions.

In this article we have discussed some of the flow features that arise when a shock interacts with a density inhomogeneity. Although we can calculate the interaction of adiabatic shocks with rigid objects with reasonable precision, we have as yet not been able to do the same thing for radiative shocks hitting clouds of finite density. Nevertheless, many of the features expected in the astrophysical flows are present in our calculations, and observations of old remnants, particularly the Cygnus Loop and IC443, are now good enough for it to be worthwhile looking for some of these effects. At least we are likely to make more progress in this way than by trying to understand all the observations in terms of steady, normal radiative shocks.

5. References.

Bryson, A.E. & Gross, R.W. 1961. J. Fluid Mech., 10, 1.
Chevalier, R.A. & Gardner, J. 1974. Astrophys. J., 192, 457.
Chevalier, R.A. & Theys, J.C. 1975. Astrophys. J., 195, 53.
Cox, D.P. 1972. Astrophys. J., 178, 143.

Cox, D.P. 1987. IAU Colloquium 101, Penticton, Canada. Eds. R.S. Roger & T.L. Landecker, p.73.

Evans, R.A. 1977. PhD Thesis, Leeds University, p.????

Falle, S.A.E.G., Garlick, A.R. & Pidsley, P.H. 1984. Mon. Not. R. astr. Soc., 208, 925.

Fesen, R.A., Blair, W.P. & Kirshner, R.P. 1982. Astrophys. J., 262, 171.

Innes, D.E., Giddings, J.R. & Falle, S.A.E.G. 1987. Mon. Not. R. astr. Soc., 226, 67.

MacLow, M.M., Norman, M.L. & McCray, R.A. 1987. IAU Colloquium 101, Penticton, Canada. Eds. R.S. Roger & T.L. Landecker, p.461.

McKee, C.F. 1987. IAU Colloquium 101, Penticton, Canada. Eds. R.S. Roger & T.L. Landecker, p.205.

McKee, C.F.& Cowie, L.L. 1975. Astrophys. J., 195, 715.

McKee, C.F., Cowie, L.L. & Ostriker, J.P. 1978. Astrophys. J., 219, L23.

Mufson, S.L., McCollough, M.L., Dickel, J.R., Petre, R., White, R. & Chevalier, R.A. 1986. Astron. J., 92, 1349.

Nittmann, J., Falle, S.A.E.G. & Gaskell, P. 1982. Mon. Not.R. astr. Soc., 201, 833.

Raymond, J.C., Davis, M., Gull, T.R. & Parker, R.A.R. 1980. Astrophys. J., 238, L21.

Shull, J.M. & McKee, C.F. 1979. Astrophys. J., 227, 131.

Tenorio-Tagle, G. & Rozycka, M. 1984. Astron. & Astrophys., 137, 276.

Tenorio-Tagle, G. & Rozycka, M. & Yorke, H.W. 1985. Astron. & Astrophys., 148, 52.

Acknowledgements.

S.A.E.G. Falle would like to thank the Volkswagen Stiftung for their hospitality during this workshop and J.R. Giddings acknowledges support from the University of Leeds during the course of this work. The computations were carried out on the Vax 8600 in the Department of Mechanical Engineering at the University of Leeds.

INTERPRETATION DIFFICULTIES OF SNR SHOCK SPECTRA

D.E. Innes
Max-Planck-Institut für Kernphysik,
Postfach 10 39 80, D-6900 Heidelberg

Abstract

Emission line ratios of unsteady radiative shocks are shown to be indistinguishable from those of steady shocks with velocities about 100 km s^{-1}, although shock parameters such as velocity and preshock density are significantly different. If the shocks are radiatively unstable, then it may be possible to observe time variations of line ratios in SNRs over a period of a few years.

1. Introduction

The gas dynamics possibly responsible for the bright optical filaments that are seen in supernova remnants (SNRs) have, to date, been seriously overlooked. The reason may be due to the partial early success of steady flow models of plane-parallel radiative shocks in explaining the observed line ratios [1,2,3]. However now that higher resolution observations at UV and optical wavelengths are available, it has become clear that non-steady flow should be considered [4,5,6,7].

The filaments in SNRs such as Cygnus Loop, IC443 or Vela SNR appear to be long thin sheets [8,2] or ropes [9,10,11] of optical emission, anti-correlated on scales of less than a parsec with patches of X-ray emission [12]. They often have an arc-like appearance, with a curvature significantly smaller than the remnant's radio shell. These arcs usually curve outwards although inwardly curved as well as straight filaments aligned along and perpendicular to the shell, are seen.

Sheet-like filamentary structure may result from the interaction of a shock wave with clouds [13,14,15,16]. When a plane shock hits a cloud a reflected shock may be produced at the front of the cloud. On passing the cloud, the shock wave is diffracted which may lead to a complex structure of shocks, sliplines and vortices in the wake of the cloud. In addition slower radiative shocks will be driven into the cloud. So far this has only been calculated for cases in which the shocks are adiabatic. If these shocks are radiative, then there is the possibility of additional shocks being formed due to unstable cooling in the postshock cooling layer [17,18,19,20,21]. Even in the planar calculation unsteady cooling produces multiple shocks and up to three distinct velocity components [22]. This leads one to conjecture that unsteady cooling behind shocks propagating in media with small density variations, without invoking cloud interaction, may be an alternative interpretation of the filamentary structure.

To compute emission line spectra appropriate to any of these cases would be rather difficult and may, as we shall see, not be very useful. Nevertheless attempts have been made. The first approach is to assume that the structure of the shocked gas is the same as that of an incomplete steady shock [23,24,7]. This model has the advantage of being straightforward to calculate and computed line ratios seem to fit the observations. Another approach is to compute the time evolution of radiative shocks. For example, calculations have recently been made for plane-parallel unsteady radiative shocks moving into uniform media [25,26]. Although the geometry and external conditions assumed in these calculations is considerably simpler than one might expect in a SNR shell, the variation in line ratios is more consistent, than results from steady flow calculations, with the scatter observed (*c.f.* [5]).

Recent analysis has shown (Falle, private communication) that the criterion for unstable cooling in the postshock gas is that the logarithmic slope of the cooling function with temperature be less than 1/2. This criterion is satisfied for shocks faster than 130 km s^{-1} assuming monatomic gas of solar abundances [26]. However, it does not mean that all shocks with line ratios matching steady shocks slower than 130 km s^{-1} are stable since, as shown in §3, the effect of the instability is to create flow configurations with the same line ratios as those predicted by steady models.

As an illustration of how one might be misled into believing that one is seeing a steady shock, we compare the spectrum of a steady 100 km s^{-1} shock with spectra of unsteady shocks. One consequence of unsteady behaviour is that for the same line ratios the face-on intensity can be up to 60 times larger than for its steady shock counterpart. Hence estimates of preshock density based on observations of surface brightness and steady shock models could be much too large.

2. Description of models

The unsteady structures are taken from the evolution of a shock with velocity 200 km s^{-1} driven by a constant pressure piston into a density 1 atom cm^{-3} and solar abundances [27]. This evolution is described in detail in [26] (their model D200A). Unsteady cooling results in the formation of secondary shocks in the postshock cooling flow. A lucid description of the secondary shock formation process is given in [17,18] and more specifically in the context of plane-parallel radiative shocks in [26]. In spite of an initial impression that the evolution was chaotic, further computations reveal cyclic behaviour [28], illustrated in Fig. 1 through the time evolution of primary shock velocity.

As stated in the introduction line ratios from the unsteady structures are compared to those of a steady 100 km s^{-1} shock. Intuitively one would expect that when the velocity of the leading shock in the unsteady evolution is about 100 km s^{-1} the line ratios would be roughly similar to the steady 100 km s^{-1} shock. We see from Fig. 1 that, in the evolution considered here, the leading shock has this velocity for at most one tenth of its cycle time. So one may argue that statistically these shocks would be quite rare. On the other hand the emission is about ten times brighter during this phase (*c.f.* [26] Fig. 10), so from the observational point of view these are probably the most relevant times.

We therefore select a few representative shock configurations from this phase in the cycle. Fig. 2 shows their temperature structures. In each case two shocks, S_1 and S_2, both propagating in the

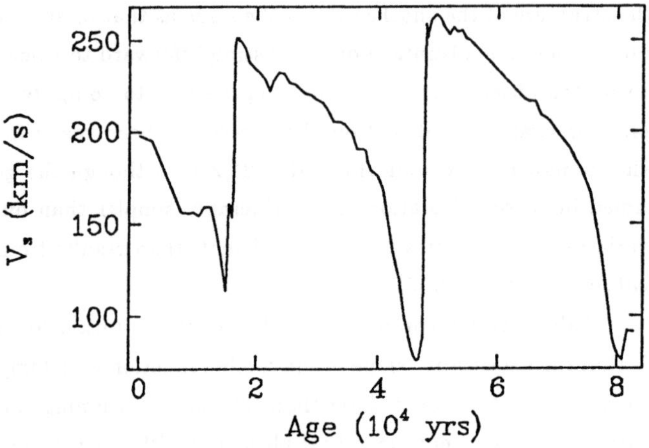

Fig. 1: Time variation of primary shock velocity during the evolution of a 200 km s^{-1} unsteady shock.

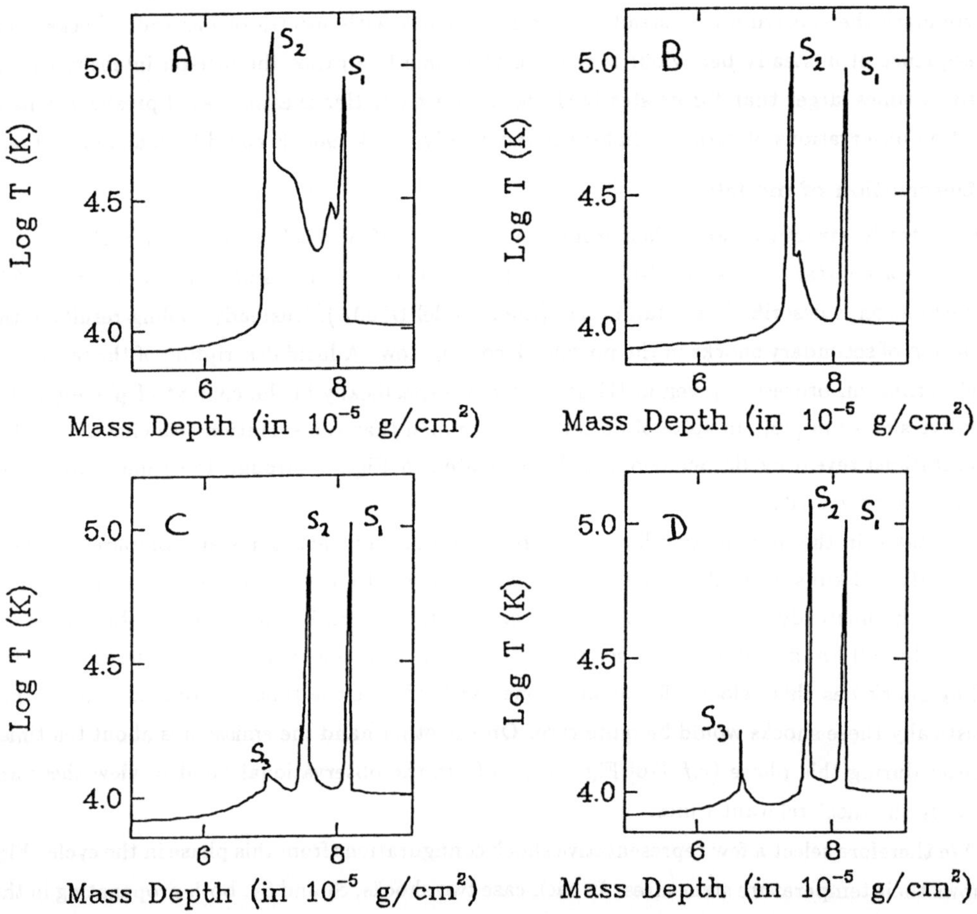

Fig. 2: Temperature structures of the unsteady shock models referred to in the text. S_1, S_2 and S_3 indicate the leading, second and backward shocks.

forward direction are clearly seen. Time 'A' is just before the gas behind the first shock, S_1, has cooled to 10^4 K. Time 'B' is soon after this gas has cooled. By time 'C' unstable cooling behind the second shock, S_2, has produced two more shocks, of which only the backward travelling shock, S_3, is visible. Finally, at time 'D' this backward shock has increased in strength and the forward shock that was formed at the same time, has reinforced S_2 causing its temperature immediately postshock to increase.

The comparison 100 km s^{-1} steady shock is computed with the same physical assumptions as in the dynamical calculations (*i.e.* no magnetic pressure but including an additional heat source at temperatures below 10^4K). For simplicity, full pre-ionization is assumed. The preshock density is 4 atoms cm^{-3} so that the density far downstream is the same for all models.

3. Line spectra

As in earlier work [26,22], line ratios are given with respect to [O II] rather than Hβ because its intensity is not as dependent on mass depth in the recombination region. Table 1 lists line ratios of the steady 100 km s^{-1}, steady 200 km s^{-1} and at various times of the unsteady shock evolution. Also given is shock velocity (V_s), the face-on intensity of [O II] and [O III], line ratios for the density sensitive [O II] and [S II] lines and finally, in the last row, the gas pressure at the position where the [O II] emissivity is a maximum. Results for the unsteady models are shown in two columns. The left hand column gives quantities for the whole structure and the right hand column for gas behind the second shock. In high resolution spectra two components may be visible if the flow direction is not normal to the line-of-sight. When there is a backward shock a third component may be seen due to emission from material upstream of the backward shock; however this is likely to be comparitively weak and for the purpose of this discussion is ignored.

Clearly, the line ratios of the unsteady models seem to be closer to the 100 km s^{-1} than the 200 km s^{-1} steady line ratios. But the remarkable thing is the level of agreement. Line ratios from the second shock in 'B' match the 100 km s^{-1} to within 25% for all lines except [O I] and He [II]. If one just looks at lines formed in doubly and triply charged ions, believed by some to be the best indicators of shock velocity [29,30], then line ratios in 'D' match to within 15%. Many 'fits' of the observations to steady models as, for example given by Raymond *et al.* in [7] can differ by more than a factor 2.

One clue that the shock is unsteady would be a second velocity component in the observed spectrum at a lower velocity. In model 'B' the [O II] intensity is comparable for both components but in 'D' it is 20 times fainter in the lower velocity component and so may not be detected. Thus the absence of a second component in the observed spectrum does not mean that the shock is steady.

Table 1 also shows that line ratios are changing very rapidly during this phase of the shock evolution. Look for example at models 'C' and 'D', where some line ratios change by more than a factor ten in only 20 years. Of course this scales with preshock density but an interstellar density of 1 atom cm^{-3} does not seem unreasonable. Variations may therefore be detectable in only a few years providing one is able to resolve individual filaments and variations are not smeared out by finite light travel times.

Table 1: Spectral characteristics of radiative shock models

	Steady		Unsteady							
Model	a	b	A		B		C		D	
Age (10^4yrs)			4.528		4.706		4.725		4.727	
V_s (km s^{-1})	100	200	82	91	82	76	87	81	87	95
I ([O II]$\lambda\lambda$3727,3730)	5.233E-05	7.818E-05	9.852E-05	7.834E-05	1.357E-04	8.535E-05	1.012E-03	9.691E-04	7.868E-04	7.480E-04
I ([O III]λ5008)	5.431E-06	5.860E-06	2.638E-05	8.497E-06	1.125E-05	9.479E-06	6.062E-05	5.880E-05	8.518E-05	8.306E-05
Hβ	4.60	4.21	4.33	5.16	3.61	4.15	3.25	3.25	6.40	6.56
He II λ1640	3.73	0.662	1.02	0.940	1.37	1.84	1.01	1.02	5.42	5.66
C II] $\lambda\lambda$2324-2339	33.0	19.7	27.5	23.8	25.5	28.9	33.9	34.7	35.4	36.5
C III] λ1908	28.5	10.7	31.0	17.8	24.5	33.4	19.9	20.3	30.0	30.9
[N II] λ6585	17.3	32.3	25.8	29.9	18.7	18.6	21.2	21.2	27.3	27.7
N III] λ1750	2.02	1.25	3.17	1.98	1.67	2.36	1.19	1.22	2.16	2.24
N IV] λ1485	0.981	1.55	1.46	1.47	0.538	0.815	0.123	0.122	1.13	1.18
[O I] λ6302	35.3	12.9	19.1	24.0	13.9	21.1	9.11	9.47	21.9	23.0
[O II] $\lambda\lambda$3727,3730	100.	100.	100.	100.	100.	100.	100.	100.	100.	100.
[O III] $\lambda\lambda$1660,1666	6.85	5.68	11.5	7.75	5.58	7.94	3.65	3.72	7.39	7.66
[O III] λ5008	10.4	7.50	26.8	10.8	8.29	11.1	5.99	6.07	10.8	11.1
O IV] λ1400	2.19	9.17	3.82	3.86	1.07	1.61	0.187	0.181	2.46	2.57
[S II]λ6718/ [S II]λ6732	1.35	1.25	1.33	1.33	1.32	1.32	0.818	0.807	0.721	0.715
[O II]λ3727/ [O II]λ3730	0.77	0.80	0.75	0.77	0.75	0.78	1.33	1.36	1.46	1.52
P_g ([O II]) (dyn cm^{-2})	8.4(-10)	8.4(-10)	6.1(-10)		8.9(-10)		1.0(-8)		2.2(-8)	

4. Estimates of preshock density

It has been suggested [7] that surface brightness measurements of Doppler shifted lines in spectra of filaments can be used to estimate preshock density, n_0, using the relationship

$$\frac{n_0}{n_{model}} = \frac{I_{obs}V_{obs}}{I_{model}V_g}$$

where I_{obs} is the observed line intensity, V_{obs} is the observed Doppler velocity, I_{model} is the computed face-on intensity for a preshock density n_{model}, and V_g is the flow velocity of the line emitting gas. The density estimate therefore depends inversely on the gas velocity and the computed model intensity.

The results in Table 1 enable us to gauge the level of uncertainty related to this density estimate. Structures at times 'B' and 'D' both give the same line ratios as the 100 km s^{-1} steady shock but in case 'B' the [O III] intensity is 1.75 times and in 'D' it is 15.3 times the value of the 100 km s^{-1} shock which has a preshock density 4 atoms cm^{-3}. These numbers are for the second shock so that the flow velocity in the emitting region is 200 km s^{-1} not 100 km s^{-1}. Therefore, if these two structures are analysed on the basis of steady shocks, one might obtain a shock velocity 100 km s^{-1} and preshock densities 14 and 122 atoms cm^{-3}, instead of an unsteady 200 km s^{-1} shock moving into a density of 1 atom cm^{-3}.

The above steady shock arguments have recently been used to estimate the nonthermal component in filaments in the Cygnus Loop [31] by comparing the estimated ram pressure to a postshock thermal pressure. A similar comparison can be made for models 'B' and 'D'. From the previous paragraph, we see that the steady shock arguments would lead one to think that the ram pressure of 'B' is 2.9×10^{-9} dyn cm^{-2} and of 'D' is 2.6×10^{-9} dyn cm^{-2}. Comparing these values with the corresponding thermal pressures given in Table 1, we see that these ram pressures are 3.3 ('B') and 1.2 ('D') times the thermal pressures. In both cases the difference is less than the factor 7 reported in [7], but it is large enough to cast some doubts on the size of the nonthermal component estimated from steady shock analyses.

5. Conclusion

It has been demonstrated that the spectra of unsteady shocks may be the same as those of slower steady shocks and that inferences based on steady shocks may substantially overestimate preshock densities and ram pressures. In an attempt to highlight an aspect of unsteady flow that was not covered in earlier work [26,22,25], we have only looked at a very limited sample of unsteady shock structures. Nevertheless, similar results would have been obtained had other unsteady shock velocities at other phases of their evolution been studied. We therefore believe that these results truthfully illustrate the large uncertainties inherent in radiative shock analyses whenever nonsteady flow is a possibility.

Since there is no obvious way of determining whether a shock is steady from spectra alone and models of unsteady shocks predict such a wide range of spectra it seems that it could be more productive to concentrate effort on understanding the dynamics of radiative shocks rather than detailed modelling of their line spectra.

Acknowledgements

I would like to thank W. Kundt who besides organizing a stimulating workshop encouraged me to write this contribution, and H.J. Völk and W.J. Markiewicz for helpful comments throughout its preparation.

References:

[1] Cox, D.P., 1972, *Astrophys. J.*, **178**, 143.
[2] Miller, J.S., 1974, *Astrophys. J.*, **189**, 239.
[3] Raymond, J.C., 1979, *Astrophys. J. Suppl.*, **39**, 1.
[4] Fesen, R.A., Blair, W.P. & Kirshner, R.P., 1982, *Astrophys. J.*, **262**, 171.
[5] Hester, J.J., Parker, R.A. & Dufour, R.J., 1983, *Astrophys. J.*, **273**, 219.
[6] Meaburn, J. & Allen, P.M., 1986, *Mon. Not. R. astr. Soc.*, **222**, 593.
[7] Raymond, J.C., Hester, J.J, Cox, D., Blair, W.P., Fesen, R.A. & Gull, T.R., 1988, *Astrophys. J.*, **324**, 869.
[8] Parker, R.A., 1967, *Astrophys. J.*, **149**, 363.
[9] Kirshner, R.P. & Arnold, C.N., 1979, *Astrophys. J.*, **229**, 147.
[10] Kundt, W., 1983, *Astron. & Astrophys.*, **121**, L15.
[11] Straka, W.C., Dickel, J.R., Blair, W.P. & Fesen, R.P., 1986, *Astrophys. J.*, **306**, 266.
[12] Hester, J.J. & Cox, D.P., 1986, *Astrophys. J.*, **300**, 675.
[13] Sgro, A.G., 1975, *Astrophys. J.*, **197**, 621.
[14] Nittmann, J., Falle, S.A.E.G. & Gaskell, P.H., 1982, *Mon. Not. R. astr. Soc.*, **201**, 833.
[15] Tenorio-Tagle, G. & Rozyczka, M., 1986, *Astron. & Astrophys.*, **155**, 120.
[16] Falle, S.A.E.G., these proceedings.
[17] Falle, S.A.E.G., 1975, *Mon. Not. R. astr. Soc.*, **172**, 55.
[18] Falle, S.A.E.G., 1975, *Mon. Not. R. astr. Soc.*, **195**, 1011.
[19] Chevalier, R.A. & Imamura, J.N., 1982, *Astrophys. J.*, **261**, 1982.
[20] Gaffet, B., 1983, *Astrophys. J.*, **273**, 267.
[21] Bertschinger, E., 1986, *Astrophys. J.*, **304**, 154.
[22] Innes, D.E., Giddings, J.R. & Falle, S.A.E.G, 1987*b*, *Mon. Not. R. astr. Soc.*, **227**, 1021, 1987
[23] Raymond, J.C., Blair, W.P., Fesen, R.A. & Gull, T.R., 1983, *Astrophys. J.*, **275**, 636.
[24] Fesen, R.A. & Itoh, H, 1985, *Astrophys. J.*, **295**, 43.
[25] Innes, D.E., 1985, *Ph.D. Thesis,* University of London.
[26] Innes, D.E., Giddings, J.R. & Falle, S.A.E.G, 1987*b*, *Mon. Not. R. astr. Soc.*, **226**, 67, 1987.
[27] Allen, C.W., 1973, *Astrophysical Quantities* (3d ed.; London: Athlone).
[28] Gaetz, T.J., Edgar, J. & Chevalier, R.A., 1988, *Astrophys. J.*, preprint.
[29] Dopita, M.A., 1977, *Astrophys. J. Suppl.*, **33**, 437.
[30] Cox, D.P. & Raymond, J.C., 1985, *Astrophys. J.*, **298**, 651.
[31] Braun, R., these proceedings.

THE ORIGIN OF KEPLER'S SUPERNOVA REMNANT

Rino Bandiera
Osservatorio Astrofisico di Arcetri
Largo E. Fermi 5, I-50125 FIRENZE (Italy)

The standard picture and its limits.

The purpose of this contribution is to put forward a scenario (Bandiera 1987) that is alternative to the standard picture for the origin of Kepler's supernova remnant (SNR).

Since the original work on this object (Baade 1943) there has always been a common belief that Kepler's supernova (SN) originated from a Population II star. The main arguments used to support this thesis are the following:

1. the light curve, as derived from the historical records, is consistent with that of a Type I SN (e.g. Baade 1943);
2. the remnant is located far from the galactic plane; its latitude is $b = 6.8°$, that, for a distance $D = 4.5\,\mathrm{kpc}$ (derived in the following, and used throughout this paper), corresponds to an elongation from the plane $z = 530\,\mathrm{pc}$;
3. this object is often regarded as a copy of Tycho's SNR (even though, as it will be shown in the next section, these two remnants are dissimilar in some aspects), which is commonly considered a typical Type I SNR.

However this picture cannot easily explain why the shock expands in a dense medium. In fact $n \approx 1\,\mathrm{cm}^{-3}$ is derived from X-ray data (White and Long 1983, Hughes and Helfand 1985), and $n \approx 10^3\,\mathrm{cm}^{-3}$ from spectroscopy of optical knots (Dennefeld 1982, Leibowitz and Danziger 1983), while the expected density for interstellar medium at a few hundred parsecs from the galactic plane is $n \leq 10^{-2.5}\,\mathrm{cm}^{-3}$ (McKee and Ostriker 1977).

The alternative scenario presented here involves a blast wave expansion into a circumstellar medium, that originated from a strong wind in a massive runaway star. Kepler's SN progenitor was probably not a very unusual object; so one can hopefully find evolutionary links with other classes of objects: Kepler's SN could have been a Type Ib SN, or a Type II linear. In addition, the expected progenitor is in many respects similar to a subclass of runaway Wolf-Rayet stars, surrounded by ring nebulae. Furthermore an event like Kepler's SN could have left behind a binary neutron star with high eccentricity; one such object has already been discovered at large z. All these classes of objects can be envisaged as different stages in the evolution of close massive binary systems (van den Heuvel 1976).

A first glance at Kepler's SNR.

Kepler's SNR is rather unconspicuous in optical; in the first plates, taken at the site of the explosion of the historical SN, "the nebulosity appears as a broken mass of bright knots and filaments covering a fan shaped area" (Baade 1943). All the brightest knots are located near the northern edge of the remnant; deeper images (e.g. D'Odorico *et al.* 1986) reveal the presence of other knots in the central region, as well as a faint arc of material connecting the knots on the northern limb.

In the first high resolution radio maps, instead, "the remnant is shown to have an irregular shell structure of 170″ arc diameter, which is brightest in the north, with a gap at the south and a peculiar dent to the east" (Gull 1975); more detailed maps have been recently presented (Matsui *et al.* 1984): apart from some fainter extensions in the SE and in the NW, the remnant is roughly circular with an angular diameter of 200 arcsec; the limb, however, is much brighter on the northern side.

High resolution X-ray images (White and Long 1983, Matsui *et al.* 1984) present a structure similar to that in radio: the northern limb is well pronounced, while no sharp limb appears on the southern side.

To summarize, at a first look one can extract the following morphological characteristics:
1. in radio and X-ray images the limb is brighter on the northern side;
2. most of the optical knots lie along the northern edge; a few of them are located near the centre;
3. no knots have been detected on the southern side, where also the X-ray limb is loosely defined.

The similarity between Kepler's and Tycho's SNRs is often pointed out. However these two objects are not at all similar, at least in some aspects. The emission from Kepler's SNR is much more asymmetric; moreover its knots are more alike the quasi stationary flocculi in Cas A than the knots in Tycho's SNR, both spectroscopically (for a discussion see Raymond 1984), and kinematically (van den Bergh and Kamper 1977; Kamper and van den Bergh 1978). Unfortunately the X-ray spectra are of little use for a comparison; the spectra in Kepler, Tycho and Cas A are, in fact, very similar to each other, in spite of the different origins of these remnants (Becker *et al.* 1980).

The kinematics of the optical knots.

The study of the knots motions, using both astrometric and spectroscopic techniques (van den Bergh and Kamper 1977), led to interesting results. The astrometric results are the following:
1. there is no evidence in favour of radial expansion; the best fitted expansion velocity is $120 \pm 140\,\mathrm{km/s}$, while the expected velocity in the case of linear expansion is $5600\,\mathrm{km/s}$;
2. there is instead evidence of a common motion for the knots pattern, the average transverse velocity being $U_T = 250 \pm 40\,\mathrm{km/s}$ (in the case of no expansion). Similar results come from spectroscopy, namely no evidence of expansion and an average radial velocity $U_R = -220 \pm 15\,\mathrm{km/s}$, that leads to a total peculiar velocity $U = 335 \pm 30\,\mathrm{km/s}$ for the knots system (all these velocity estimates have been obtained using $D = 4.5\,\mathrm{kpc}$, and correcting for galactic rotation).

The nature of the knots.

On the basis of the observational evidence given above, one can speculate on the nature of the optical knots:
1. they are not ejecta; in fact their expansion velocities should be much larger than actually measured; moreover, the abundances inferred spectroscopically are rather standard, in contrast with those expected for processed material;
2. they are not interstellar; at such a large z it is very unlikely to find matter as dense and as clumped as that observed; moreover, an interstellar origin cannot provide an easy explanation for the high average velocity for the knots;

3. they should therefore be circumstellar; in favour of this hypothesis are also their optical spectra, consistent with a slight nitrogen overabundance (Dennefeld 1982, Leibowitz and Danziger 1983).

However two objections are usually raised against the circumstellar nature:
1. the northern limb, in X-rays, is sharper than that produced by a blast wave expanding in a r^{-2} density distribution;
2. the mass loss required to produce such a circumstellar medium is stronger than that usually associated with an old population progenitor (as the standard picture requires).

We shall see how the model presented in the following sections can naturally overcome these objections.

The origin of the knots.

Let us accept that:
1. the knots have a circumstellar origin;
2. the average knots velocity reflects the progenitor's motion.

Thus the progenitor should have been a runaway star subject to strong mass loss. This view is strengthened by the fact that the progenitor's velocity points away from the galactic plane. By approximating its trajectory with a linear motion it follows that it must have left (or crossed) the plane only $3 \cdot 10^6$ yr ago. On the other hand it is very unlikely that Kepler's SN progenitor was a old star; in fact a star moving through the Galaxy with high velocity (close to the escape velocity) has a rather small probability of being within only 500 pc from the plane, unless the star only recently has been ejected from the plane.

While modelling the wind from a runaway star one should take into account the effects of its interaction with the interstellar medium. The circumstellar matter must feel an interstellar ram pressure from the opposite direction to that of the stellar motion. That pressure produces a distortion in the wind bubble and, when a steady situation is eventually reached, a bow shock is formed on the front side, where matter accretes onto a dense layer; probably, due to instabilities during this phase of condensation, knots are formed in that region. If the wind is supersonic, the internal flow remains spherical, with an r^{-2} density profile.

Let us now imagine how a SNR looks if the central star explodes and a blast wave moves through that density distribution:
1. when the blast wave moves in the innermost circumstellar region the remnant still looks symmetric;
2. but when the blast wave reaches the bow shock, the matter reached in the direction of the stellar motion is much denser: therefore the limb must appear much brighter on that side;
3. when eventually the blast wave goes beyond the bow shock, it interacts only with thin interstellar medium, and the emission from the remnant becomes very weak.

The asymmetry in the emission from Kepler's SNR well fits the second phase of the evolution drawn above. In fact the direction of the average transverse motion U_T coincides with that of the brightest limb. Therefore we can conclude that the present blast wave radius is almost equal to the size of the bow shock. We shall see later, while giving a more detailed model, that the assumption of being in this intermediate phase is not very stringent; this assumption will turn out to be very useful in determining the parameters of the wind that produced the knots now visible in optical.

The X-ray radial profile.

Models for the X-ray emission have been computed, to fit the X-ray radial profiles obtained by Einstein. White and Long (1983) considered both the case of free expansion and that of Sedov expansion for a shock wave moving into a homogeneous gas, under the assumption of an equilibrium ionization. Hughes and Helfand (1985) extended that work to the more general case of a time dependent ionization.

However, according to the scenario presented for Kepler's SNR, the northern bright limb is mainly due to the interaction of the blast wave with individual high density knots, rather than to a propagation in a dense, homogeneous medium. Therefore a model in spherical symmetry should not be used to fit that limb.

The southern side, instead, can be more safely modelled, since there the shock is expanding in a circumstellar medium, whose density does not deviate considerably from the standard r^{-2} profile. Models should then be recomputed with that density profile to fit the southern limb. In the free expansion regime, however, the emission radial profile must still present a sharp limb; the only way to fit the observed profile, that shows a loosely defined limb, is to consider a Sedov phase.

A simple model has been developed to simulate the X-ray southern side. A Sedov solution is adopted for the dynamics, as well as an equilibrium ionization emissivity, computed by White and Long (1983) for a gas with abundances consistent with those derived from the X-ray spectra (Becker *et al.* 1980).

In Fig. 1 the measured profiles (White and Long 1983) are compared with simulations of a Sedov expansion in homogeneous gas and in a r^{-2} density profile, respectively. Both models assume a supernova energy of 10^{51} erg; fixing the SNR age and distance, the models do not have free parameters. While the former model requires a bright limb, the latter one fits rather well both the shape of the observed profile and its intensity (within a factor of 2). This agreement gave us confidence that the blast wave is now expanding, in a Sedov regime, into circumstellar material.

The wind of the progenitor.

Let us assume that the present circumstellar density distribution was produced during a phase of steady wind. As explained above, a bow shock is formed at a distance R from the star at which the ram pressure of the wind is equal to that of the interstellar medium, namely when $\rho_{ism}U^2 = \dot{M}w/4\pi R^2$, where $\dot{M}$ and w are mass loss rate and wind velocity, respectively: this allows one to evaluate $\dot{M}w$. Moreover, from the Sedov solution one derives $\dot{M}/w = 6Et^2/R^3$. Taking the energy of the explosion $E = 10^{51}$ erg, also $\dot{M}/w$ can be evaluated. Combining these two results, one gets the following wind parameters: $\dot{M} = 5 \cdot 10^{-5}\, M_{\odot}/\mathrm{yr}$; $w = 10\,\mathrm{km/s}$.

To obtain a steady situation the wind must have reached a distance R from the star, namely it must have lasted more than $2 \cdot 10^5$ yr; therefore the mass lost as wind is larger than $10\,M_\odot$; on the other hand, the requirement that the Sedov regime is valid fixes the mass ejected by the supernova to less than $10\,M_\odot$. The resulting scenario is that the progenitor was originally rather massive, but lost most of its mass during a phase of strong, low velocity wind.

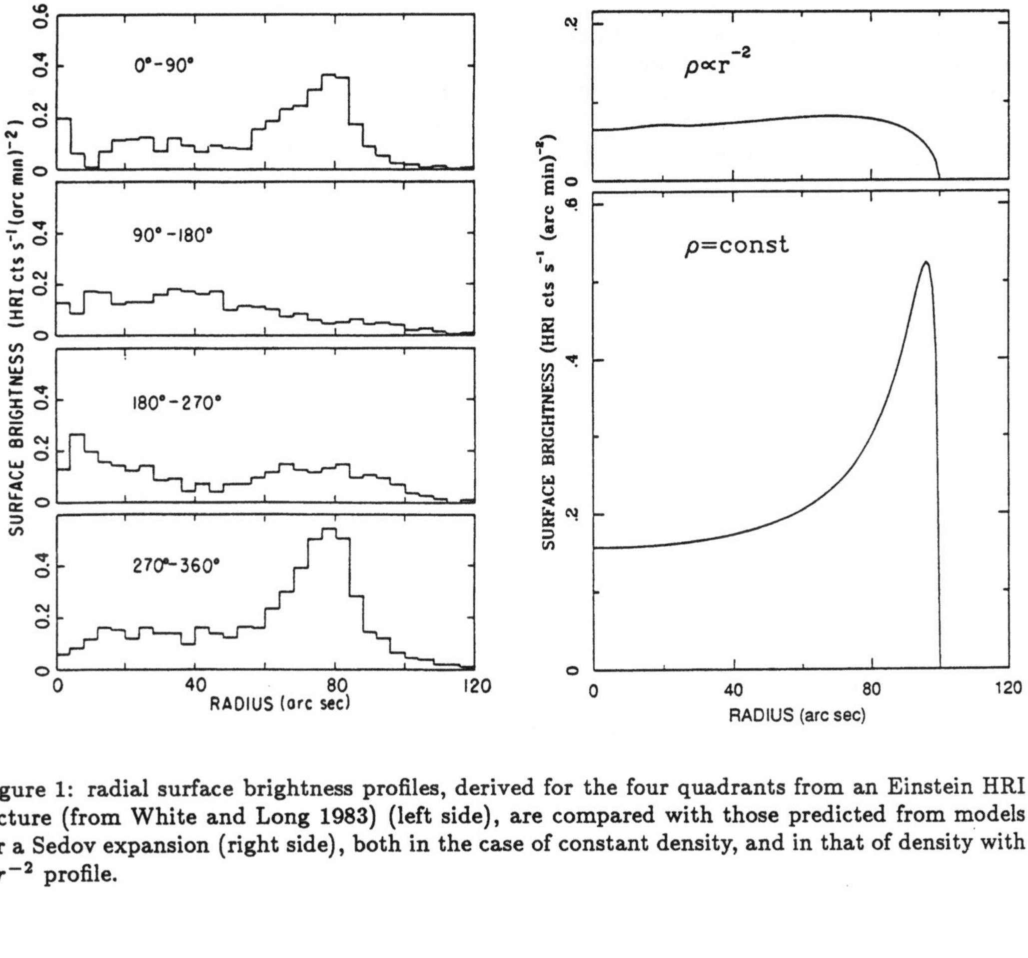

Figure 1: radial surface brightness profiles, derived for the four quadrants from an Einstein HRI picture (from White and Long 1983) (left side), are compared with those predicted from models for a Sedov expansion (right side), both in the case of constant density, and in that of density with a r^{-2} profile.

The distance of the remnant.

If one relies on the bow shock model presented above there is a direct way to derive the distance of Kepler's SNR. If the dense knots are formed on the bow shock they are confined to a nearly paraboloidal sheet; the knots seen in optical are those which have been recently triggered by the impact on the (spherical) blast wave: then they must lie on the intersection of these two surfaces, namely on an annulus.

Its projection on the sky is an ellipse, whose axis ratio is equal to the ratio between the radial (U_R) and total ($\sqrt{U_R^2 + U_T^2}$) velocity of the progenitor star: therefore it should be possible to recognize such an elliptical pattern in the knots distribution on the sky. We assumed above that the progenitor motion is well represented by the mean motion of the optical knots; spectroscopy gives U_R, while astrometry gives U_T/D: therefore the equation for the axis ratio allows one to separate transverse velocity U_T and distance D of the object.

While fitting the pattern of the brightest knots by an ellipse, one can give some constraints using the velocity measurements as well as a model for the bow shock (Huang and Weigert 1982). In this way there are only 2 free parameters: the distance, and the time elapsed (the latter quantity will be discussed in the next section). In Fig. 2 (from Bandiera 1987) the solid line ellipse represents the present position of the intersection between the bow shock and the blast wave, as fitted to the outer edge of the knots pattern: this fit gives a distance of the remnant $D = 4.5$ kpc.

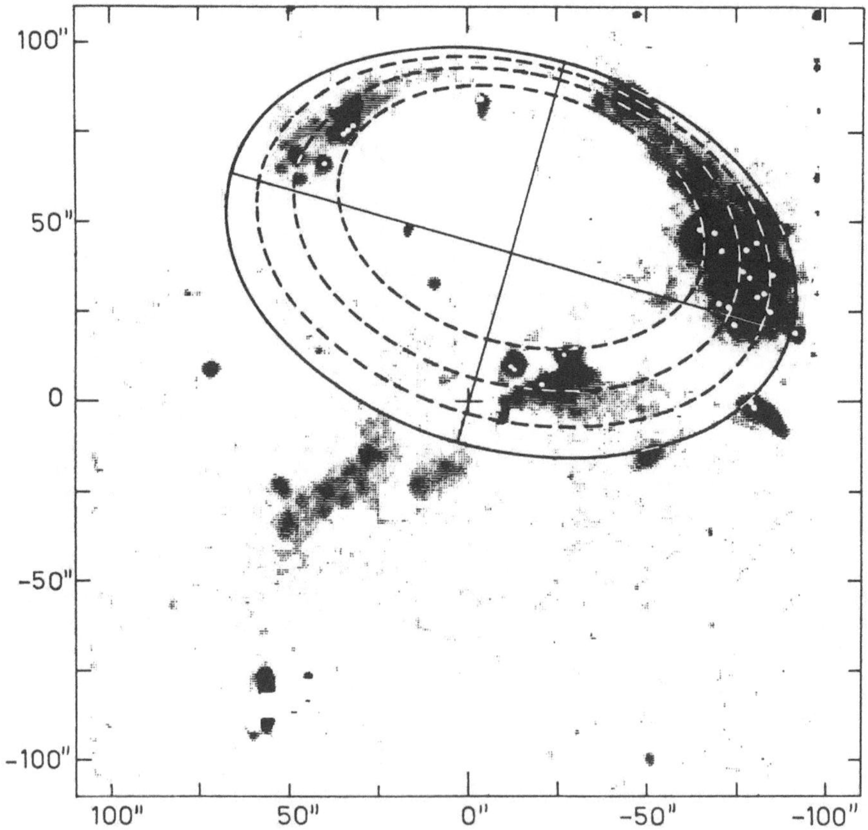

Figure 2: overlay of the model for the positions of optical knots to an image of Kepler's SNR. A central cross indicates the SNR centre. The solid line ellipse represents the present location of the intersection between the bow shock and the blast wave; while dashed line ellipses give the position of that intersection 20, 40 and 60 yr ago (from outside inwards).

Various estimates of Kepler's SNR distance have appeared in the literature, often giving contrasting values: recent estimates range from 3 to 9 kpc. Most of them rely on XVII century measurements of the SN brightness, and on the assumption that it was a standard Type I SN. A crucial point is the derivation of the visual absorption in the direction of the SN. On the basis of reddening of field stars van den Bergh and Kamper (1977) derived $A_V = 2.2 \pm 0.7$; more recently, Danziger and Goss (1980) put forward a higher value ($A_V = 3.5 \pm 0.2$), based on the Balmer decrement measurements in the optical knots.

The latter absorption estimate is often used to determine Kepler's SNR distance (see, e.g., Strom in this meeting), but it is probably not correct; the Balmer decrement is found to vary from knot to knot, leading to a variation in A_V of more than 2 mag over a few arcsec. Therefore a large fraction of the absorption is inside, or very close to the emitting knots, and the minimum measured Balmer decrement should be used to give an upper limit to the interstellar absorption; Danziger and Goss, instead, used the average Balmer decrement to estimate the interstellar absorption.

For this reason the estimate made by van den Bergh and Kamper will be used here, but in the opposite direction to that normally taken: given the distance of the remnant, it will serve to determine the absolute magnitude of the SN, which turns out to be −18.5, more than a magnitude fainter than that of a normal Type I SN: this result will be of use in the following to speculate on the nature of the SN event.

The evolution of the remnant.

From the model for the interaction of the blast wave with the bow shock, presented in the previous section, one can derive the expected time evolution of the pattern of optical knots: as soon as the blast wave reached the apex of the bow shock optical knots started to appear, at first in a very confined region, and then evolving to an annulus whose size increased with time. From the fit to the knots pattern the second free parameter, namely the time elapsed from the moment at which the blast wave reached the bow shock apex, turns out to be nearly 100 yr. Therefore the assumption we originally made, namely that blast wave radius and bow shock size are nearly equal, does not require fine tuning on the remnant age: the present phase, that began a century ago and will still last for more than another one, must be compared with the total age, namely 380 yr.

The dashed line ellipses shown in Fig. 2 represent the positions of the intersection of blast wave and bow shock 20, 40 and 60 yr ago. From the width of the region with optical knots one can argue that knots continue emitting for nearly 60 yr after the passage of the blast wave. According to the model, new knots should gradually appear near the zone indicated by the solid line ellipse: new knots have been actually discovered (van den Bergh and Kamper 1977) near that line on the NW region; the model predicts that, if new knots appear near the centre, they should be more south than those already existing.

Another prediction about the evolution of the radio size is the following: since the blast wave is in Sedov expansion in a circumstellar medium, one expects the radius to evolve as $t^{-2/3}$. However this expansion law should be obtained only for the southern side of the remnant. On the northern side most of the radio emission is probably related to the knots; therefore it should follow the expansion rate of the knots pattern, given in Fig. 2: in this case different expansion times should be derived along different directions, e.g. 200 yr towards NE, 700 yr northwards and 600 yr towards NW.

The progenitor star.

The scenario presented above is consistent with a rather massive progenitor. But the evidence that Kepler's SN was a Type I event is in favour of an old population progenitor: how do we reconcile these two facts? It could have been a Type Ib SN (Panagia 1985); furthermore, Doggett and Branch (1985) showed that the historical records are consistent also with a Type II linear SN. In fact the absolute magnitude at the SN maximum, derived above, is consistent with both choices, while it is fainter than that expected for a classical Type I SN.

If the radio emission from SNe is interpreted as due to the interaction of the blast wave with a circumstellar medium (Chevalier 1984), one can derive the quantity $\dot{M}/w$ for the wind of the progenitor: both Type Ib and Type II L SNe require typically $\dot{M}/w \approx 5 \cdot 10^{-7}\,(M_{\odot}/\mathrm{yr})/\,(\mathrm{km/s})$, a factor 10 smaller than found for Kepler's SN progenitor. This discrepancy can be explained as due to a change in the wind parameters of the progenitor: for instance, the wind velocity could have been increased in the last stages before the explosion.

One may ask what class of objects Kepler's SN progenitor belonged to; they must be runaway stars with a high original mass, most of which has been lost as wind; moreover, since Kepler's SN appeared only a few centuries ago, our belief, or probably prejudice, is that the SN was not a very unusual event, and that various other objects like Kepler's SN progenitor are present in our Galaxy. Even though the wind that produced the bow shock was that typical of a red supergiant, the star did not explode as a red supergiant, otherwise the light curve would have been different from that observed. Interesting classes of massive runaway stars are instead OB runaway and Wolf-Rayet (WR) runaway stars.

Nebulae similar to that of Kepler's SN progenitor.

Some WR runaway stars are in fact associated with "ring nebulae" that present most of the characteristics needed for the circumstellar matter surrounding Kepler's SN progenitor, namely:

1. a brighter limb in the direction of the stellar motion, when it is known; otherwise, a brighter limb on the side further from the galactic plane;
2. a highly clumped structure, with densities up to $n \approx 10^3\,\mathrm{cm}^{-3}$;
3. the star displaced towards the brighter limb (the bow shock apex);
4. a stellar radial velocity equal to that of the brighter limb (since the bow shock apex is almost stationary with respect to the star).

A few such objects have been already observed, namely:

1. S308: an almost circular nebula, with a brighter limb on the NW side, and the WR star slightly displaced toward that limb. In the direction opposite to the brighter limb there is a young open cluster, NGC 2362; if the star originated from that cluster, it must have moved faster than $150\,\mathrm{km\,s^{-1}}$ to reach its present position. The almost circular shape probably means that the bow shock did not develop completely.
2. NGC 3199: with an almost paraboloidal shape, indicating a well developed bow shock, with the star close to the apex, that is also the brightest part of the nebula. On the opposite direction there is the cluster NGC 3293, from which the WR star possibly originated.

3. NGC 6888: with an elliptical shape and a developed filamentary structure; the star is located near the brighter side, pointing in the NW direction; if this is the direction of the stellar motion, the star is moving away from the galactic plane.
4. M1-67: a highly inhomogeneous nebula (with densities up to more than $10^3\,\mathrm{cm}^{-3}$), associated with the WR star 209 BAC, famous for being the fastest known runaway WR, with a lower limit to its velocity of $178\,\mathrm{km\,s}^{-1}$. Spectroscopic observations of the nebula revealed two components; the red-shifted one is also the brightest, and has a radial velocity almost equal to that of the star. Apart from the different perspective (the bow shock points away from us), this nebula is the most similar to that surrounding Kepler's SN progenitor.

The evolutionary scenario.

Runaway WR stars are typically single-line spectroscopic binaries, namely they have an unseen companion, probably a neutron star, orbiting with a period of only a few days. This class of stars is normally considered as a phase in the evolution of massive binary systems (van den Heuvel 1976); in a close binary system the primary star evolves more rapidly, transfers part of its mass to the companion, and eventually explodes, leaving a stellar remnant, namely a neutron star. The system receives a kick, and becomes a runaway system; but it is not disrupted, and after some time it will appear as a massive X-ray binary, and then as a runaway WR star.

A SN like Kepler's event corresponds to the explosion of the survived star; if the SN event is actually due to a WR star explosion, it should appear as a Type I (a Type Ib) SN. From the study of Kepler's SNR it follows that the star was subject for some time to a typical red supergiant wind; a similar wind is also needed to produce the above mentioned ring nebulae. If also the second SN produces a neutron star and the system remains bound, it will eventually evolve into a binary neutron star with a high eccentricity orbit; one such object, namely PSR 2303+46, has been actually discovered at 500 pc from the galactic plane; it is possible that one such object is imbedded in Kepler's SNR.

References.

Baade, W. 1943, *Astrophys. J.*, **97**, 119.
Bandiera, R. 1987, *Astrophys. J.*, **319**, 885.
Becker, R.H., Boldt, E.A., Holt, S.S., Serlemitsos, P.J., and White, N.E. 1980, *Astrophys. J.*, **237**, L77.
Chevalier, R.A. 1984, *Astrophys. J.*, **285**, L63.
Danziger, I.J., and Goss, W.M. 1980, *M. N. R. A. S.*, **190**, 47P.
Dennefeld, M. 1982, *Astron. Astrophys.*, **112**, 215.
D'Odorico, S., Bandiera, R., Danziger, I.J., and Focardi, P. 1986, *Astron. J.*, **91**, 1382.
Doggett, J.B., and Branch, D. 1985, *Astron. J.*, **90**, 2303.
Gull, S.F. 1975, *M. N. R. A. S.*, **171**, 237.
Huang, R.Q., and Weigert, A. 1982, *Astron. Astrophys.*, **116**, 348.
Hughes, J.P., and Helfand, D.J. 1985, *Astrophys. J.*, **291**, 544.
Kamper, K.W., and van den Bergh, S. 1978, *Astrophys. J.*, **224**, 851.
Leibowitz, E.M., and Danziger, I.J. 1983, *M. N. R. A. S.*, **204**, 273.
Matsui, Y., Long, K.S., Dickel, J.R., and Greisen, E.W. 1984, *Astrophys. J.*, **287**, 295.
McKee, C.F., and Ostriker, J.P. 1977, *Astrophys. J.*, **218**, 148.

Panagia, N. 1985, in *Supernovae as Distance Indicators*, ed. N. Bartel (Berlin: Springer-Verlag) p. 14. Raymond, J.C. 1984, *Ann. Rev. Astron. Astrophys.*, **22**, 75.

van den Bergh, S., and Kamper, K.W. 1977, *Astrophys. J.*, **218**, 617.

van den Heuvel, E.P.J. 1976, in *IAU Symposium 73, The Structure and Evolution of Close Binary Systems*, ed. P. Eggletos, S. Mitton, and J. Whelan (Dordrecht: Reidel), p. 35.

White, R.L., and Long, K.S. 1983, *Astrophys. J.*, **264**, 196.

Discussion.

van den Bergh: If you find a shell in the galactic halo there are two possible interpretations: one is that the progenitor star was a halo star, and the other that it was a runaway star in the disk. Could you summarize the arguments that lead you to believe that this is a runaway star?
Bandiera: From a halo star I do not expect a mass loss as large as that required in this case, namely $10\,M_{\odot}$.
van den Bergh: The number of 10 solar masses is of course model dependent. Can you give us an estimate of the mass that is actually seen in the knots?
Bandiera: It is difficult to get a direct estimate of the total mass ejected; the value given before relies on the self-consistence of the model.
Biermann: If your scenario is correct then in every case where the wind bubble is ellipsoidal or hyperboloidal you should have the same kind of topology, giving rise to a ring. This particular topology should then be fairly common.
Bandiera: Yes, but only for very young SNRs.

CTB80: THE SUPERNOVA REMNANT WITH (ALMOST) EVERYTHING

Richard G. Strom
Netherlands Foundation for Radio Astronomy
Postbus 2, 7990 AA Dwingeloo

Summary: The diversity of properties found in the peculiar supernova remnant CTB80 is briefly reviewed. A generic connection between the pulsar-driven flat spectrum component and the larger scale emission seems likely, but the exact nature of their relationship remains unclear.

1. Introduction

If one were to make a list of items which we might expect, or even hope, to encounter in a supernova remnant (SNR), then it would certainly include: galactic object; radio emission of a nonthermal nature; optical nebulosity; X-ray component; infrared emission. In the majority of cases we would not be surprised to find an indication of shell structure, albeit with some irregularity; less usual would be a flat spectrum (Crab Nebula-like) component, which if we are lucky might contain a neutron star visible as a pulsar. Finally, we might hope for an SNR young enough that its initiating supernova might have been recorded in historical chronicles - that it was seen as a "guest star", to use the Oriental terminology.

From such a shopping list we would be able to choose the elements required for constructing most known SNRs. In the majority of cases, depending upon the amount and quality of observational material available, we would not make an exhaustive selection from the list. Unusual in this respect, then, is the galactic SNR CTB80, in which essentially all the items listed above have been found to be present. Although some of its more unusual properties were apparent at a fairly early stage (e.g., Velusamy, Kundu and Becker, 1976; Angerhofer, Strom, Velusamy and Kundu, 1981), it has only recently become apparent to what extent one and the same object can encompass such a diversity of remnant qualities.

2. Large scale structure

CTB80 consists of several arcs of radio emission which emanate from a central high-brightness plateau, stretching over nearly 1° (Fig. 1). Its radio spectrum is largely nonthermal, with the notable exception of the high brightness flat spectrum component, centrally located at the western end of the plateau. The central component has associated optical nebulosity (Angerhofer et al., 1981), as do the arcs where they are nearest to the central plateau (Blair, Krishner, Fesen and Gull, 1984).

Infrared emission from CTB80 can be seen in IRAS survey data at 60 and 100 μ. The spatial coincidence with the arc which stretches to the north and then east is particularly striking (Fig. 1), but the infrared emission then appears to turn south (where no radio emission is present) and continues on, making a nearly complete circle encompassing most of the radio emission from CTB80. The spectral characteristics of this emission are similar to what has been found in the infrared for typical

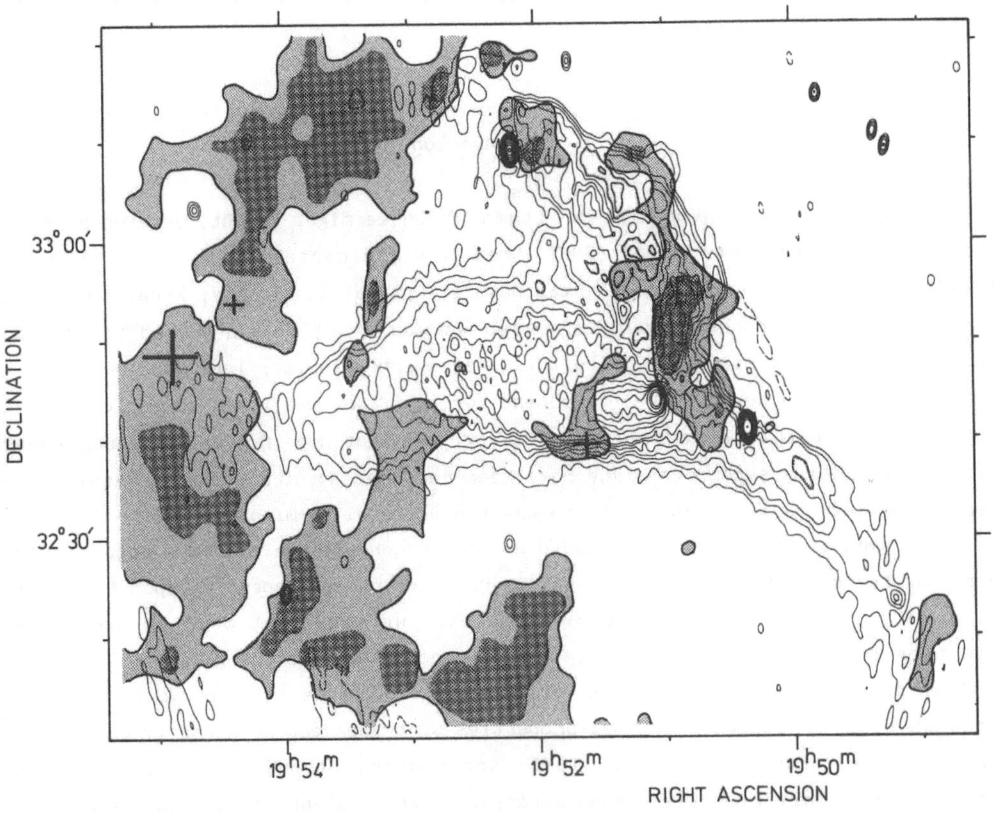

Figure 1. Contours showing the 49 cm brightness distribution of CTB80 observed with the Westerbork Telescope, with the distribution of infrared emission from warm, presumably shock-heated, dust superimposed (gray shading). Emission from bright stars (crosses) has been suppressed. The distribution of infrared emission has been derived from the IRAS survey.

shell SNR such as IC443 (Braun and Strom, 1986a) and the Cygnus Loop (Braun and Strom, 1986b).

X-ray emission from CTB80 has been mapped by the Einstein Observatory (Becker, Helfand and Szymkowiak, 1982; Wang and Seward, 1984). It is dominated by an unresolved component at the western edge of the flat spectrum radio feature, although Wang and Seward also find X-ray emission extending eastwards along the radio plateau. The compact X-ray feature was the first direct evidence for the possible presence of a neutron star in CTB80.

To round off this section, let us briefly consider the distance to CTB80. Angerhofer, Wilson and Mould (1980) concluded that the best estimate, based upon both radio and optical evidence, was 3±1 kpc, while Blair et al. (1984) revised this slightly downwards to 2.5 kpc. An HI spectrum obtained with the Westerbork Telescope (WSRT) shows deep absorption of emission from the high brightness flat spectrum central component in CTB80 over the velocity range 0 to +14 km s^{-1}. In this direction of the galaxy (l=69°), the velocity of gas in circular motion relative to the sun is expected to increase to a maximum of just under 20 km s^{-1} (at the tangential point), and then decrease, with negative values for gas beyond the solar circle. The absorption spectrum of an extragalactic source some 10' arc from the flat spectrum component shows features from nearly -100 km s^{-1} to a maximum of +19 km s^{-1}. The fact that the CTB80 component shows no absorption at velocities greater than +14 km s^{-1} (and none at negative velocities) suggests a distance of 2 kpc. Finally, it is worth noting that the dispersion measure (DM) of the pulsar associated with CTB80 (see below) would, on the face of it, imply a distance of only 1.4 kpc.

3. Flat spectrum source

The high brightness, flat spectrum radio feature near the geometrical center of CTB80 has been observed with increasing resolution (Angerhofer et al., 1980; Strom, Angerhofer and Velusamy, 1980; Strom, Angerhofer and Dickel, 1984; Velusamy and Kundu, 1983; Strom, 1987). The highest resolution maps presented in these papers show an egg-shaped emission region, the brightest part of which is about 30" in size. There is some evidence for limb-brightening along the western and southern edges, with a striking hot spot in this rim near the southwest apex.

The component coincides with much optical nebulosity (Strom et al., 1980). While Hα filaments delineate the radio emission, [OIII]λ5007 emission is strikingly seen only on the western half (Fesen and Full, 1985). An Hα image taken in supurb seeing shows many fine filaments both along the edges and protruding into the interior of the component (Van den Bergh and Pritchet, 1986). A proper motion study may have found evidence for expansive motions (Strom and Blair, 1985).

Recently, Strom and Greidanus (in preparation) have analyzed data on the kinematics of Hα in the central component of CTB80 using TAURUS on the Isaac Newton

Telescope. Most of the emission is concentrated in a ring which shows velocities of up to 100 km s^{-1}, and which are predominantly negative. If they are interpreted in terms of a uniformly expanding shell (which must be an over-simplification, since the ring is not circular), then the implied expansion rate is 100-200 km s^{-1}.

3.1. Discovery of a pulsar

In a high resolution, 20 cm VLA map, a polarized, unresolved source was found 2" arc east of the hot spot in the southwest rim (Strom, 1987). The steep radio spectrum construed for this source strongly suggested a pulsar, and pulsations were quickly discovered (Clifton et al., 1987) with a period of 39.5 ms. The relatively low value of $\dot{P}$, P being the pulse period, found for PSR1951+32 (as it was called), implies a weak magnetic field, and this has consequences for field generation and decay in neutron stars (Fruchter et al., 1988).

From the morphology, Strom (1987) argued that the neutron star is probably moving rapidly to the west, and interpreting the hot spot on the radio limb as the stagnation point in the bow shock, derived a speed of about 400 km s^{-1}. He also noted the similarity between the hot spot and Wisp 1 in the Crab Nebula (toward which its pulsar, PSR0531+21, is also moving). The interpretation of rapid motion for PSR0531+21 appears to have been confirmed by scintillation measurements (Fruchter et al., 1988) which suggest a speed of about 300 km s^{-1}, in satisfactory agreement with Strom's estimate based upon pressure balance. These values are also in agreement with the expansion speed found for the optical nebula (Strom and Blair, 1985).

Strom (1987) showed that, by analogy with the Crab Nebula, the X-ray luminosity of the CTB80 nebula can be used to constrain the properties of its pulsar, and in particular that $\dot{P}/P^3$ (which is proportional to the rate of loss of rotational energy) should be much smaller than the value found for PSR0531+21. Clifton et al. (1987) noted with some surprise the low value of $\dot{P}$ found for PSR1951+32, but in view of its period and the low X-ray luminosity this was to be expected.

We can compare the X-ray luminosity of extended emission associated with various neutron stars with the rate at which they are losing rotational energy, $\dot{E}$ (assuming a moment of inertia of 10^{45} g cm^2). If the X-ray emission is predominantly of synchrotron origin, then the relatively short radiative lifetimes of the emitting particles should ensure that the observed luminosity is related to the rate at which energy is being deposited in the ambient medium. The correlation, shown for a number of objects like the Crab Nebula and CTB80, as well as several "normal" radio pulsars (Helfand, 1983), is shown in Fig. 2. If the energy conversion occurred with the same

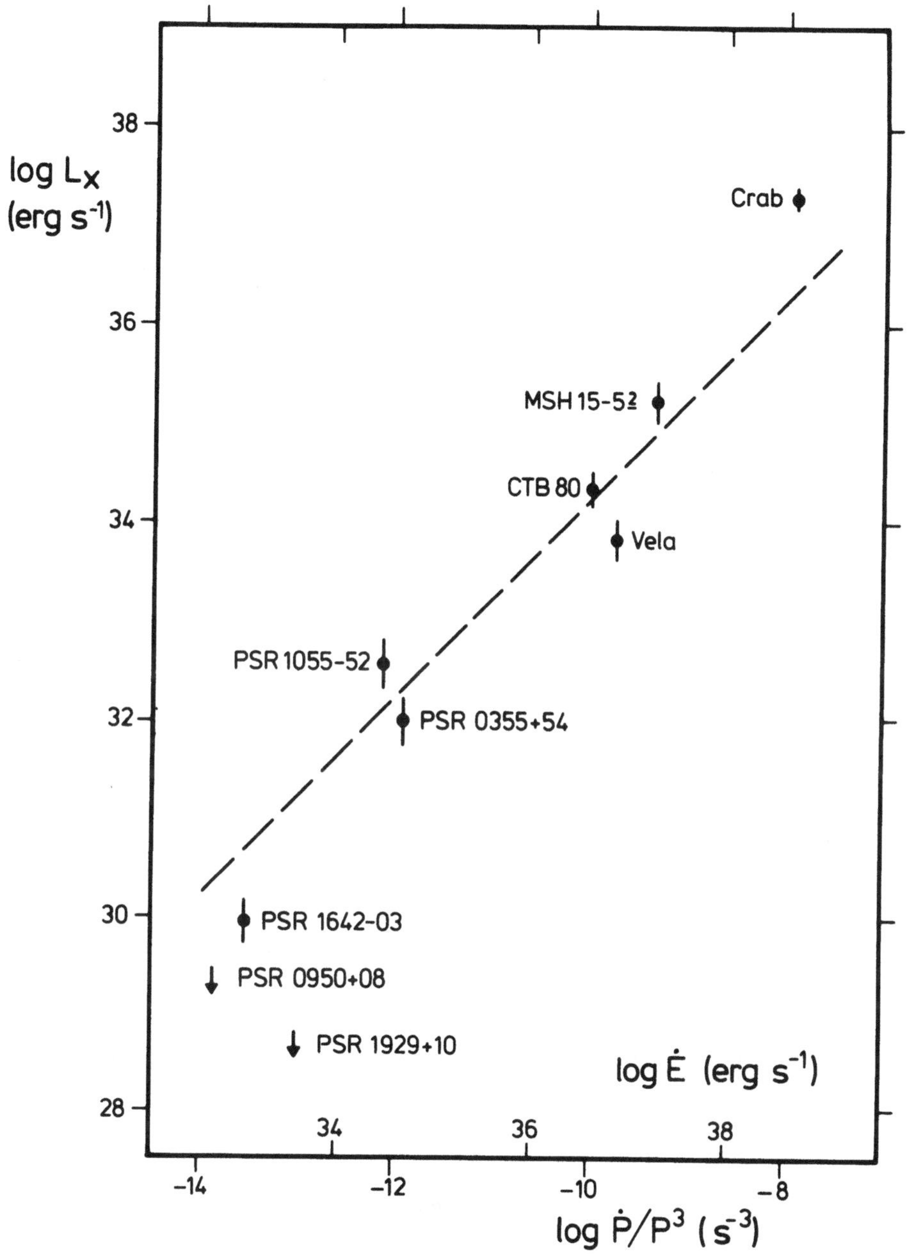

Figure 2. The X-ray luminosity, L_x, of extended emission associated with various neutron stars (indicated by either nebulae or pulsar name), shown as a function of $\dot{P}/P^3$, which corresponds to the $\dot{E}$ shown for an assumed moment of inertia (see text).

efficiency everywhere, then we would expect to find luminosity $\propto \dot{E}$ (as indicated by the dashed line). There is clearly some correlation, although the relationship appears to be steeper than that suggested above, and there is some scatter in the points.

4. Conclusions

Like the proverbial anecdote about an animal constructed by a committee (was it the wildebeest?) whose disparate anatomical elements seem irreconcilably anachronistic, so the surfeit of exotica in CTB80 deepens rather than resolves its mystery. The pulsar, it is true, helps explain the presence of the flat spectrum component, but given the existence of the latter (and an X-ray point source to boot), an associated neutron star seemed all but unavoidable. The real question is, how do we explain the presence of the neutron star, particularly in view of the large scale structure of CTB80? It seems clear that some of the more extended radio emission - specifically the plateau stretching eastwards from the pulsar - must be generically related to the flat spectrum component, for the spectrum gradually steepens along its length (Angerhofer et al. 1981). This interpretation is strengthenend by the extended X-ray emission coming from the plateau.

If we look at the remaining large scale emission, the radio arc stretching north and east has the semblance of a (partial) shell, and the associated infrared emission (Fig. 1) has the characteristics of shock-heated dust. The infrared map shows, moreover, a nearly complete ring encircling most of the radio emission, so plausibly CTB80 also has a shell component. The eccentric location of the flat spectrum component suggests a largely westward motion (if we are to link its origins to an event near the center of the shell), agreeing with my own conclusions about the motion of the pulsar with respect to the flat spectrum component itself (Strom, 1987). So it must be that PSR 1951+32 bears some generic relationship to the event which produced all of the large scale emission in CTB80.

Acknowledgements

The Netherlands Foundation for Radio Astronomy is financially supported by the Netherlands Organisation for Scientific Research (NWO).

References

Angerhofer, P.E., Strom, R.G., Velusamy, T., Kundu, M.R., 1981. Astron. Astrophys. 94, 313.

Angerhofer, P.E., Wilson, A.S., Mould, J.R., 1980. Astrophys. J. 236, 143.

Becker, R.H., Helfand, D.J., Szymkowiak, A.E., 1982, Astrophys. J. 255, 557.

Blair, W.P., Kirshner, R.P., Fesen, R.A., Gull, T.R., 1984. Astrophys. J. 282, 161.

Braun, R., Strom, R.G., 1986a. Astron. Astrophys. 164, 193.

Braun, R., Strom, R.G. 1986b. Astron. Astrophys. 164, 208.

Clifton, T.C., Backer, D.C., Foster, R.S., Kulkarni, S.R., Fruchter, A.S., Taylor, J.H., 1987. IAU Circular No. 4422.

Fesen, R.A., Gull, T.R., 1985. Astrophys. Letters 24, 197.

Fruchter, A.S., Taylor, J.H., Backer, D.C., Clifton, T.C., Foster, R.S., Wolszczan, A., 1988. Nature 331, 53.

Helfand, D.J., 1983. Proc. IAU Symp. 101 (eds. J. Danziger, P. Gorenstein), D. Reidel, p. 471.

Strom, R.G., 1987. Astrophys. J. 319, L103.

Strom, R.G., Angerhofer, P.E., Dickel, J.R., 1984. Astron. Astrophys. 139, 43.

Strom, R.G., Angerhofer, P.E., Velusamy, T., 1980. Nature 284, 38.

Strom, R.G., Blair, W.P., 1985. Astron. Astrophys. 149, 259.

Van den Bergh, S., Pritchet, C.J., 1986, PASP 98, 448.

Velusamy, T., Kundu, M.R., 1983. J. Astrophys. Astr. 4, 253.

Velusamy, T., Kundu, M.R., Becker, R.H., 1976. Astron. Astrophys. 51, 21.

Wang, Z.R., Seward, F.D., 1984. Astrophys. J. 285, 607.

G316.3-0.0 AND G332.4+0.1 - TWO SUPERNOVA REMNANTS WITH BLOWOUTS

D.K. Milne, J.L. Caswell, M.J. Kesteven, and R.F. Haynes
Division of Radiophysics, CSIRO, PO Box 76, Epping, N.S.W. 2121, Australia
and
R.S. Roger
Dominion Radio Astronomical Observatory, Penticton, B.C., Canada

Summary High-resolution 843 MHz observations of the supernova remnants G316.3-0.0 and G332.4+0.1 exhibit forms which suggest that the spherical shells have blown out over part of their surface into adjacent lower-density regions. This view is reinforced by observations of the polarization, and hence the magnetic field directions, in these objects.

1. Introduction

Recent models for supernova remnant (SNR) evolution have invoked a "blowout" from spherically symmetric shells into adjacent cavities of low-density presumably swept clear of gas by intense stellar wind from nearby and perhaps pre-existing stars (Braun and Strom 1986; Pineault et al. 1987). Milne (1987) suggested that three or four of the remnants in his atlas of SNR magnetic fields might exhibit such departures from spherical symmetry; in this paper we discuss two of these: G316.3-0.0 and G332.4+0.1.

Little has been published on the supernova remnant G316.0-0.0 (MSH 14-57). In the 5 GHz map of Milne and Dickel (1975) it appears to have a central peak with an extension to the west and south-west. There is 5 GHz polarization in the south-west which does appear to be associated with the remnant; however, polarization along the western edge is possibly unrelated.

G332.4+0.1 (Kesteven 32) is also included in Milne and Dickel (1975) and is seen there as a double-peaked source with significant polarization in the direction of the western peak. Milne and Dickel note an abrupt rotation of the E vector in this direction. Recently Roger et al. (1985) published an 843 MHz Molonglo Observatory Synthesis Telescope (MOST) map of G332 which exhibits a jet and plume seemingly emitted from the SNR. A Parkes 8.4 GHz map was also included in that paper. In this paper we present an 843 MHz MOST map of G316 together with 8.4 GHz total-power and polarization maps of G316 and G332.

2. MOST 843 MHz Observations

SNR observations with the University of Sydney MOST have been described by us (e.g. Roger et al. 1986; Milne et al. 1986). Basically the MOST consists of two steerable, cylindrical, east-west arrays which synthesize a field up to 70' arc in diameter at a resolution of 43" arc (E-W) by 43" cosec Dec. (N-S). Because of the gap between the eastern and western arms of the array the instrument is not sensitive to structural scales >20' arc. This array has been used to map ~40 galactic SNRs in a program described by Milne et al. (1985).

The maps of G316 and G332 were made over two nights in August 1983 and September 1984 respectively; they have been CLEANed to remove a ~8% sidelobe. The 843 MHz map of G316 is shown in Figure 1. As mentioned above, MOST observations of G332 (and also a Parkes 8.4 GHz total-power map) were included in Roger et al. (1985) and in this paper we reproduce this MOST map as a grey-scale .

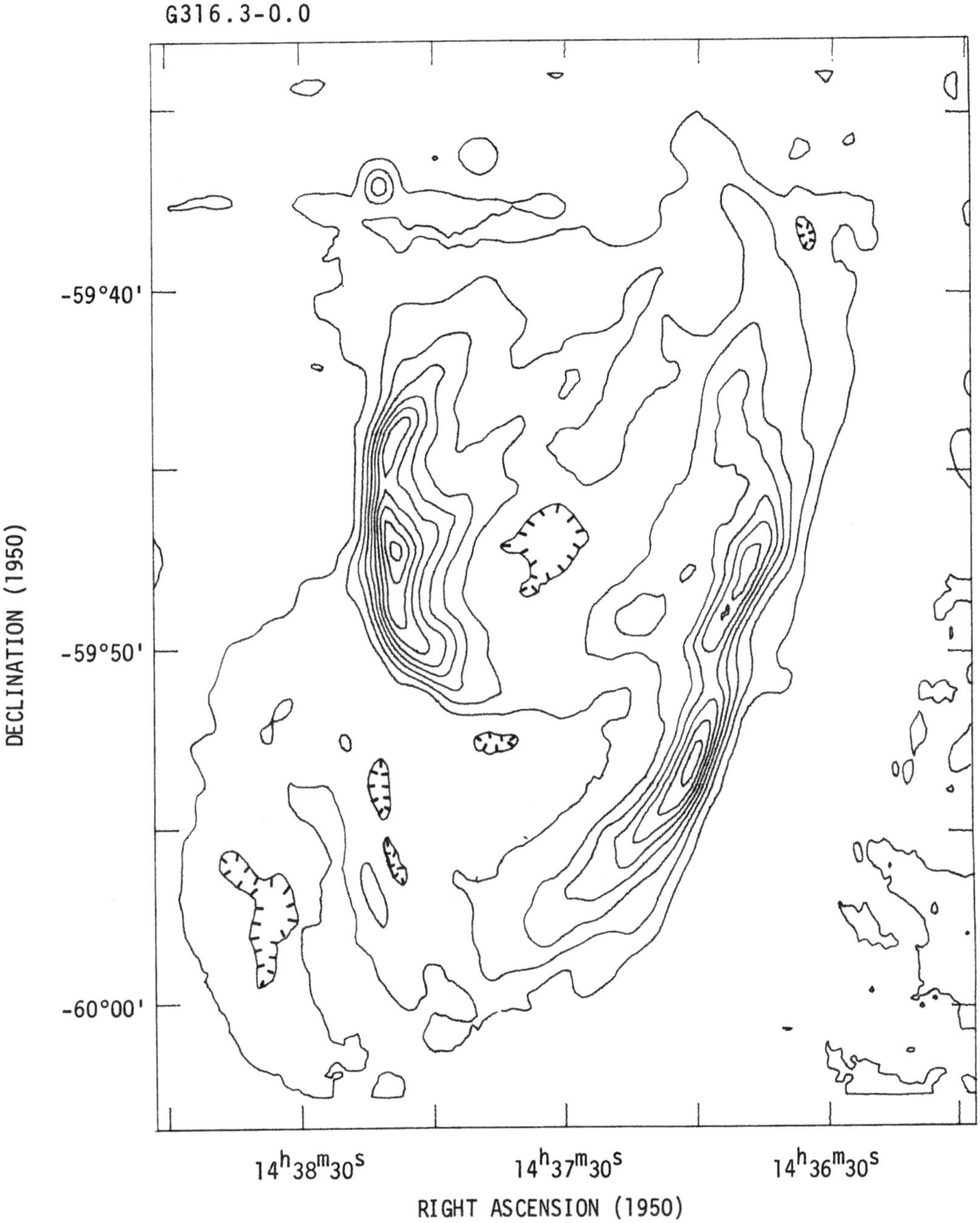

Fig. 1 Intensity contours of the 843 MHz CLEANed radio map of the SNR G316.3-0.0. The contour interval is 25 mJy per beam area. Fringed contours indicate local intensity minima. The beam size is 43" arc EW by 49" arc NS.

3. Parkes Polarization Observations

A description of the observing techniques for polarization and total-power mapping of SNRs is given by Milne and Dickel (1975), and more recent details can be found in Roger et al. (1986). The receiver was switched between orthogonal linear polarizations and then phase-detected; the half-power beamwidth at 8.4 GHz is 3'.0 arc.

The polarization (and total-power) maps were made from telescope scans at four feed angles: 0°, 45°, 90°and 135°. The scans were baselined to remove the background as much as possible; these two remnants are close to the galactic plane and unfortunately suffer from considerable polarization confusion. The 8.4 GHz results for G316 and G332 are presented in Figures 2 and 6 respectively. The scans extended some distance north and south of the bounds of these figures; tying-in scans were made in right ascension.

Whilst both of these remnants were included in Milne and Dickel's (1975) study, new 5 GHz maps were made at Parkes to supplement these for comparison with the 8.4 GHz results. These 5 GHz maps were made in the same manner as described at 8.4 GHz; the half-power beamwidth is 4'.6 arc. They are similar to the Milne and Dickel maps but have improved sensitivity and spatial coverage. The G332 map confirms the sudden change in polarization near the western peak noted by Milne and Dickel.

4. Results

(a) G316.3-0.0

The MOST map (Fig. 1) shows a typical shell source in the northern part of G316, in that there are two opposing arcs of emission that embrace a roughly circular region. However, the western arc is extended towards the south as a linear, tail-like protrusion beyond the otherwise spherical shell. In addition this tail encompasses an area of low-level emission to the east and south-east. Altogether its appearance suggests that the remnant plasma has burst out to the south of the main shell.

In the 8.4 GHz map (Fig. 2) G316 is clearly delineated in polarization against a relatively strong polarized background. There are features visible in this background polarization, particularly the strong, uniformly directed polarization to the north-east of the remnant. Polarization can be seen to the south-east, in the suggested blowout. The distribution of percentage polarization at 8.4 GHz is 2-3% over most of the shell and 7-8% on the western rim and tail and rises to >10% in the blow-out.

We have compared the polarization directions at 8.4 GHz (convolved to 4'.6 arc resolution) with those obtained at 5 GHz to obtain the Faraday rotation measure and hence the intrinsic polarization and the magnetic field directions (the latter shown in Fig. 3). With polarization at only two frequencies it is not possible to rule out an additional 180° Faraday rotation between them; this would imply an additional 1322 rad m^{-2} in the rotation measure. In our assessment of rotation measure (Fig. 4) the lowest absolute value consistent with the observations has been adopted.

The vectors shown in Figure 3 indicate direction of the projected magnetic field; the magnitudes represent the geometric mean polarization intensity at the two frequencies used and is a measure of the reliability of the field determination. Figures 3 and 4 have been overlaid with selected contours from the (higher-resolution) MOST map (Fig. 1). In this figure we see non-random field directions over the whole area both on and off G316, and this is consistent with the closeness of the object to the galactic plane and indeed to the galactic centre; compare this with the observations on G7.7-3.7 (Milne et al. 1986) or G291.0-

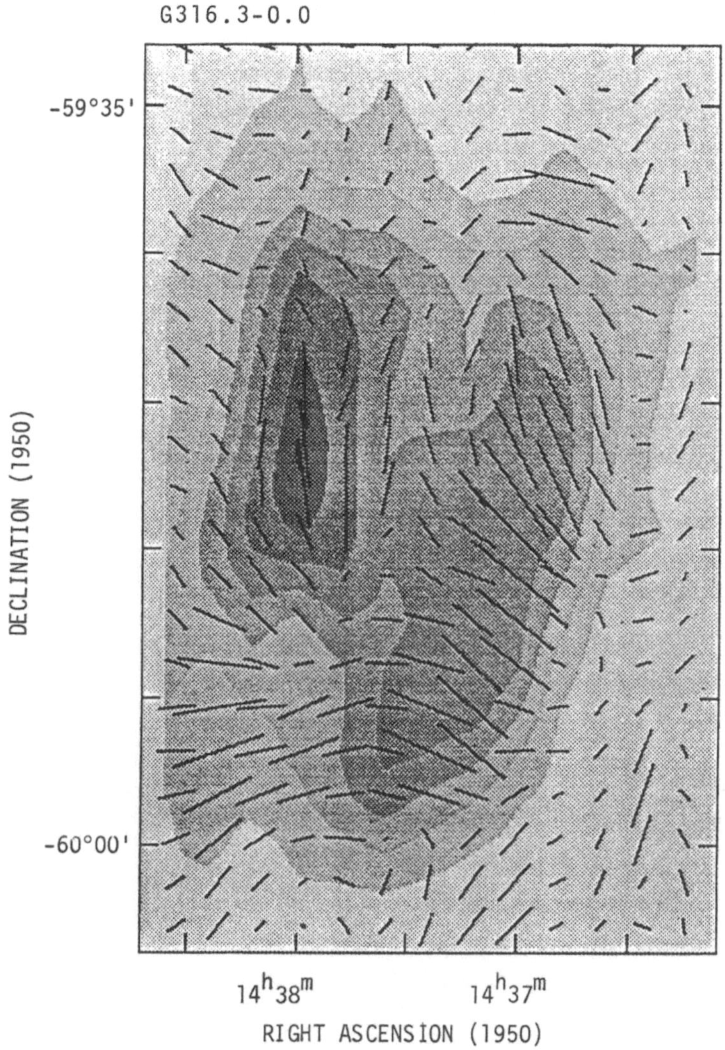

Fig. 2 Total power grey-scale levels and polarization **E** vectors at 8.4 GHz of the SNR G316.3-0.0. The grey-scale interval is 118 mJy per beam area, the maximum polarization is 140 mJy per beam area (at R.A. $14^h38^m12^s$, Dec. -59°58') and the beamwidth is 3".0 arc.

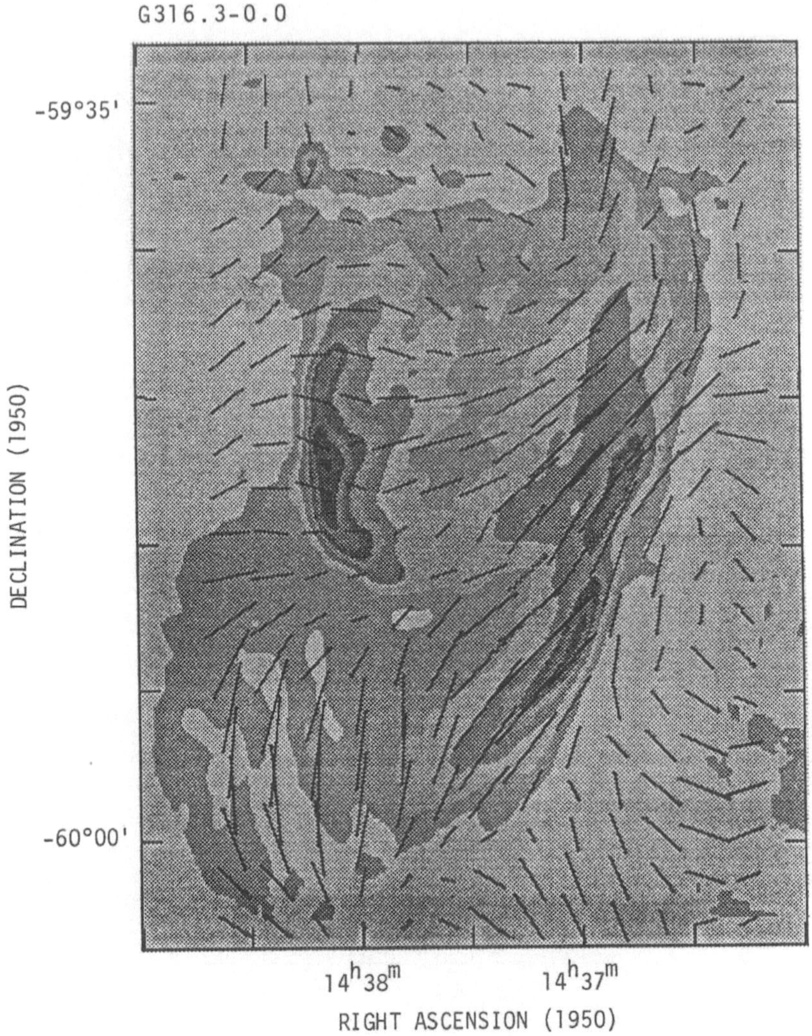

Fig. 3 Direction of the projected magnetic field in G316.3-0.0 (resolution 4'.6 arc) superimposed on a grey-scale representation of the higher-resolution 843 MHz map (Fig.1). The length of the vectors are proportional to the geometrical mean of the 8.4 and 5 GHz polarization intensities.

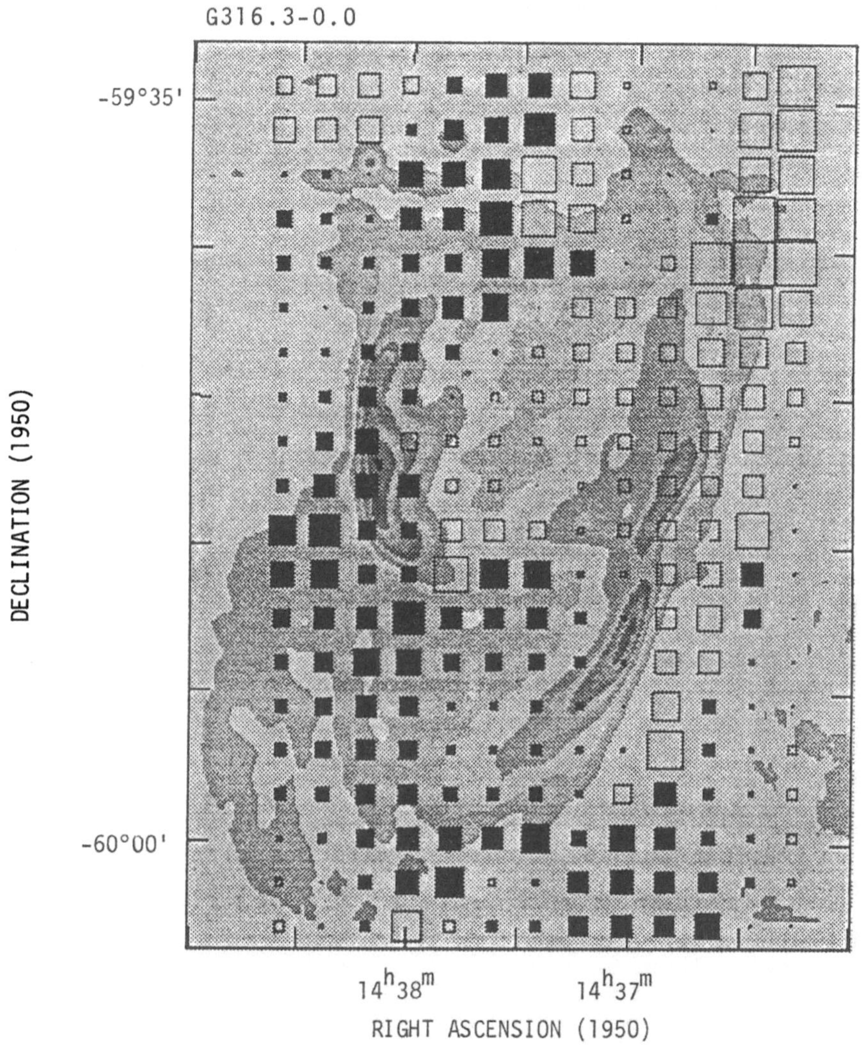

Fig. 4 Distribution of rotation measure over G316.3-0.0 (resolution 4'.6 arc) superimposed on the 843 MHz grey-scale map as in Figure 3. Positive and negative rotation measures are indicated by filled and unfilled squares respectively. The size of each square is a linear representation of the rotation measure in that direction. The scale in rad m^{-2} is: ■ 200 ■ 400 ■ 600

0.1 (Roger et al. 1986), where these remnants are either sufficiently far off the galactic plane or away from the galactic centre to minimize the polarized background. However, G316 does seem to have an influence on the direction of the field, perhaps not so obvious in the northern shell but more so in the suggested blowout to the south, where it appears to be drawn out in that direction.

There is considerable variation in the rotation measure (Fig.4) over the field with relatively moderate values, of around -200 rad m^{-2}, in the region occupied by G316. The rotation measure increases to over twice that figure (negative in the west and positive in the east) around its periphery, where perhaps the remnant is interacting with the ambient magnetic field. The negative values seen within the northern shell extend along the bright tail but become negative further east in the blowout.

The depolarization ratio (percentage polarization at 8.4 GHz/percentage at 5 GHz) is displayed in Figure 5 for the region where the total power is >10% of its peak value at both frequencies. Here we see very high depolarization (>10) around the eastern rim of the shell coincident with the sign change in rotation measure. Relatively high depolarization is also found where other sign changes in rotation measure occur. The bulk of the values over the remainder of G316 lies between 1 and 3.

Integrated flux densities measured at 843 MHz and at 5 and 8.4 GHz are respectively 19.0±3.0 Jy, 14.7±2.0 Jy and 11.6±1.5 Jy. Because the MOST is not sensitive to structural scales greater ~20' arc this 843 MHz flux density should be regarded as a lower limit. The spectral index obtained from these and previously published flux densities (Shaver and Goss 1970; Milne and Dickel 1975) is -0.35±0.07.

(b) G332.4+0.1

As remarked earlier, the MOST map of this remnant has been discussed by Roger et al. (1985) and no satisfactory explanation for the jet and plume was advanced. In this paper we draw attention to the more general appearance of the shell. It seems, as in G316, that there is a "normal" spherical shell to the west with a breakout into the region in the east *plus* the jet and plume.

The 8.4 GHz total-power grey-scale map and the polarization E vectors in the direction of G332 are shown in Figure 6. As in G316 there is well-defined polarization off the SNR; this field is also imbedded in the galactic plane, but G332 is clearly polarized: ~12% and ~5% on the western and eastern peaks respectively. There is polarization in the direction of the jet which increases in the plume to the north-east of Figure 6; unfortunately this part of the field was not mapped at 5 GHz.

The direction of the projected magnetic field and the Faraday rotation are shown in Figures 7 and 8 respectively. These figures are limited to directions where the polarization was greater than one-tenth of the relevant peak polarization. We see a projected magnetic field which runs across the north of the bright western peak and looping around in the west, presenting essentially a tangential field in the spherical western part of the SNR. In the east, where we believe that the remnant has broken away from its spherical form, the magnetic field is radial and in the north-east it mostly runs parallel to the jet towards the plume. The field directions perpendicular to the jet near its base seem to be a contradiction to this pattern and a third frequency must be used to clarify this.

The rotation measure distribution shown in Figure 8 and used to construct Figure 7, represents the lowest absolute values that give consistency in the field directions and rotation measure distribution. The Faraday rotation is negative over the entire remnant,

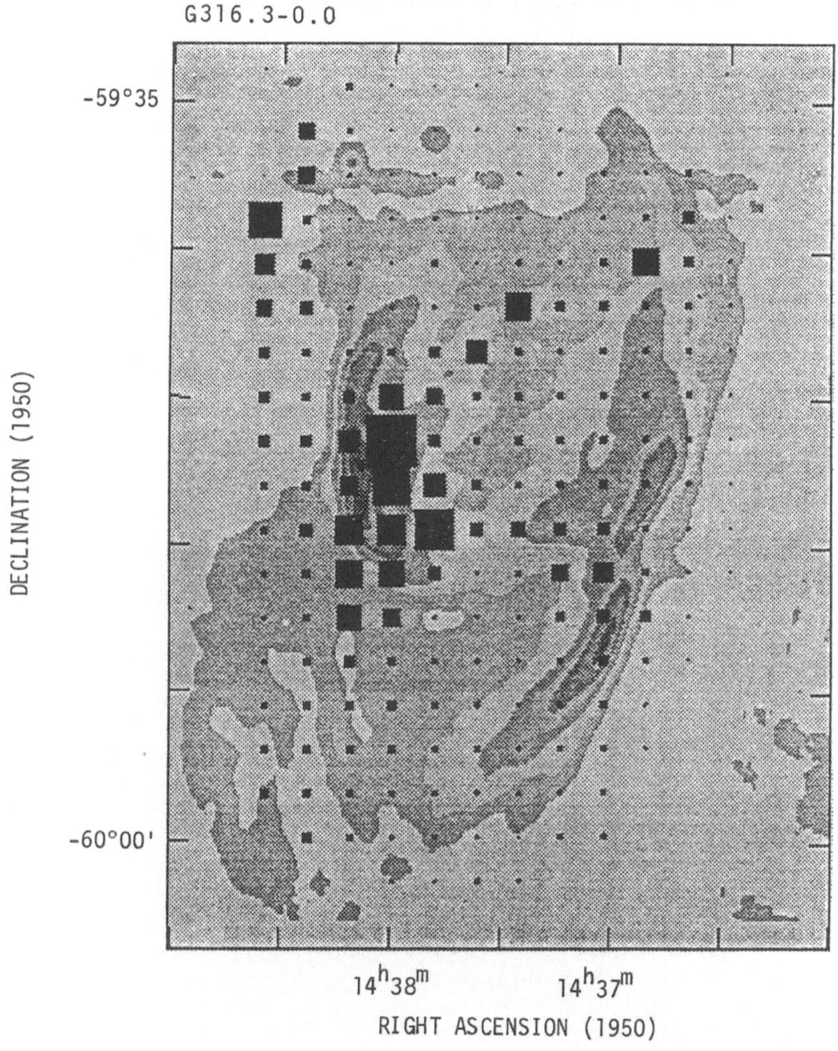

Fig. 5 Distribution of the 8.4/5 GHz depolarization ratio over the SNR G316.3-0.0 (resolution 4'.6 arc) for the region where the polarization intensity is greater than one-tenth of the relevant peak value at each frequency. The linear size of each square is a representation of the depolarization in that direction. The maximum depolarization shown here is 24; however, the bulk of the values shown in Figure 5 are around one-tenth of that value. The superimposed grey-scale map is the same as in Figure 3.

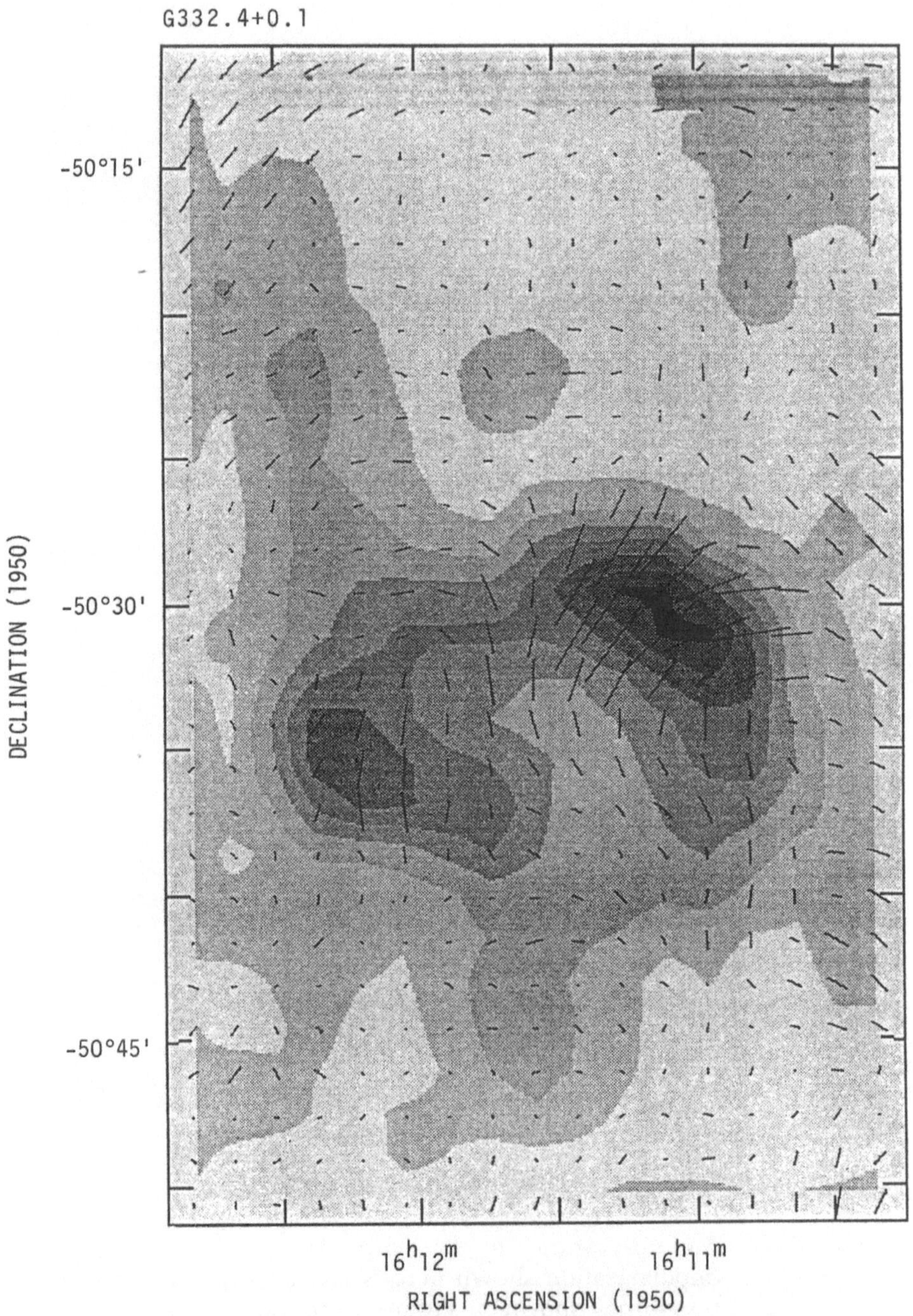

Fig. 6 Total-power grey-scale levels and polarization **E** vectors at 8.4 GHz of the SNR G332.4+0.1. The grey-scale interval is 175 mJy per beam area, the maximum polarization is 262 mJy per beam area (at R.A. $16^h11^m20^s$, Dec. -50°30') and the beamwidth is 3'.0 arc.

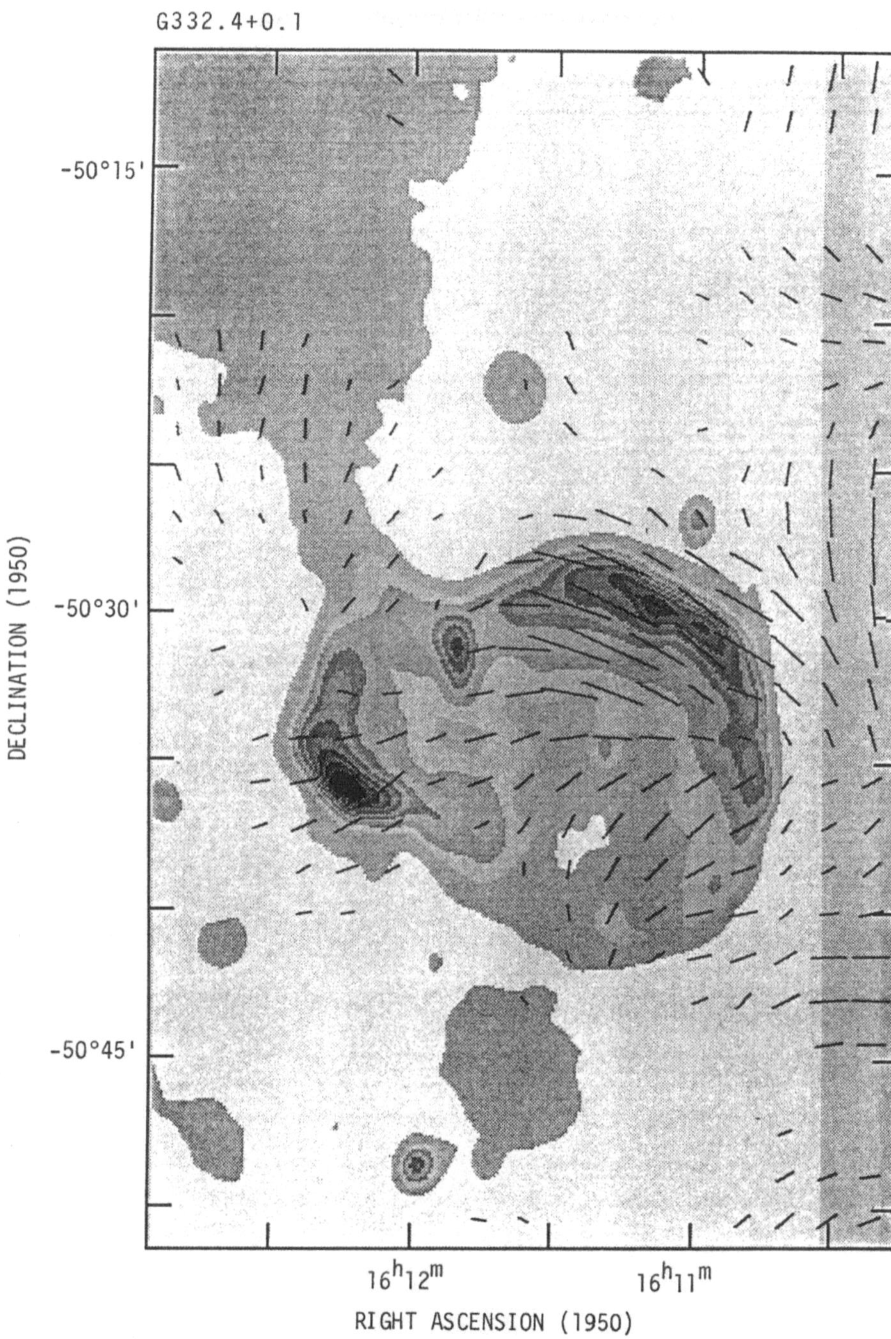

Fig. 7 Direction of the projected magnetic field in G332.4-0+0.1 (resolution 4'.6 arc) superimposed on a grey-scale representation of the higher-resolution 843 MHz map (Fig. 7). The mean 8.4 and 5 GHz polarization intensity is represented by the "vector" magnitude. Data are not shown in Figures 7, 8 and 9, where the 5 and 8.4 GHz polarization is less than one-tenth of the relevant peak polarization.

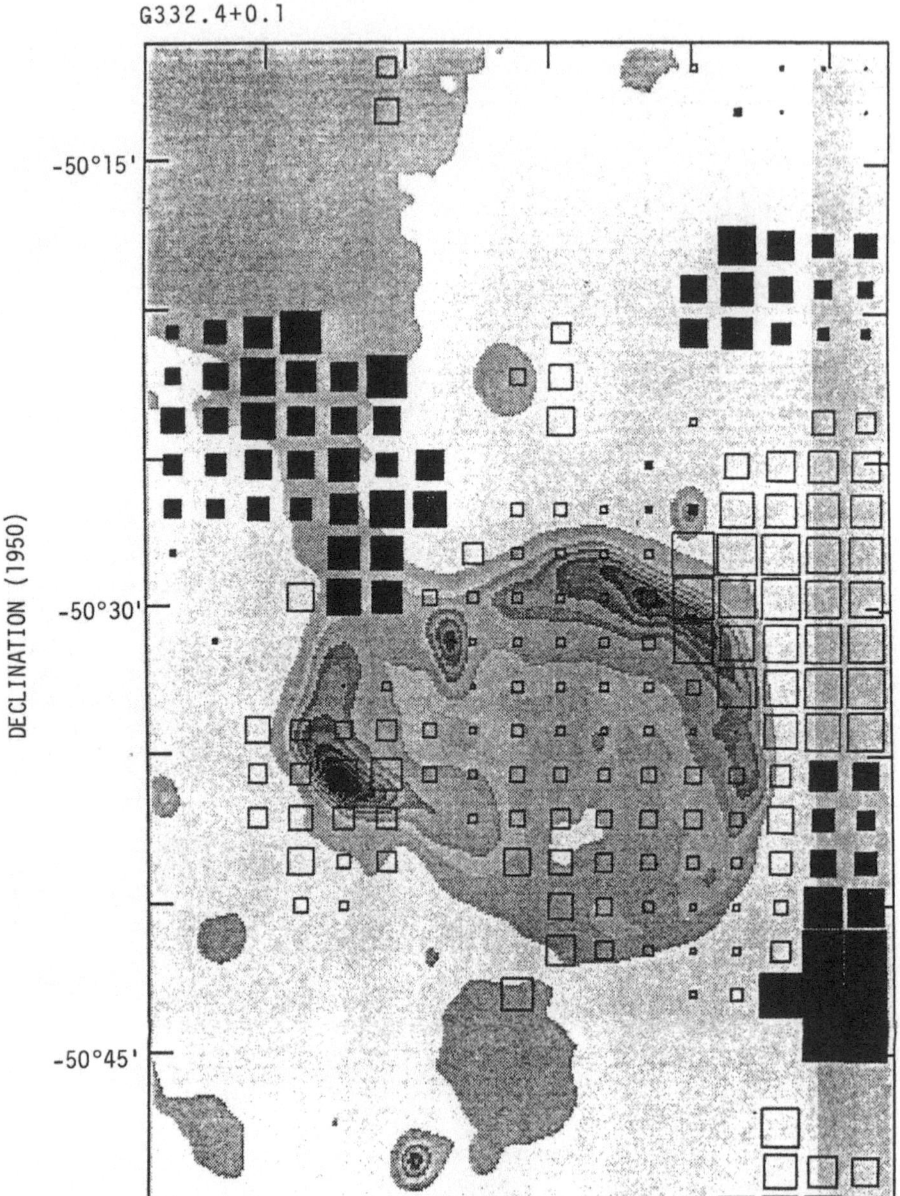

Fig. 8 Distribution of rotation measure over G332.4-0+0.1 (resolution 4'.6 arc) superimposed on the 843 MHz grey-scale map as in Figure 7. Other details are as given for Figure 4.

with some positive values away from the remnant and in particular in the direction of the jet and plume.

The 8.4/5 GHz depolarization (Fig. 9) is generally fairly low, between 1 and 3, with higher values at the two peaks; the depolarization is low in the direction of the jet.

The integrated flux densities are 37.0±3.0, 16.5±2.0 and 13.0±1.5 Jy at 843 MHz, 5 GHz and 8.4 GHz respectively. These are all a little higher than previously published values (see Milne 1979) and yield a spectral index of -0.45±0.05. The spectral index of the jet from 843 MHz to 8.4 GHz lies between -0.3 and -0.4, allowing for uncertainties in the baselevels.

5. Discussion

G316 and G332 both exhibit a circular region flanked by extensions to the south and east respectively. There is some evidence in the direction of the magnetic field that plasma motions have drawn the field into the extensions. This also can be seen in the Cygnus Loop, as noted in Milne (1987). The most likely explanation for this form of structure is that the remnant, expanding spherically, has met a low-density cavity, most likely a wind-driven bubble, into which it has expanded drawing the magnetic field along with it. (This view is contrary to the two-arc model proposed by Manchester (1987) for G316, and perhaps the presence of magnetic field directed into the blowout is the best evidence for the cavity model.)

An alternative explanation for G316 is suggested by the rotation measure map (Fig. 4), where we see evidence for a magnetic field directed away in the northern spherical shell and towards us in the southern extension. Is it that the field is toroidal, or does it merely indicate a tilt in the blowout towards us? No great differences in Faraday rotation is seen across G332 except that the jet may be directed a little towards us. Polarization at another frequency would clarify some problems associated with the Faraday rotation over G332.

No mention was made in Roger et al. (1985) of the point source in G332 at R.A. $16^h11^m18^s.8\pm0^s.6$, Dec. -50° 34' 34"±5" (1950.0). This source is very close to the centre of the suggested spherical shell and may indeed be the compact remnant. Unfortunately, G332 has not been observed in X-rays, which might have been able to detect a stellar remnant. A search for a pulsar in G332 was unsuccessful to a limit of 1 mJy at 1.4 GHz (Manchester et al 1985).

Acknowledgements

The Molonglo Observatory is funded by the Australian Research Grants Committee, the University of Sydney Research Committee and the Science Foundation for Physics within The University of Sydney. We are grateful to Professor B.Y. Mills and his staff, especially Dr J.M. Durdin, for generously making these facilities available and assisting us.

References

Braun, R., Strom, R.G., 1986. *Astron. Astrophys.*, **164**, 208.

Manchester, R.N., 1987. *Astron. Astrophys.*, **171**, 205.

Manchester, R.N., D'Amico, N., and Tuohy, I.R., 1985. *Mon. Not. R. Astron. Soc.*, **212**, 975.

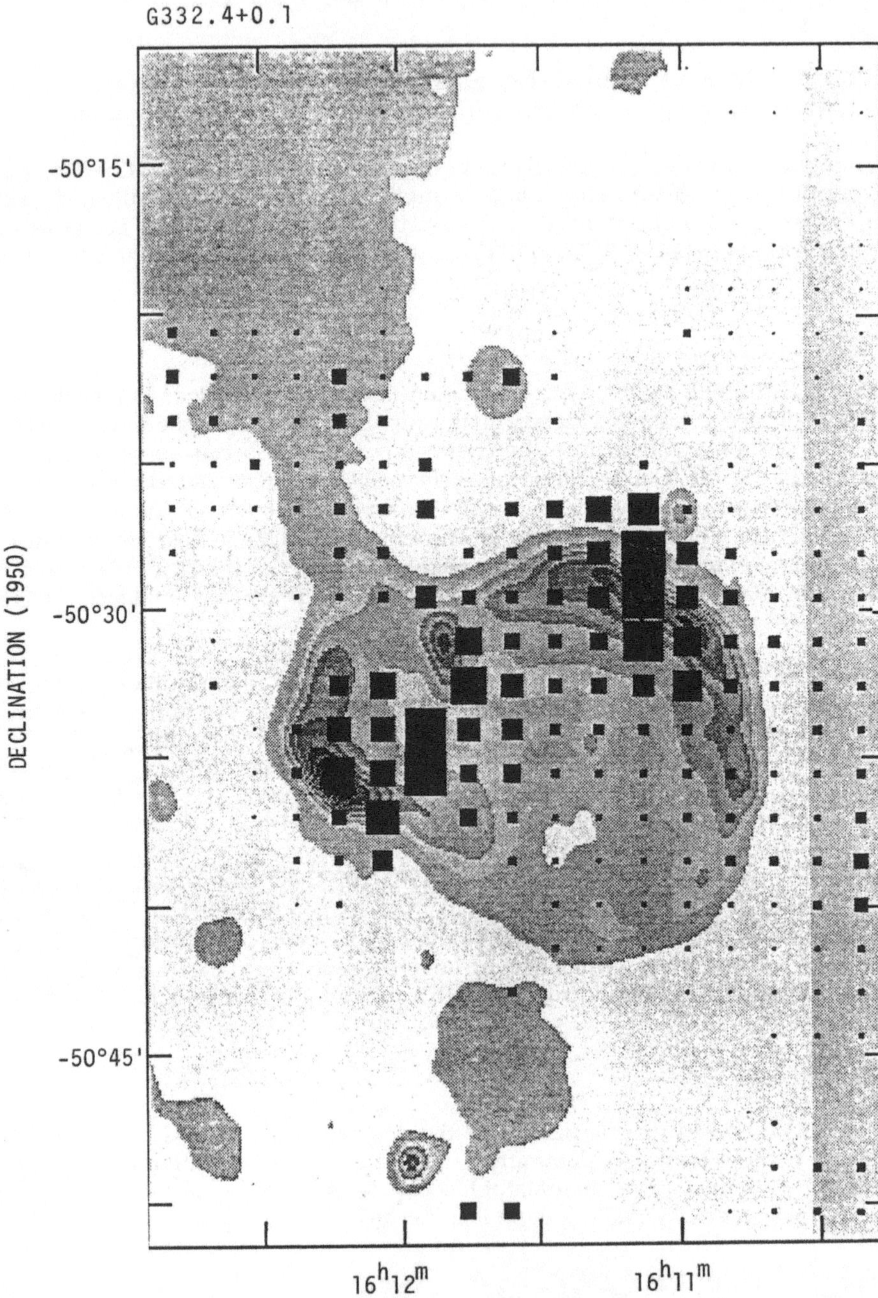

Fig. 9 Distribution of the 8.4/5 GHz depolarization ratio over the SNR G332.4-0+0.1 (resolution 4'.6 arc). Other details are as given for Figure 5. The maximum depolarization ratio shown here is 9; however, the bulk of the values shown in Figure 9 lie between 2 and 3. The superimposed grey-scale map is the same as in Figure 7.

Milne, D.K., 1987. *Aust. J. Phys.*, **40**, 771.

Milne, D.K., 1979. *Aust. J. Phys.* , **32**, 83.

Milne, D.K. and Dickel, J.R., 1975. *Aust. J. Phys.*, **28**, 209.

Milne, D.K., Caswell, J.L., Haynes, R.F., Kesteven, M.J., Roger, R.S. and Bunton, J.D.,1985. *Proc. Astron. Soc. Aust.*, **6**, 78.

Milne, D.K., Roger, R.S., Kesteven, M.J., Haynes, R.F., Wellington, K.J. and Stewart,R.T., 1986. *Mon. Not. R. Astron. Soc.*, **223**, 487.

Pineault, S., Landecker, T.L. and Routledge, D., 1987. *Astrophys. J.*, **315**, 580.

Roger, R.S., Milne, D.K., Kesteven, M.J., Haynes, R.F. and Wellington, K.J., 1985. *Nature*, **316**, 44.

Roger, R.S., Milne, D.K., Caswell, J.L. and Little, A.G., 1986. *Mon. Not. R.Astron. Soc.*, **219**, 815.

Shaver, P.A. and Goss, W.M., 1970. *Aust. J. Phys. Astrophys. Suppl.*, No. 14, 77.

Question

Kundt: How sure are you that linear (radio) polarization "sticks out" beyond the edges of the remnant shell (as in your magnetic field map of IC443)? If this is so then it would indicate either a larger shell or a strong dependence on the ambient field.

Milne: Firstly, each of the diagrams shown in my atlas (Milne 1987) was constructed from the highest-resolution total-power map available, overlaid with magnetic fields which were almost invariably of lower resolution. Consequently the field does stick out past the high-resolution shell (this point was noted in my atlas). Secondly, there is polarization of the galactic background, and this can be seen in the map of G316.3-0.0 presented here. This background polarization is much less confusing at 8.4 GHz than at 5 GHz, indicating a much steeper (i.e. more non-thermal) spectrum for the background polarization than for the remnant. Thirdly, I have gained the impression over the years that the background polarization is strongest around SNRs, and if this is true it does suggest an influence well out into the ambient medium.

PUPPIS A AND ITS ENVIRONMENT AS REVEALED BY INFRARED OBSERVATIONS

Richard G. Arendt
Astronomy Department, University of Illinois
349 Astronomy Building, 1011 W. Springfield Ave.
Urbana, IL 61801, USA

Eli Dwek
Lab. for Astronomy & Solar Physics
NASA Goddard Space Flight Center, Code 685
Greenbelt, MD 20771, USA

Robert Petre
Lab. for High Energy Astrophysics
NASA Goddard Space Flight Center, Code 666
Greenbelt, MD 20771, USA

Abstract. We present observations of the supernova remnant (SNR) Puppis A obtained by the Infrared Astronomical Satellite (IRAS). The infrared (IR) data can be used to study the structure of the interstellar medium (ISM) in which the SNR expands, to constrain physical parameters of the SNR and the surrounding medium, and to investigate gas - dust interactions in the shocked x-ray emitting gas and the nature of the IR emitting dust.

I. Introduction

The work presented here uses the infrared analysis of Puppis A to demonstrate what can be learned through a detailed investigation of individual SNRs. Puppis A is relatively bright in the 25μm and 60μm IRAS data. It is large enough to be fairly well resolved by the IRAS, yet not so large that it covers widely varied background emission. Its location well away from the galactic plane places it in a relatively unconfused region. Puppis A has also been observed in detail at other wavelengths, particularly in the x-ray regime.

In section II we show how the IR data can be used to identify the structure of the (ISM), and how that structure affects the evolution and morphology of the SNR. Section III demonstrates how the IRAS data can be used to evaluate the global properties (e.g. density, temperature, mass, etc.) of the SNR and the surrounding ISM. In section IV, we show how detailed modelling of the IR spectrum can yield information on the size distribution of dust grains and the destruction of dust in a hot plasma.

II. Structure of the ISM

The general brightening of the High Resolution Imager (HRI) image of Puppis A toward the northeast, or toward the galactic plane, has been attributed to a moderately smooth increase in the ambient density by a factor of ~4 in that direction (Petre et al. 1982). The scale height of this gradient is significantly smaller than that expected from the general density gradient of the galactic disk. The true nature of this feature and its full extent are revealed by the IRAS data. Slices through the region of Puppis A at constant galactic longitude reveal that the

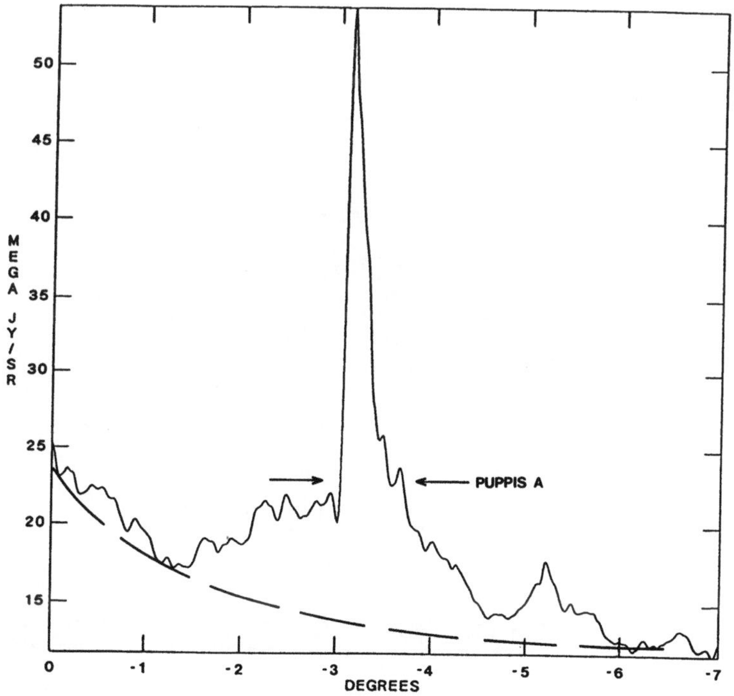

Figure 1.
A plot of 60μm surface brightness vs. galactic latitude. This slice shows that Puppis A lies near the peak of a ridge of emission ($-1^\circ\!.5 > b > -4^\circ\!.5$) which is separate from the general emission of the galactic plane (approximately indicated by the dashed line).

SNR sits near the peak of a prominent "ridge" of emission which is distinct from the emission of the galactic disk. A slice at galactic longitude l = 260°.37 is shown in figure 1. Large scale maps indicate that the "ridge" extends several degrees in galactic longitude. The location of Puppis A atop this feature qualitatively agrees with the enhanced density gradient inferred by Petre et al. (1982).

A specific example of the interaction of Puppis A and the local ISM is revealed at the location of the "eastern knot." This knot is the brightest feature of the SNR in both the x-ray and the infrared regimes. From the HRI data, Petre et al. (1982) demonstrated that this feature is due to a cloud with a density of 10 - 30 cm^{-3} which is being enveloped by the SNR. The x-ray morphology of the knot and its thermal structure, as revealed through optical observations, were explained as indications that the interaction is a recent event. The IRAS data clearly show a cloud which lies to the east of Puppis A (see Figure 2). This cloud abuts the SNR at the position of the eastern knot. Thus, the infrared data reveal that the eastern knot is only the tip, or an outlying portion, of a larger cloud which will eventually be enveloped by the SNR. This cloud has been recently identified by Dubner and Arnal (1988) from its HI and CO emission. They derive a density for the

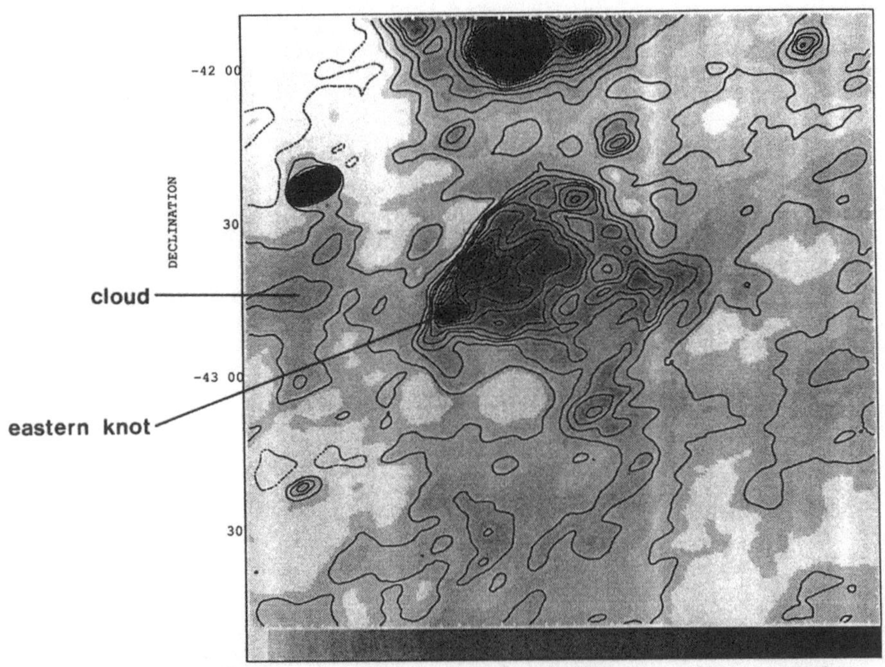

Figure 2.
Puppis A at 60μm. The eastern knot and the center of the cloud (discussed in section II) are indicated. The contour levels are: -5.5, 0, 5.5, 11, 16.5, 22, 33, 44, 55, 66, 77, 88, 99 x 10^{-8} W/m^2/sr.

cloud of ~12 cm^{-3}, in good agreement with the value derived from the x-ray data (Petre et al. 1982). The HI morphology of the cloud is very similar to that observed in the infrared.

III. Physical properties of Puppis A

Only a small fraction of the IR emission of Puppis A can be attributed to synchrotron and free-free emission processes. Lack of correlation between the optical and IR morphologies suggests that line emission is also unimportant. These facts and the excellent correlation between the IR and x-ray emission, lead us to conclude that most of the IR emission from Puppis A is thermal emission from shock-heated dust. The IR emission can therefore be used to set constraints on the temperature and density of the x-ray emitting gas and to estimate the mass of the IR emitting dust. These may then be used to derive various SNR parameters (Dwek 1987).

The flux densities measured for Puppis A, after removal of a planar background level, are listed in Table 1. Also shown in Table 1 are the color corrected flux densities (IRAS Explanatory Supplement, 1985), and the dust temperatures derived from the ratios of the flux densities at the given and the next shorter wavelengths. The color correction factors and dust temperatures were calculated for a blackbody source modified by a $\lambda^{-1.5}$ emissivity law. The dust temperature listed for the ratio of the 12μm to the 25μm flux densities was derived from only the region of the eastern knot, since only this region shows discernable emission at 12μm.

Table 1

Wavelength (μm)	Flux Density (Jy)	Color Corrected Flux Density (Jy)	Dust Temperature (K)
12	<100	...	
25	245	291	120
60	1195	1250	65
100	1265	1250	44

We first make some simplifying assumptions, which will be relaxed in a forthcoming paper (Arendt et al. 1988):

1) We adopt an effective grain size of 0.1μm.

2) The gas temperature is $T_g = 10^7$K. This value is characteristic of the smooth, x-ray emitting regions of the SNR, as determined by non-equilibrium ionization models

(field N of Szymkowiak 1985) and by the hotter component of two temperature models of the x-ray spectrum (Zarnecki et al. 1978, Szymkowiak 1985). Lower temperatures characterize dense knots, but these knots are atypical of the medium which influences the overall expansion of Puppis A, and the knots provide only a fraction of the total IR emission.

3) No significant amount (by mass) of grain destruction has taken place. This assumption is supported by the relatively young ages derived for Puppis A by various methods (see below).

4) The SNR is approximated as being spherically symmetric and undergoing Sedov phase expansion in a uniform medium. These approximations may be improved in the future as the structures of Puppis A and its environment are better defined.

From the 60μm to 100μm flux density ratio we derive a dust temperature of 44K. We find an angular radius of ~28′ for Puppis A from the IR maps.

The gas density (n) of the hot plasma is found from the gas and dust temperatures as demonstrated in figure 7 of the work by Dwek (1987). The density of the ambient medium (n_0) is the idealized, adiabatic ratio of one quarter of the gas density found behind the shock front of the SNR. The radius (R) is determined from the ambient density, the angular radius, the gas temperature, and the total IR flux of the SNR through the standard Sedov solution (see equation 23b of the work by Dwek (1987)). The distance (d) to the SNR is a simple geometric relation between the actual and angular radii of the SNR. The mass (M) of swept up material is found from the product of the SNR's volume and the ambient density. The infrared luminosity (L) of the SNR is determined from the total flux and the distance. Next, if we assume that Puppis A is in the Sedov phase, we can derive the initial energy (ϵ_0) of the explosion and the age (t) of the SNR from the standard Sedov relations for the radius and temperature. The energy and age are thus found as functions of the radius, gas temperature, and ambient density. The values derived for all these parameters are listed in table 2. Also listed are the results obtained by using a slightly lower gas temperature and a slightly higher dust temperature.

Table 2

T_d (K)	T_g (K)	n (cm^{-3})	n_0 (cm^{-3})	R (pc)	d (kpc)	M ($M_\odot$)	L ($L_\odot$)	ϵ_0 (10^{51} ergs)	t (yrs)
44	10^7	0.58	0.14	12	1.5	34	5600	0.38	5500
44	$8x10^6$	0.79	0.20	7.6	0.9	12	2200	0.10	3900
45	10^7	0.66	0.17	9.2	1.1	18	3000	0.20	4200

From x-ray data, other estimates of these parameters include T_g = 3 - 10x10^6 K, n_0 = 0.1 - 1.0 cm^{-3} (Szymkowiak 1985), and $n_0 \approx 1$ cm^{-3} (Petre et al. 1982). From radio Σ-D relations Milne et al. (1983) derive R = 12.5 or 16.5 pc, d = 1.8 or 2.4 kpc, and t ≈ 5000 yrs, though an age of ~2000 yrs is suggested by comparison to other young SNRs. Observations of optical filaments (Winkler and Kirshner 1985) indicate a kinematic age of $\leq$ 10000 yrs.

To summarize Table 2, our preferred values (T_g = 10^7K, T_d(0.1μm grains) = 44K) favor an ambient density (n_0 = 0.14 cm^{-3}) and distance (d = 1.5 kpc) at the low end of the range of previous estimates, and an age (t = 5400 yrs) a little toward the high end of the range. It is not possible to simultaneously bring all three of these parameters into closer agreement with previous estimates by altering T_g, T_d, the total IR flux, or the assumed grain size.

Our results are very strongly dependent on the derived dust temperature, which is uncertain by ~20% due to uncertainties in the IRAS flux densities, and somewhat less strongly on the assumed grain size. However this also means that the dust temperature is tightly constrained by assuming that the parameters calculated should not fall greatly outside the range of previous estimates.

IV. Gas - Dust Interactions and the Nature of the Emitting Dust

Detailed modelling of the x-ray spectra of selected regions of the SNR (Szymkowiak 1985) provide strong constraints on the gas temperature and density in these regions. These constraints can be used in conjunction with the IRAS data to infer the size distribution of the dust particles and the effect of grain destruction. The x-ray spectra were obtained by the Einstein satellite's Solid State Spectrometer (SSS), which had an aperture with a diameter of 6'. This compliments the IRAS data which have a resolution of ~3'x5'. For four spectrally distinct regions, Szymkowiak calculated non-equilibrium ionization models parameterized by T_g, the gas temperature, and η, the product of the initial energy and the square of the ambient density. The infrared images were integrated over the same regions as observed by the SSS to obtain the IR spectra to be modelled.

The grain size distribution can be characterized by the exponent of a power law spectrum of sizes, and by the maximum and minimum grain sizes which may differ for silicate and graphite grains. There are too many free parameters and the data are insufficient in both quantity and quality to attempt to find a unique grain size distribution from the IR spectrum and the gas parameters. Therefore, our approach is to model the IR data starting with a standard grain size distribution and then find what modifications, if any, are needed to provide a better fit. The standard initial grain size distribution is an MRN distribution (Mathis et al. 1977) extended to smaller grain sizes using the same power law. The extension to smaller grain sizes is suggested by the IR spectra of diffuse cirrus clouds in the ISM (Draine and Anderson 1985; Weiland et al. 1986; and references therein). The adopted distribution had a power law index α = -3.5, maximum grain size a(max) = 0.25μm for both graphite and silicate grains, and minimum grain size a(min) = 0.0003μm and 0.0026μm for graphite and silicate grains respectively. The procedure used to calculate the IR spectrum from the grain size distribution and the plasma parameters is explained in detail in the work by Dwek (1986). Grains smaller than 0.05μm were found to be stochastically heated. Larger grains were effectively at their equilibrium temperatures.

The first modification made to improve the fit of the models was to include the effect of sputtering on the grain size distribution. This resulted in a distribution which was relatively unchanged at the large end where the fractional reduction in grain size was small, and flattened at the small end where the reduction in size was a significant fraction or as large as the initial size of the grains. The effect on the IR spectrum was to reduce the relative amount of emission at 12μm, 25μm and to a lesser extent at 60μm, and also results in an increase in the inferred mass of IR emitting dust. The emission at these shorter wavelengths is primarily due to the stochastically-heated, small grains which radiate most of their energy at higher temperatures.

Figure 3 shows models of the IR emission from Szymkowiak's region "C," which contains the eastern knot. The light curve is the model using the initial grain size distribution. The heavy curve shows the spectrum from the initial distribution after modification by sputtering for ~1800 years (Draine and Salpeter 1979). For other regions we find that a similar amount of sputtering also produces improved

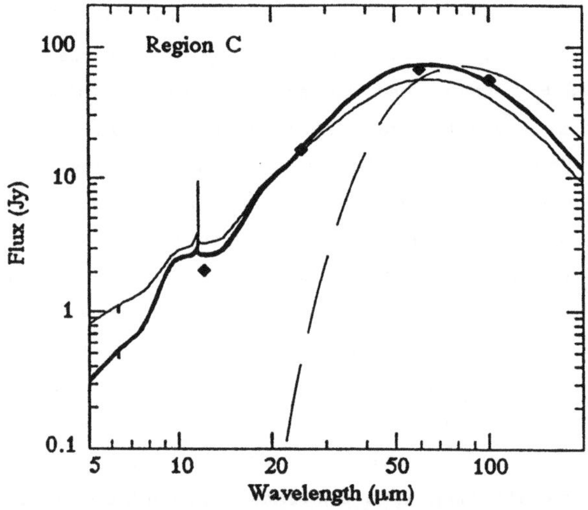

Figure 3
Model IR spectra for a 6′ diameter region of Puppis A. The light curve is modelled from the standard grain size distribution. The dark curve is modeled from the same distribution after sputtering acts for ~1800 yrs to reduce the grain sizes. The dashed curve shows the spectrum of a grain distribution which lacks small, stochastically-heated grains. (The spike at 11.5μm is a graphite resonance feature (see e.g. Draine and Lee 1984) which is too narrow to contribute a significant amount of flux to the IRAS 12μ band.)

fits to the IR data at 12μm and 25μm. For two of the other regions however the fit is still poor unless the maximum or minimum grain sizes (depending on the region) are raised.

Also shown in Figure 3 is the spectrum produced from a model distribution which lacks small, stochastically-heated grains (dashed curve). Such models produce spectra which underestimate the 12μm and 25μm flux densities by at least an order of magnitude. Thus, small stochastically-heated grains must be included to obtain reasonable model spectra.

V. Conclusion

From this preliminary study of Puppis A, we find evidence that distinct structure in the ISM influences both the large and small scale morphology of the SNR. We estimate a number of parameters (density, mass, radius, etc.) of Puppis A from the IR data. Finally we demonstrate the importance of sputtering and small, stochastically-heated grains in modelling the observed IR spectrum of Puppis A.

Acknowledgement: We thank J. Dickel for valuable comments on earlier versions of this text.

References:

Arendt, R. G., Dwek, E., Petre, R., and Dickel, J. R. 1988, in preparation.
Draine, B. T., and Salpeter, E. E. 1979, Ap. J., 231, 438.
Draine, B. T., and Lee, H. M. 1984, Ap. J., 285, 89.
Draine, B. T., and Anderson, N. 1985, Ap. J., 292, 494.
Dubner, G. M. and Arnal, E. M. 1988, in Supernova Remnants and the Interstellar Medium, eds. R. S. Roger, and L. Landecker, (Cambridge: Cambridge University Press), p. 249.
Dwek, E. 1986, Ap. J., 302, 363.
———, 1987, Ap. J., 322, 812.
Dwek, E., Dinerstein, H. L., Gillett, F. C., Hauser, M. G., and Rice, W. L. 1987, Ap. J., 315, 571.
IRAS Catalogs and Atlases, Explanatory Supplement 1985, edited by Beichman, C. A., Neugebauer, G., Habing, H. J., Clegg, P. E., and Chester, T. J., (Washington D. C.: U. S. Government Printing Office).
Mathis, J. S., Rumpl,. W., and Nordseick, K. H. 1977, Ap. J., 217, 425.
Milne, D. K., Goss, W. M., and Danziger, I. J. 1983, M. N. R. A. S., 204, 237.
Petre, R., Canizares, C. R., Kriss, G. A., and Winkler, F. P. 1982, Ap. J., 258, 22.
Szymkowiak, A. E. 1985, "X-ray Spectra of Supernova Remnants," NASA Technical Memo. 86169.
Weiland, J. L., Blitz, L., Dwek, E., Hauser, M. G., Magnani, L., and Rickard, L. J. 1986, Ap. J. (Letters), 306, L101.
Winkler, P. F., and Kirshner, R. P. 1985, Ap. J., 299, 981.
Zarnecki, J. C., Culhane, J. L., Toor, A., Seward, F. D., and Charles, P. A. 1978, Ap. J. (Letters), 219, L17.

HIGH RESOLUTION RADIO OBSERVATIONS OF G11.2−0.3

D A Green
National Research Council, Herzberg Institute of Astrophysics,
Dominion Radio Astrophysical Observatory,
P.O. Box 248, Penticton, B.C. V2A 6K3, Canada.

Summary: High-resolution (3 arcsec) radio observations of the relatively bright 'shell' SNR G11.2−0.3 made with the VLA at 1.46 and 4.76 GHz are presented. These reveal a clumpy shell of emission, which resembles an evolved version of Cassiopeia A, rather than the smooth shell of emission that is seen from the remnants of Tycho's and Kepler's SN.

1 Introduction

G11.2−0.3, a non-thermal radio source in the Galactic plane, has appeared in catalogues of supernova remnants (SNRs) for many years and has been proposed (Clark & Stephenson 1977) as the possible remnant of the supernova (SN) of AD386, one of only eight historical supernovae recorded over the last two millenia. Downes (1984) made radio observations of G11.2−0.3 with the VLA at 1465 MHz with a resolution of 20×25 arcsec^2 which showed an apparently sharply bounded, almost perfectly circular shell ≈ 4 arcmin in diameter, brightest in the southeast. The structure resembled that seen in the remnants of both Tycho's and Kepler's SN, with little evidence for the clumps or extended plateau seen outside the main ring of emission in the case of Cassiopeia A. Downes argued that its circular appearance and its relatively high radio surface-brightness indicate that G11.2−0.3 is a young remnant, but with the relatively poor resolution of the observations then available it was not possible to distinguish between two likely alternatives: either that G11.2−0.3 is similar to Kepler's and Tycho's remnants (and therefore 300–500 yr old and, presumably, the result of a 'type I' SN), *or*, that it is an evolved version of Cassiopeia A (and possibly the result of the SN of AD386, a 'type II' SN explosion of a massive star). Here I present radio observations of G11.2−0.3 with sufficiently high resolution (3 arcsec) to show that its radio structure is a clumpy shell, as expected for an evolved version of that of Cassiopeia A, rather than a sharply bounded shell seen in the case of both Tycho's and Kepler's SNRs.

2 Observations and Results

G11.2−0.3 was observed with the Very Large Array (VLA) at frequencies in the L and C bands in various configurations between 1984 July and 1985 May. The A-array data in C band, and to a lesser extent those in L band, had poor signal-to-noise ratios, so it was decided to make graded maps at both frequencies with a resolution corresponding to that of the good data in L band (*i.e.*, ≈ 3 arcsec). The data were calibrated by standard procedures, with the flux density scale based on 3C286. In each band the data were combined in the visibility plane, provided valid data were available for all four channels (*i.e.*, formal Stokes parameter I), with data from larger baselines given less weight (using a gaussian that gave a grading 0.3

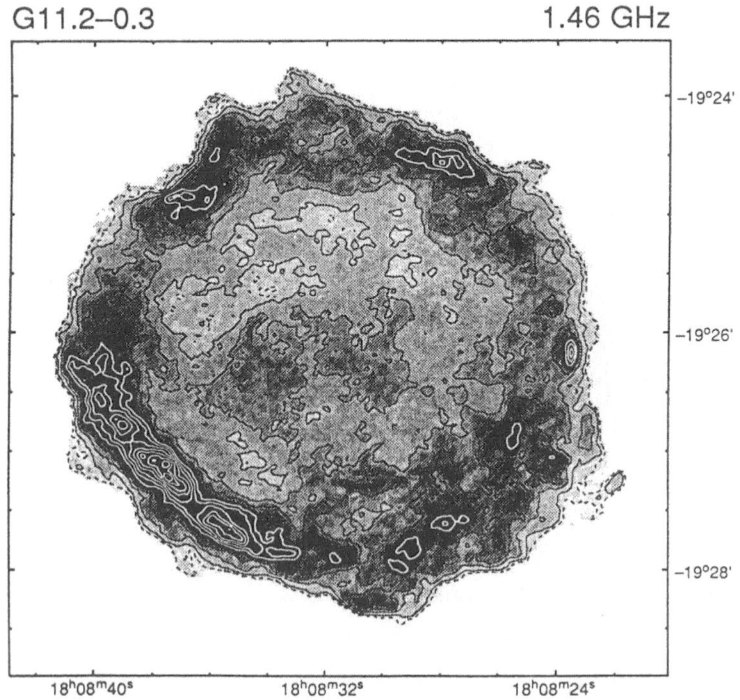

Figure 1: MEM map of G11.2–0.3 at 1.46 GHz convolved with a gaussian of 3 arcsec HPBW. The greyscale, white to black, is 0 to 5 mJy beam^{-1}. Contours are: 0.5 (dashed black), 1, 2, 3, 4 (continuous black), and 5, 6,... (continuous white) mJy beam^{-1}.

at 74 kλ). 'Dirty' maps and beams were then produced and deconvolved using the Maximum Entropy Method (MEM) with a featureless initial model (the noise levels were 0.21 and 0.16 mJy beam^{-1} at 1.46 and 4.76 GHz respectively). The MEM maps were then convolved with a 3-arcsec (HPBW) gaussian beam to give the maps shown in Figs 1 and 2.

Comparison of Figs 1 and 2 for a study of the variation of spectral index across G11.2–0.3 is limited by the lack of data from small interferometer baselines. This is particularly a problem at C band, and accounts for the relative faintness of the emission from the middle of the remnant in Fig.2 compared with Fig.1.

3 Discussion

Although Figs 1 and 2 confirm the already-known basic features of the radio emission from G11.2–0.3, they reveal for the first time its clumpy nature, and the lack of a sharp outer boundary. There are several clumps of emission *outside* the boundary of the main emission, most notably those near $18^h08^m22^s$, $-19°27'30''$ (1950.0), which are very reminiscent of, but apparently somewhat larger than the knots of emission seen in Cassiopeia A (*e.g.*, Bell, Gull & Kenderdine 1975; Tuffs 1986).

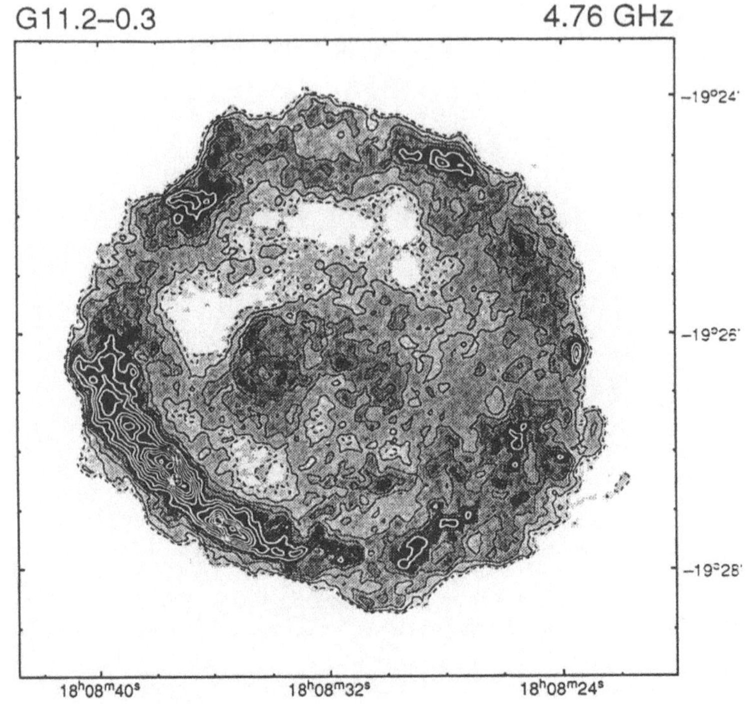

Figure 2: MEM map of G11.2−0.3 at 4.76 GHz convolved with a gaussian of 3 arcsec HPBW. The greyscale, white to black, is 0 to 2 mJy beam^{-1}. Contours are: 0.2 (dashed black), 0.4, 0.8, 1.2, 1.6 (continuous black), and 2.0, 2.4,... (continuous white) mJy beam^{-1}.

Tycho's and Kepler's SNRs both have 'smooth' shells of radio emission, which is thought to be the result of having swept-up a considerable amount of the interstellar medium and being in the Sedov phase (see Strom, Goss & Shaver 1982 or Tan & Gull 1985 for radio expansion observations of Tycho's SNR), with magnetic fields and relativistic particles generating radio emission at the outer shock of the blast-wave. The dominant radio emission from Cassiopeia A, on the other hand, is from the bright ring that represents the zone of Rayleigh-Taylor instability (*e.g.*, Gull 1975) at the interface between the swept-up interstellar medium and the SN ejecta. This is expected for remnants relatively early in their evolution, before the swept-up mass dominates the dynamics of the remnant. The bright ring is surrounded by a weak 'plateau' of emission, the edge of which marks the outer blast wave. The fast optical knots associated with Cassiopeia A have an expansion timescale of ≈ 300 yr (Kamper & van den Bergh 1976), and since the optical knots are essentially undecelerated their expansion timescale is the age of the remnant, which makes Cassiopeia A the youngest known Galactic shell SNR. This view is consistent with expansion timescale of the *bulk* of the radio emission from Cassiopeia A (Green 1988), rather than that given by the compact radio knots (≈ 1000 yr, Tuffs 1986) which contribute only a small fraction of the total radio emission from the remnant. As Cassiopeia A sweeps up more material it will presumably tend towards a Sedov phase of expansion, in which case the radio emission would then be dominated, as in Tycho's and Kepler's SNRs, by a sharply bounded, smooth shell (provided that the surrounding medium is uniform on the appropriate scales); the knots of emission being of decreasing importance as the dense clumps of ejecta are decelerated and dissipated. The structure of G11.2−0.3's radio emission (Figs 1 and 2) is, therefore, that to be expected from an evolved version of Cassiopeia A.

A distance estimate of > 26 kpc is available for G11.2−0.3 from an H I absorption spectrum made by Becker, Markert & Donahue (1985), who regard as significant the weak absorption that is seen out to -30 km s^{-1}. At this distance G11.2−0.3 would have a diameter of > 30 pc, which is difficult to reconcile with the apparent youth of the remnant inferred from its high radio surface brightness and its structure. There is, however, no strong absorption seen in Becker *et al.*'s spectrum of G11.2−0.3 for velocities between $+45$ km s^{-1} and the tangent point at $+120$ km s^{-1}, which is unlikely if the remnant is at 26 kpc or more, beyond the solar circle. An alternative interpretation of Becker *et al.*'s H I spectrum is that G11.2−0.3 is at the near distance of ≈ 5 kpc corresponding to $+45$ km s^{-1}, with the weak absorption at negative velocities being due to unusual motions in local gas. At this distance, the diameter of G11.2−0.3 is ≈ 6 pc, which is more easily reconciled with the apparent youth of the remnant. However, an association of G11.2−0.3 with the SN of AD386 is not easy, as the small size for the then presumed age of the remnant implies that, unless the ejection velocity was quite small, the SN ejecta have been strongly decelerated, which is not what is inferred from the radio structure revealed in Figs 1 and 2.

Acknowledgements

This work was done in collaboration with S.F. Gull, S.M. Tan and A.J.B Simon at the Mullard Radio Astronomy Observatory, Cambridge, U.K., during which time I was supported by a Junior Research Fellowship from Churchill College, Cambridge. The VLA is a facility of the National Radio Astronomy Observatory which is operated by Associated Universities Inc., under contract with the U.S. National Science Foundation.

References

Becker, R.H., Markert, T. & Donahue, M., 1985. *Astrophys. J.*, **296**, 461.
Bell, A.R., Gull, S.F. & Kenderdine, S.K., 1975. *Nature*, **257**, 463.
Clark, D.H. & Stephenson, F.R., 1977. *The Historical Supernovae*, (Pergamon, New York, U.S.A).
Downes, A.J.B., 1984. *Mon. Not. R. astr. Soc.*, **210**, 845.
Green, D.A., 1988. In: *Supernova Remnants and the Interstellar Medium*, (IAU Colloquium 101), p.51, eds. Roger, R.S. & Landecker, T.L., (Cambridge University Press, England).
Gull, S.F., 1975. *Mon. Not. R. astr. Soc.*, **171**, 263.
Kamper, K.W. & van den Bergh, S., 1976. *Astrophys. J. Suppl.*, **32**, 351.
Strom, R.G., Goss, W.M. & Shaver, P.A., 1982. *Mon. Not. R. astr. Soc.*, **200**, 473.
Tan, S.M. & Gull, S.F., 1985. *Mon. Not. R. astr. Soc.*, **216**, 949.
Tuffs, R.J., 1986. *Mon. Not. R. astr. Soc.*, **219**, 13.

G70.7 + 1.2: SUPERNOVA REMNANT?

M. de Muizon[1,2], R.G. Strom[3], M.J.A. Oort[1], J.J. Claas[4], R. Braun[5]

[1] Sterrewacht Leiden, Postbus 9513, NL-2300 RA Leiden, The Netherlands
[2] Observatoire de Paris, Section de Meudon, F-92190 Meudon, France
[3] Netherlands Foundation for Radio Astronomy, Postbus 2, NL-7990 AA Dwingeloo, The Netherlands
[4] Laboratory for Space Research, Postbus 9504, NL-2300 RA Leiden, The Netherlands
[5] National Radio Astronomy Observatory, P.O. Box 0, Socorro, NM 87001, USA

Summary: Recent data in several wavebands on the peculiar galactic object G70.7+1.2 is reviewed. A new VLA map clearly shows the radio shell. Possible interpretations of the object include supernova remnant, nova shell, and nebula resulting from mass loss from a giant or supergiant, although each of these interpretations has its own difficulties.

1. Introduction

G70.7+1.2 has long been known as a prominent but compact nebulosity (e.g. Minkowski, 1948), which can be readily seen on the Palomar Sky Survey (Fig. 1). This fact, combined with its low galactic latitude, place it in the Milky Way. It is also a source of rather weak radio emission (peak flux density of slightly more than 1 Jy), which Green (1985) found to be in the form of an irregular shell some 20" arc in diameter. Its radio spectrum has not been terribly well-defined, largely as a result of the weakness of its emission and the effects of confusion from galactic features, but it is relatively flat.

There have consequently been doubts about its nature. Green (1985), largely on the basis of its radio spectrum, thought it was probably an HII region, while Reich et al. (1985) concluded that there was evidence for a nonthermal spectrum and that G70.7+1.2 is therefore a supernova remnant (SNR). Moreover, its compactness and consequent high surface brightness suggest a young object, hence one resulting from a recent supernova (SN). This very fact, however, led Green (1986) to argue that a SN origin was unlikely, since its proximity should have ensured a prominent, if not spectacular, optical event which was apparently never recorded.

Here we will concern ourselves with recent radio, infrared, optical and X-ray observations which throw new light on this still puzzling object.

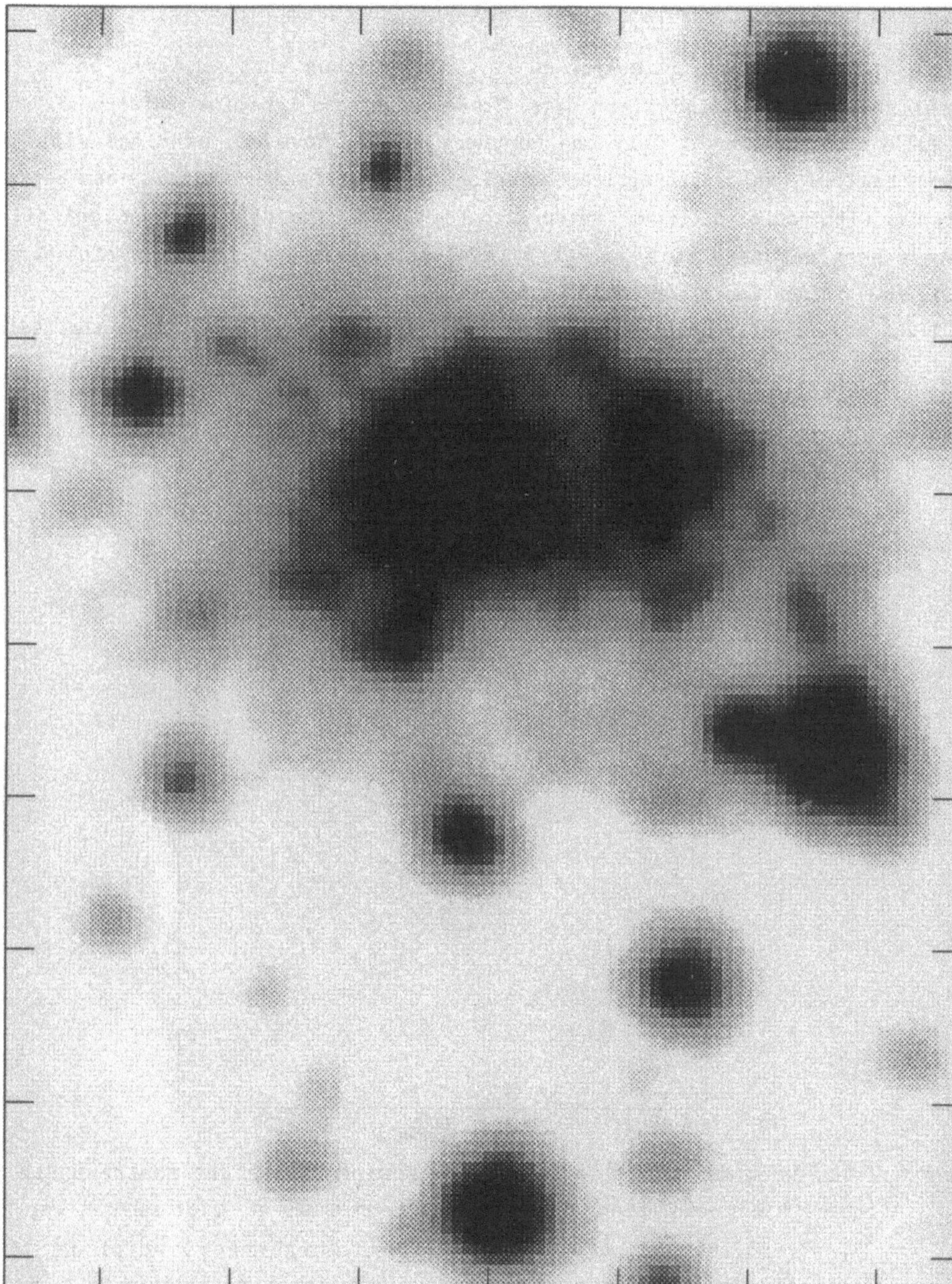

Figure 1. Red image of G70.7+1.2 from the PSS 'E'-plate. The emission is mainly from Hα and [NII]$\lambda\lambda$6548, 6583. Notice the dust lane bifurcating the brightest nebulosity.

2. New observations

A WSRT 6 cm observation (Muizon et al., 1988) shows that in addition to the partial shell along the northern edge, lower brightness irregular emission extends some 30" arc to the south. Only the northern shell, however, overlaps with the western half of the bright optical emission, although faint nebulosity can be seen coinciding with radio continuum features to the south. The fact that polarization may have been detected at 6 cm, at a level similar to that found by Reich et al. (1985), is evidence for a nonthermal component.

A VLA map recently obtained with the A-array at 20 cm (Fig. 2) shows the basic

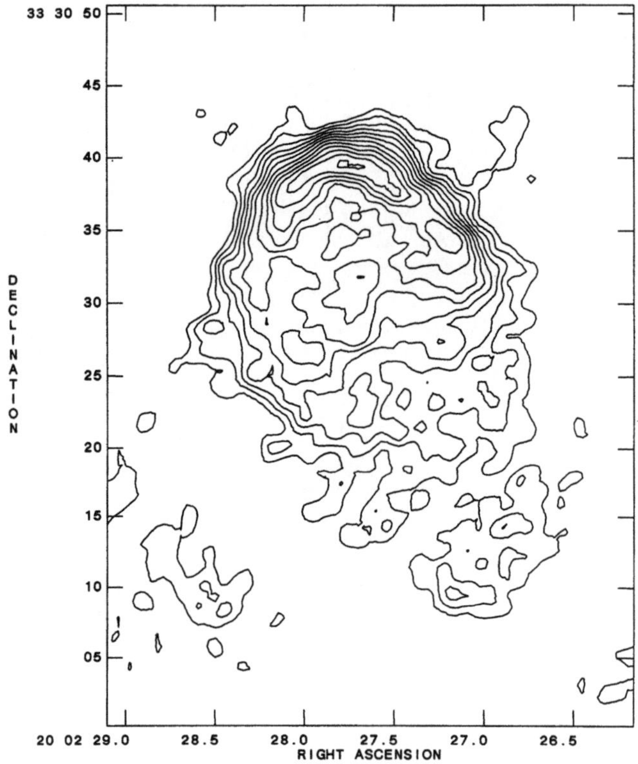

Figure 2. A VLA 20 cm map of the radio shell, showing the bright northern rim at about 1" arc resolution. Contours have been drawn at 3, 6, 10, 15, 21, 28, 37, 47, 58, 70, 83, and 97% of the peak brightness, which is 17.6 mJy/beam. The feature southeast of the main emission is unreal, being caused by weak grating lobes.

structure in still greater detail. Evidence for limb brightening in the southern extension is weak at best, although it is possible to trace the eastern rim southwards, to a point where it has curved substantially to the west. Additional observations were done with the VLA (D-Array) at 6 and 2 cm to search for recombination line emission. Nothing could be detected to limits of 1% (at 6 cm) and 5% (2 cm), levels at which recombination lines have been readily detected in compact HII regions. The nondetection argues against a substantial thermal radio component, as does the spectral index found for the bright northern shell based upon a comparison between the WSRT (6 cm) and VLA (20 cm) measurements: $\alpha = -0.6$ (where flux density $\propto$ frequency$^{\alpha}$).

Strong infrared emission was detected from G70.7+1.2 with IRAS, both the survey instrument and the LRS. Spectra in the 8 to 23 μ range reveal the 11.3 μ feature believed to be associated with PAH-type molecules (e.g. Léger and Puget, 1984), but do not detect other lines typical of HII regions (such as those at 12.8 and 18.7 μ of [NeII] and [SIII]). Follow-up observations with an infrared spectrophotometer on UKIRT readily detected the 3.3 μ line usually associated with the 11.3 μ feature, in addition to a strong continuum. Measurements with various apertures reveal that both line and continuum are extended, and that they peak near a stellar feature in the eastern half of G70.7+1.2. Near infrared photometry suggests that it is a red, late-type star.

Two long-slit spectra obtained with the IDS on the Isaac Newton Telescope cover the wavelength range from about 5000 to 6800 Å. In one the slit was aligned north-south, the other was east-west, and both (partially) include the stellar object referred to above. The most interesting lines found fall in the range from [OI]λ6300 to [SII]λ6731, although the north-south scan was slightly shifted in wavelength and did not include the latter. Briefly summarizing the main results, [OI]λλ6300, 6363, [NII]λλ6548, 6583 and [SII]λλ6716, 6731 are all quite strong compared to Hα on the western half of the source (with both components of the [SII] doublet equalling or exceeding Hα in intensity in many places). At the position of the stellar object, the Hα intensity increases by about two orders of magnitude, and remains the dominant line throughout the eastern half of the nebula. The line ratios are thus qualitatively indicative of shock-excited gas on the west, and radiative ionization on the east. The [SII] doublet ratio (λ6716:λ6731 = 0.73) suggests electron densities of about 5×10^3 cm^{-3} and a temperature near 2×10^5 K.

The spectra of the stellar object on the east of G70.7+1.2 show, in addition to the strong Hα, emission lines of numerous quadrupole transitions of FeII superimposed upon a rather red continuum.

A 30,000 s observation of G70.7+1.2 with the X-ray satellite EXOSAT failed to detect any emission in the 0.05 to 2 keV band.

3. Discussion

The radio data indicate that the emission from G70.7+1.2 is predominantly if not entirely nonthermal. The optical spectra of the nebulosity which coincides with the radio shell indicate shock excitation, and are hence consistent with the radio. However, the spectra of the eastern nebulosity suggest a compact HII region which has not been detected in any of our radio measurements. The infrared data, on the other hand, seem to contradict this with the nondetection of typical compact HII region diagnostics such as [NeII] and [SIII]. All these facts, together with a rather low radio luminosity and undetectable X-ray emission, have to be satisfactorily accounted for if a model is to be acceptable.

Let us first consider the major elements, without regard to their possible origins. In the red image (Fig. 1), there appears to be a dust lane running diagonally across the nebulosity from northeast to southwest. To the east of the dust lane we see a roughly circular region which has been radiatively ionized. The dust lane suggests that this eastern nebula lies in front of the shock excited emission to the west, and that it is associated with dust, which qualitatively agrees with the substantial infrared emission observed. To explain the relatively flat radio spectrum, in particular the apparent low frequency cutoff (Green, 1986), we now suppose that the plasma frequency of the eastern nebula is such as to produce free-free absorption at frequencies below 300 MHz.

For a cutoff frequency of 300 MHz, we require an electron density of about 10^3 cm^{-3}, a value similar to that suggested by the [SII] doublet ratio in the shock excited gas. Are such densities ruled out by other considerations? The total flux density radiated by free-free emission from the eastern nebula would amount to about 100 mJy at 6 cm, considerably less than the 600 mJy observed from the entire source. While this may be an acceptable proportion, it is somewhat disturbing that no radio emission has been detected to the east of the shell. The total mass, given a density of 10^3 cm^{-3}, amounts to only 0.1 $M_\odot$ which is certainly no unreasonable value. The high densities inferred agree with the line ratio estimate made above, which probably refers to the collisionally excited gas. This suggests that both components of the nebulosity are in close proximity.

4. Conclusions

G70.7+1.2 clearly comprises both thermal and nonthermal components. The thermal component may be a compact HII region, although the question then arises, why have no continuum and recombination line emission been detected from it? How significant the present limits are is a question requiring further examination. The stellar object could be providing the required ionizing radiation, although it is also possible that it may be a background object having nothing to do with the rest of the complex. The numerous FeII lines are found in several types of objects including W-R stars, novae, and η-Car. They could also originate in a B[e] supergiant.

It seems quite probable that the nonthermal and thermal components are in proximity. The main mystery concerns the former, and we see three possible explanations for its origin. They, along with major problems relating to them, are:

Supernova remnant. This could certainly explain the nonthermal emission, although the radio luminosity appears to be one to two orders of magnitude low compared to other SNR. While this objection might be surmounted by assuming it to be a young SNR, the question of why the supernova itself was never observed must be accounted for.

Nova shell. Nonthermal radio emission has been detected from one nova shell (Reynolds and Chevalier, 1984), and if the stellar object should indeed show other characteristics of a nova, then the candidate progenitor is also present. In the case of a nova, one must wonder however whether the radio luminosity (some three orders of magnitude greater than that found by Reynolds and Chevalier for GK Per) and mass estimates for G70.7+1.2 are not too high. However, it would seem that a nova releases sufficient energy to power the shell, so such an explanation cannot be ruled out.

Stellar shell. Even a red giant or supergiant in the last phases of mass loss releases enough energy to power the nonthermal shell, although it would clearly have to be an unusual object since no other stellar shells are known to be associated with nonthermal radio sources. Here too, the density and mass may prove to be too high.

Acknowledgements

The Netherlands Foundation for Radio Astronomy is financially supported by the netherlands Organisation for Scientific Research (NWO). The VLA is a facility of the National Radio Astronomy Observatory which is operated by Associated Universities Inc., under contract with the U.S. National Science Foundation.

References

Green, D.A., 1985. Mon. Not. Roy. Astron. Soc. 216, 691.

Green, D.A., 1986. Mon. Not. Roy. Astron. Soc. 219, 39P.

Léger, A., Puget, J.-L., 1984. Astron. Astrophys. 137, L5.

Minkowski, R., 1948. Publ. Astron. Soc. Pacific 60, 386.

Muizon, M. de, Strom, R.G., Oort, M.J.A., Claas, J.J., Braun, R., 1988. Astron. Astrophys. 193, 248.

Reich, W., Fürst, E., Altenhoff, W.J., Reich, P., Junkes, N., 1985. Astron. Astrophys. 151, L10.

Reynolds, S.P., Chevalier, R.A., 1984. Astrophys. J. 281, L33.

A MOLECULAR CLOUD IN THE DIRECTION OF G70.68+1.20

W. Reich, N. Junkes, E. Fürst

Max-Planck-Institut für Radioastronomie
Auf dem Hügel 69
D-5300 Bonn 1
Federal Republic of Germany

Abstract

We have discovered a small diameter molecular cloud of high density ($\gtrsim 10^3$ cm^{-3}) in the direction of the non-thermal shell-type source G70.68+1.20. The peak CO intensity of the molecular cloud is displaced only a few arcseconds from the apparent centre of G70.68+1.20, strongly suggesting a physical relation between both objects. The kinematic distance of the molecular cloud is most likely 5.5 kpc and the diameter of G70.68+1.20 is about 0.55 pc. Its luminosity is nearly the same as that of SN 1006, suggesting that G70.68+1.20 is a very young supernova remnant expanding in a dense molecular cloud.

1. Introduction

In the course of high-resolution radio observations of compact radio sources in the Galactic plane, GREEN [1] discovered the shell-type object G70.68+1.20 (apparent diameter 20"). REICH et al. [2] investigated the radio spectrum of this source. They found a non-thermal spectral index of $\alpha = -0.55$ ($S_\nu \sim \nu^\alpha$) at frequencies above 1 GHz and identified G70.68+1.20 as a shell-type supernova remnant (SNR), based on the radio spectrum and morphological structure. The associated optical and infrared emission of G70.68+1.20 was recently spectroscopically studied by DE MUIZON et al. [3]. They find evidence for shock excitation, which supports the SNR identification although they could not rule out other interpretations.

The distance to G70.68+1.20 is unknown. If we assume a location inside the Galaxy the distance should be < 15 kpc corresponding to a linear diameter of < 1.5 pc. A SNR of this size must be very young ($\approx$ 100 years) if the ambient density is about 5 cm^{-3} [2]. Alternatively, the supernova explosion may have occurred in a dense molecular cloud. In order to investigate this possibility we searched for CO gas towards G70.68+1.20 with the IRAM 30-m telescope.

2. CO Observations and Results

We observed ^{12}CO spectra towards several positions near G70.68+1.20 (marked in Figure 1) with the IRAM 30-m telescope. The HPBW of the telescope at ν = 115 GHz is

22". The pointing accuracy was 5". We used a filterbank spectrometer and applied a position switching procedure. The r.m.s. noise was found to be 0.7 K T_R^* for a velocity resolution = 0.52 km s^{-1}. Linear baselines were subtracted and the absolute temperature scale was established by comparison with a single sideband spectrum of DR21. Errors in the absolute calibration are estimated to be less than 20%.

We integrated the emission in the velocity interval 0 km s^{-1} to 8 km s^{-1}. No emission exceeding the 3 x r.m.s. level was detected outside this interval. A contour map of the integrated emission is displayed in Figure 1. The peak ^{12}CO intensity at the position (15",-15") is T_R^* = 35 K.

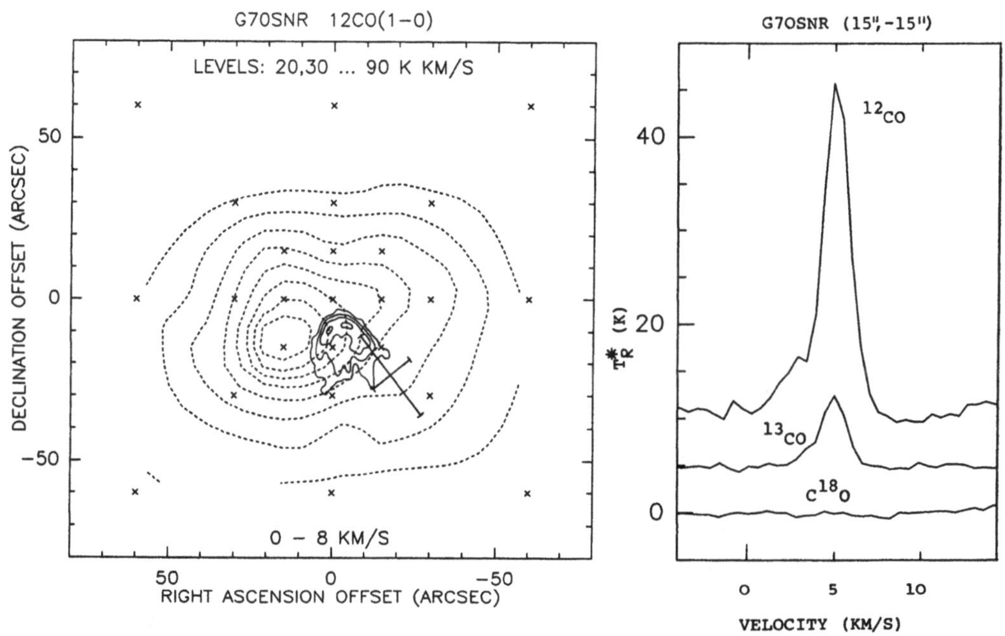

Fig. 1: Contour map of integrated ^{12}CO centred on RA(1950) = $20^h02^m28^s$, DEC(1950) = 33°30'48" (dashed contours). Superposed are some contours (solid lines) of the 5 GHz continuum map as observed by GREEN [4]. The location of the IRAS point-like source 200226.8+333025 is marked by a cross indicating the positional error. Spectra of ^{12}CO, ^{13}CO, and C^{18}O obtained at the peak position of the ^{12}CO temperature map are also shown.

We also observed ^{13}CO ($\nu \approx$ 110.2 GHz) and ^{12}C^{18}O ($\nu \approx$ 109.8 GHz) towards the positions (0",0") and (15",-15") with the same equipment but 6 min integration time per position. The spectrum towards the position (15",-15") is shown in Figure 1. Using formulae given by DICKMAN [5], the optical thickness τ for ^{13}CO was calculated to be $\tau \approx 0.2$ at both positions. Assuming $N_{H_2} = 5\ 10^5\ N_{^{13}CO}$ (for LTE [5]), we obtain column densities of molecular hydrogen of $N_{H_2} = 7\ 10^{21}\ cm^{-2}$ at position (0",0") and $N_{H_2} = 1.7\ 10^{22}\ cm^{-2}$ at position (15",-15").

The diameter of the cloud corresponding to half the peak intensity is about 50". The full extent exceeds 200", for even at the observed position most distant from the peak, integrated emission was detected at a low (< 10 K km s^{-1}) but still significant level.

At the galactic longitude of $\ell = 70°$ the kinematic distance is either 0.5 kpc or 5.5 kpc for a central velocity of $v_{LSR} = 5$ km s^{-1}. In case of a physical relation of G70.68+1.20 and molecular cloud, the distance is most likely 5.5 kpc. At the smaller distance its linear diameter D would be only ≈ 0.05 pc, which is very unlikely in view of the available data.

For a distance of 5.5 kpc, we may estimate the density distribution of molecular hydrogen in the cloud. Assuming spherical symmetry we obtain a radial density decrease with $n_{H_2} > 3000\ cm^{-3}$ at the centre of the cloud and $n_{H_2} < 400\ cm^{-3}$ at a distance of 45" (1.2 pc). However, clumpiness and locally higher density is possible.

The total mass inside a sphere of radius 1.2 pc is of order 400 $M_{\odot}$, about ten times the Jeans' mass.

3. Concluding Remarks

According to the available observations G70.68+1.20 is a nonthermal shell-type radio source expanding inside a molecular cloud. At the kinematic distance of 5.5 kpc the linear diameter of G70.68+1.20 is 0.55 pc. Its spectral luminosity at 5 GHz is $\sim 2 \cdot 10^{22}$ erg s^{-1} Hz^{-1}, which is the same as that of SN 1006. This supports its identification as a SNR. However, this luminosity is more than one order of magnitude lower than that of other historical SNRs like Kepler and Tycho [2], but this is expected for the large number of missing historical SNRs, because of the completeness of current SNR catalogues for bright objects.

If sweeping is assumed the mass swept up by G70.68+1.20 is about 10 $M_{\odot}$ assuming an ambient density of ≈ 3000 cm^{-3}. This swept-up mass is comparable to the mass of ejecta of a type II supernova. G70.68+1.20 is likely at the end of the free expansion phase. Its expansion velocity is unknown, but if we assume for instance an average expansion velocity of 2000 km s^{-1}, the age is about 135 years. There is no historical record of a supernova at the position of G70.68+1.20. However, the distance is large and the extinction by the dust contained in the molecular cloud may be several magnitudes; therefore the event may have escaped detection (see also [3]). The present extinction may be lower, since weak optical emission is seen near G70.68+1.20. This may indicate that the supernova has disrupted the molecular cloud.

Acknowledgements

We thank K. Menten, Dr. H. Steppe and H. Hein for help with the CO observations.

References

1. Green, D.A.: 1985, Monthly Notices Roy. Astron. Soc. 216, 691
2. Reich, W., Fürst, E., Altenhoff, W., Reich, P., Junkes, N.: 1985, Astron. Astrophys. 151, L10
3. de Muizon, M., Strom, R.G., Oort, M.J.A., Claas, J.J., Braun, R.: 1988, Astron. Astrophys. 193, 248
4. Green, D.A.: 1986, Monthly Notices Roy. Astron. Soc. 219, 39P
5. Dickman, R.L.: 1978, Astrophys. J. Suppl. 37, 407

DETECTION OF FOUR SUPERNOVA REMNANTS IN THE POLARIZED EMISSION FROM THE EFFELSBERG 11 CM SURVEY

N. Junkes, E. Fürst, W. Reich

Max-Planck-Institut für Radioastronomie
Auf dem Hügel 69
D-5300 Bonn 1
Federal Republic of Germany

1. Introduction

The Effelsberg 11 cm survey [1] is the most sensitive radio continuum survey of the Galactic plane in this wavelength range. Observations were made with a three-channel receiver recording the total intensity and Stokes parameters U and Q simultaneously. The latter were used to produce a polarization survey [2], giving linearly polarized intensity and the polarization position angle.

The whole area was searched for polarized sources, these are good candidates for supernova remnants. Four of the newly found polarized sources were established as supernova remnants by additional information (radio spectral index, infrared-to-radio flux ratio). We found 50% of the previously known supernova remnants significantly polarized at 11 cm wavelength. The lack of polarization for the rest of them may be due to depolarization.

2. The Survey

The first part of the Effelsberg 11 cm survey covers a 3° broad stripe along the Galactic plane from the Galactic centre to the Cygnus region ($357^{\circ}.4 < l < 76^{\circ}$, $|b| < 1^{\circ}.5$). Some basic parameters of the survey are given in Table 1. The polarization data (Stokes U,Q) were corrected by subtracting relative baselevels before combining them to maps of polarized intensity and polarization angle [3]. Due to this procedure, large-scale regular foreground polarization was removed [4]. Observation and data reduction procedures are described in former publications [1,2].

Table 1. Basic properties of the Effelsberg 11 cm survey

frequency	2695 MHz
bandwidth	80 MHz
T_{SYS}	60 K
HPBW	4.27 '
T_B/S_ν	2.51
instrumental polarization	0.7±0.3 %
σ_{PI} (rms-value)	8 mK

The data include instrumental polarization. The amount of instrumental polarization is ~0.7%, determined by unpolarized calibration sources (e.g. 3C295). Inspection of strong well-known HII regions supports this value.

The polarization survey omits the Galactic center region (l < 5°) because of ground radiation effects at very low elevations [4].

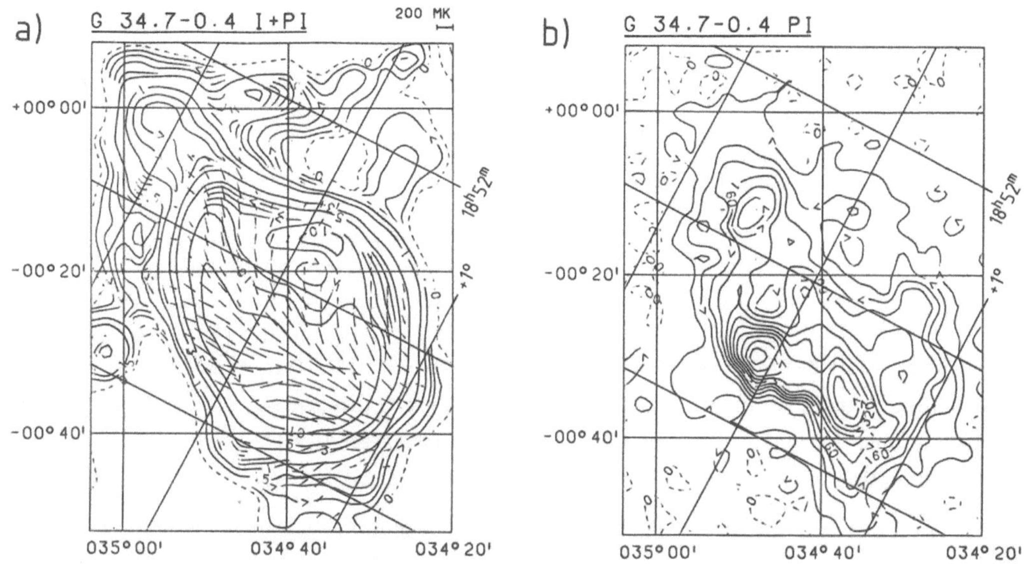

Fig. 1: Polarization properties of the supernova remnant W44 at 2695 MHz. a) Total intensity contours [Kelvin] and superposed polarization E-vectors. b) Contours of polarized intensity, labelled in [mK].

3. The Sources

The most prominent polarization features in the survey are due to supernova remnants. The source G34.7−0.4 (W44) is shown as an example in Fig. 1a,b. Fig. 1a is an overlay with contours in total power distribution (in K T_B) and superposed bars of linearly polarized intensity in E-field direction. In Fig. 1b the polarized intensity is separately given in contours above 40 mK T_B. The Galactic background radiation was subtracted in Stokes I, U, Q to give a new relative zero-level. The integrated polarization value across the source is 1.8% in accordance with former observations [5].

Fig. 2 shows G20.0−0.2 which has been identified as a plerionic SNR by Becker and Helfand [6]. The integrated polarization is found to be 0.9%. This is near the limit of instrumental polarization, but the source contains a clearly polarized component which is distinct from the maximum emission in total power. It is not sufficient to obtain polarization information from integrated flux values; often it is necessary to check the degree of polarization pointwise.

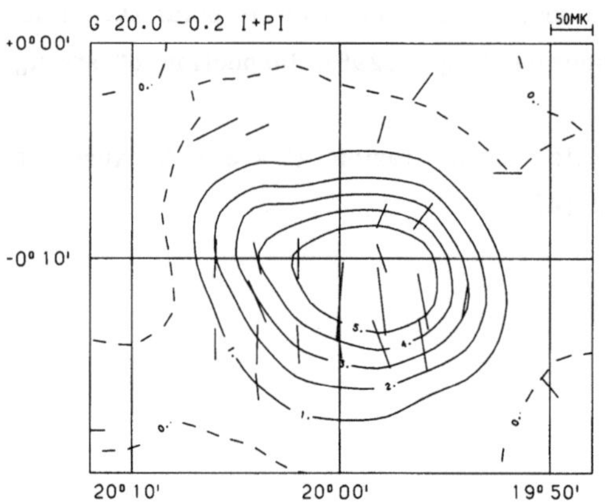

Fig. 2: Supernova remnant G20.0−0.2. Total intensity contours [K] and superposed polarization E-vectors at 2695 MHz.

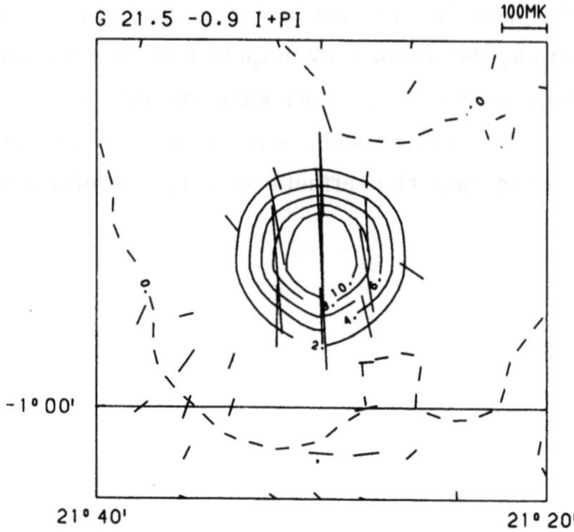

Fig. 3: SNR G21.5−0.9 at 2695 MHz. Contours and vectors as in Fig. 2.

Fig. 3 shows G21.5−0.9, a plerionic SNR with a size of ≈1'. Our data show a uniform polarization feature in north-south direction with an integrated value of 3.2%. High resolution observations, however [7], detect a circumferentially oriented polarization field with values up to 30% at an angular resolution of ≈15".

We checked the polarization structure of all known supernova remnants in the survey area. The results are given in Table 2. Only 50% of the remnants were found to be significantly polarized at 11 cm wavelength. The rest is near or below the limit of instrumental polarization or in confused regions where it is not possible to distinguish emission of different source components (e.g. W51−complex).

Table 2. Polarization of SNRs in the survey area

degree of polarization	nr. of sources
< 1%	28
1% - 5%	21
5% - 10%	4
> 10%	3

Moreover we found some candidates for new remnants from the polarized emission of the 11 cm survey. Four are confirmed to be SNR by further observations:

<u>G6.1+1.15</u> is an extended source (30' · 26') not far away from the Galactic center. An overlay of total power contours and bars of polarized emission is shown in Fig. 4. Two peaks of polarized emission are visible, both having values up to 60%. The integrated value is 23%. Combining flux values of λλ21 cm, 11 cm and 6 cm wavelength, we get a

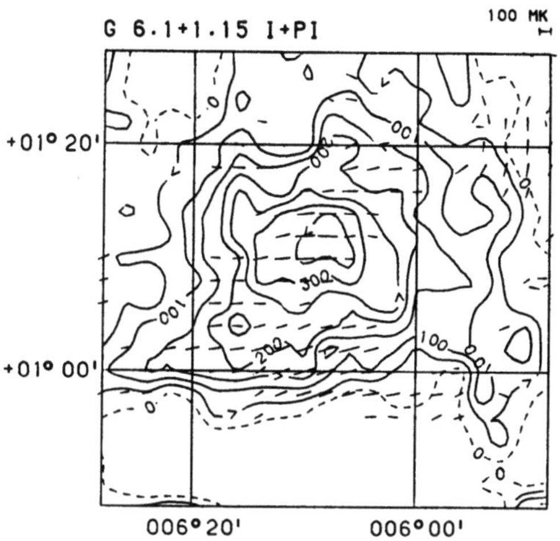

Fig 4: SNR G6.1+1.15 at 2695 MHz (cf. Fig. 2). The total intensity contours are labelled in [mK].

rather flat spectral index of $\alpha \approx 0.3$ ($S_\nu \sim \nu^{-\alpha}$). After correcting for the background there is no infrared emission at $\lambda 60\mu$ wavelength near the position of G6.1+1.15. According to Fürst et al. [8] this gives strong evidence for the source being a SNR. G6.1+1.15 is most likely a plerionic supernova remnant.

G59.7+1.2 is another polarized source in the survey area. Fig. 5a gives an overlay of contours of total power emission with polarized intensity bars, Fig. 5b gives contours of the polarized intensity. The polarization feature at $\lambda 11$ cm wavelength is clearly connected with the source giving an integrated polarization value of 4%. The spectral index is $\alpha \sim 0.5$. A comparison of infrared and radio flux densities giving a value of $S_{60\mu}/S_{11cm} < 200$ supports the identification as a supernova remnant.

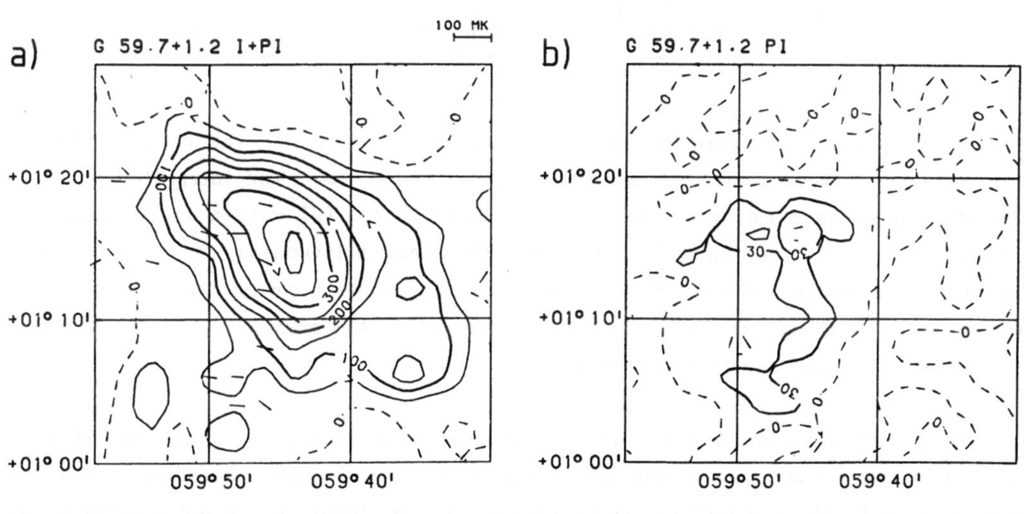

Fig. 5: SNR G59.7+1.2 at 2695 MHz. a) Total intensity contours [mK] and superposed polarization E-vectors. b) Contours of polarized intensity [mK].

G68.6−1.2 is a very weak, extended radio source, first detected in the 11 cm survey. The source (Fig. 6) is highly polarized at this wavelength (≈ 35%). Comparison with 21 cm observations gives a spectral index value of ≈0.0. The $S_{60\mu}/S_{11cm}$ ratio is about 30. According to these results, G68.6−1.2 is probably a plerionic SNR.

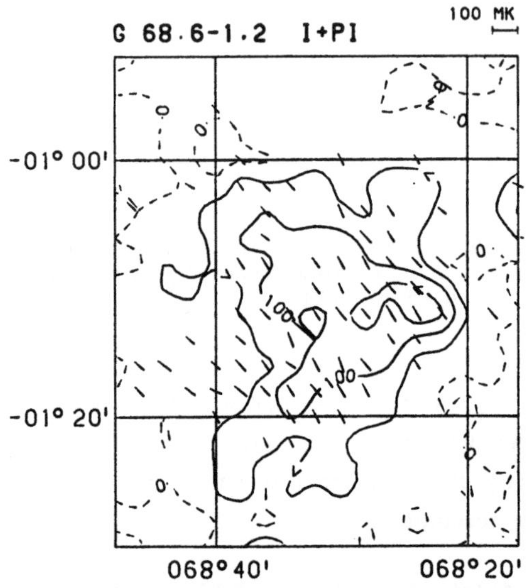

Fig. 6: SNR G68.6−1.2 at 2695 MHz. Contours and vectors are shown as in Fig. 5a.

Table 3. Properties of new-found SNRs

Source	11cm-flux [Jy]	polarization [%]	spectral index ($S_\nu \sim \nu^{-\alpha}$)
G6.1+1.15	3.1 ± 0.6	23 ± 13	0.3?
G59.7+1.2	1.1 ± 0.2	4.0 ± 2.0	0.5
G68.6-1.2	0.8 ± 0.3	32 ± 13	0.0?
G69.7+1.0	0.7 ± 0.15	6.4 ± 3.7	0.8

G69.7+1.0 is a slightly extended source (16' diameter) near the W58 complex. Fig. 7a shows 11 cm contour lines with superposed polarization bars. The integrated value is 6%. Fig. 7b shows 6 cm observations of the same source [9] with a strong polarized component in the upper right half (> 15%). The weak polarization at 11 cm is again due to depolarization effects. Flux values at λλ11 cm and 6 cm give a spectral index of 0.8, there is no source feature present in the λ60μ IRAS map. We identify G69.7+1.0 as a new shell-type supernova remnant.

Altogether we have found four new supernova remnants from their polarization properties. Some basic parameters are compiled in table 3 (spectral indices are given according to [10]). A list of compact polarized sources found in the survey, similar to those shown in fig. 3, will serve as reference for high resolution observations.

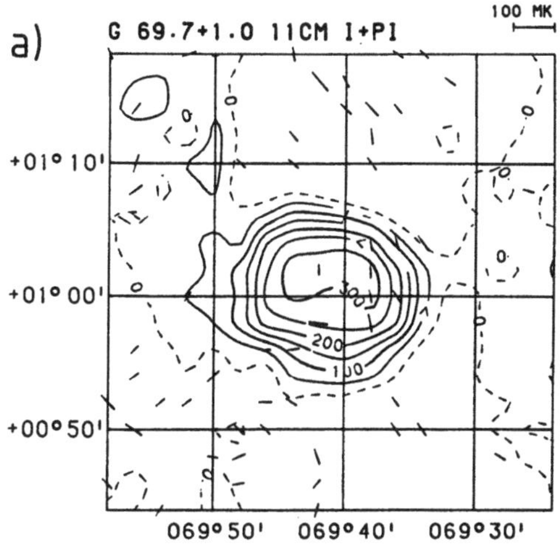

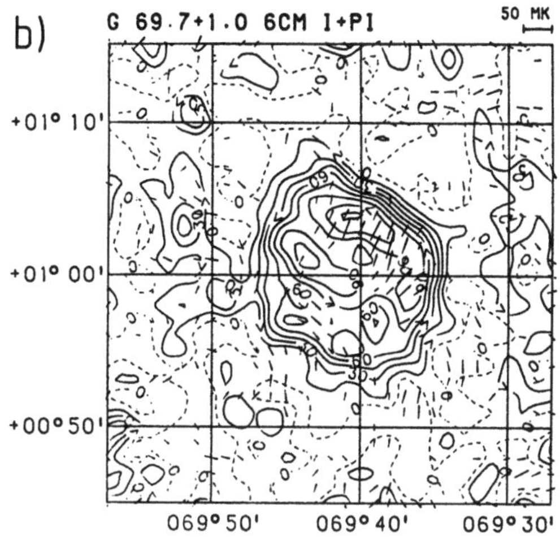

Fig. 7: SNR G69.7+1.0. Total intensity contours and superposed polarization E-vectors at 2695 MHz (a) and at 4650 MHz (b). The contours are labelled in mK.

References

1. Reich, W., Fürst, E., Steffen, P., Reif, K., Haslam, C.G.T.: 1984, Astron. Astrophys. Suppl. Ser. 58, 197
2. Junkes, N., Fürst, E., Reich, W.: 1987, Astron. Astrophys. Suppl. Ser. 69, 451
3. Wardle, J.F.C., Kronberg, P.P.: 1974, Astrophys. J. 194, 249
4. Junkes, N., Fürst, E., Reich, W.: 1987, in *Interstellar Magnetic Fields,* eds. R. Beck and R. Gräve, Springer, Berlin, p. 115
5. Baker, J.R., Preuss, E., Whiteoak, J.B.: 1973, Astrophys. Lett. 14, 123
6. Becker, R.H., Helfand, D.J.: 1985, Astrophys. J. 297, L25
7. Becker, R.H., Szymkowiak, A.E.: 1981, Astrophys. J. 248, L23
8. Fürst, E., Reich, W., Sofue, Y.: 1987, Astron. Astrophys. Suppl. Ser. 71, 63
9. Seiradakis, J., Han Tie: 1985, private communication
10. Reich, W., Fürst, E., Reich, P., Junkes, N.: 1988, in *Supernova Remnants and the Interstellar Medium,* IAU Coll. 101, eds. R.S. Roger and T.L. Landecker, Cambridge University Press

KINEMATICS OF OPTICAL FILAMENTS IN THE CYGNUS LOOP.

H. Greidanus, Sterrewacht Leiden, The Netherlands.
R.G. Strom, Netherlands Foundation for Radio Astronomy.

Abstract: We describe the kinematical character of optically emitting gas in a small area of the Cygnus Loop, as seen with moderately high resolution in Hα and [OIII]. In both lines a number of different components contributing to the emission are recognized. The [OIII] emission indicates that the gas may lie in a thin sheet, while from the Hα emission this indication is much less clear.

1. Introduction.

We have observed a number of fields in the evolved supernova remnant the Cygnus Loop in the Hα 6563A and [OIII] 5007A lines, in order to study the kinematics of the optically emitting gas. The instrument used was TAURUS, an imaging Fabry-Perot interferometer, with an IPCS as 2-dimensional detector, at the 2.5 m Isaac Newton telescope at La Palma. The way TAURUS works is decribed in detail in Atherton et al. (1982). After proper calibration, a 3-dimensional matrix of intensities is produced, the axes of which are two spatial coordinates and radial velocity; i.e., for each point on the sky within the field of view a spectrum is obtained. For the observations presented here, the fieldsize is 7 arcmin, and the range in radial velocity is 375 km/s for Hα and 285 km/s for [OIII]. After smoothing to improve the signal to noise, an FWHM resolution of 6" spatially and 25 km/s in velocity is obtained. The absolute accuracy of the radial velocity determination is estimated to be 2 km/s.

2. Results.

We will describe the character of the emission for one of the observed fields here. This particular field is centered on $\alpha = 20^h45^m.9$, $\delta = 31°31'.0$, which is in the upper part of the so-called carrot region of the Cygnus Loop. An image of the Hα and [OIII] intensities is displayed in Figure 1. The continuum has been subtracted, and

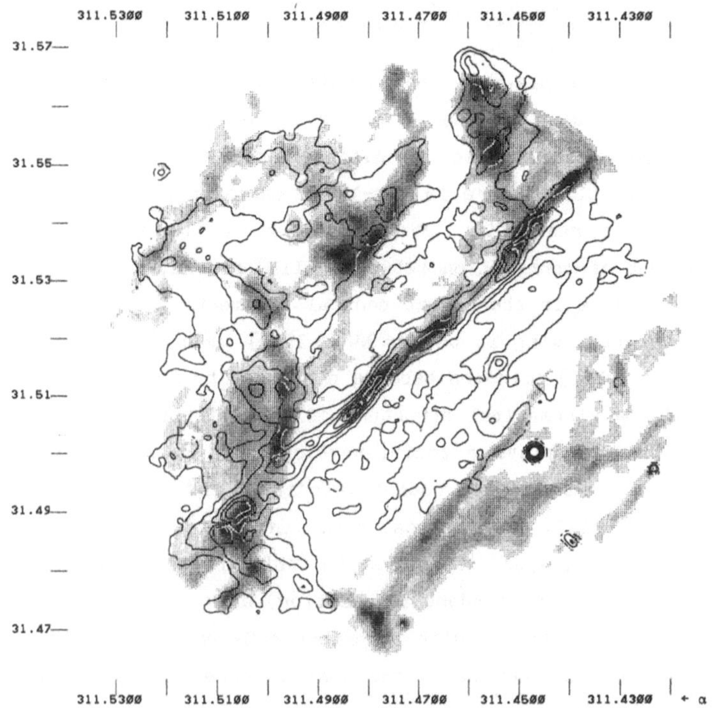

Figure 1. Intensity in the Hα line (contours) overlaid over intensity in the [OIII] λ5007 line (grayscale). Contours and gray levels have a linear increment.

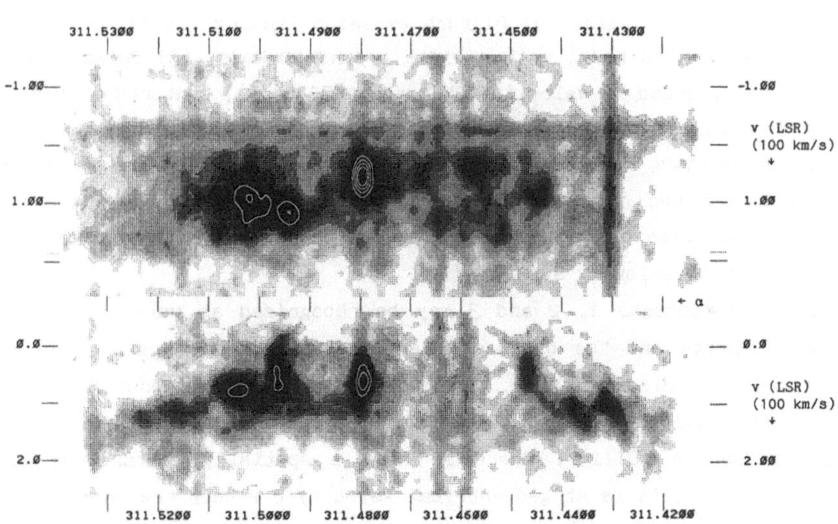

Figure 2. Intensity as a function of right ascension and velocity, for declination 31.511°. The grayscale shows the lowest intensities with a logarithmic increment, while the contours show higher intensities with a linear increment; the lowest two contours are not drawn. The top figure is in Hα, the bottom one in [OIII] λ5007.

the only stellar-like images remaining (cf. at 311.447°,31.500°) are ghosts produced by the interference filter. From a comparison of the stellar continuum emission, we estimate that positional errors in the Hα/[OIII] overlay due to geometrical distortions in the detector are less than 1". Only the brightest emission is shown; this consists of filaments surrounded by diffuse emission of lower intensity. Actually, for both Hα and [OIII] there is low-level emission over the entire field which is amorphous at this resolution. From Figure 1, it is clear that although there is a certain amount of correlation between [OIII] and Hα, the [OIII]/Hα ratio is far from constant. Notable is the bright central filament: around 311.49°,31.50° it has only Hα emission; following it to the NW, [OIII] turns on. But then, at the centre of the image, the [OIII] emission becomes displaced by up to 4" from the Hα.

An impression of the kinematical structure of the emission may be obtained by inspecting Figure 2, which displays a position-velocity diagram. The intensity of the emission is shown as a function of radial velocity v and right ascension, at a constant declination of 31.511°. The vertical bars (intensity independent of v at a certain position) are the images of stars in this representation. The weak horizontal features (intensity independent of position) at v = -49 km/s and v = -100 km/s in the Hα plot may be identified with atmospheric OH emission at λ = 6553.7 Å and λ = 6569.0 Å (these are not imaged in the same interference order as Hα). The bright knot at α = 311.480° is the cross section of the bright central filament.

From an inspection of these profiles for the entire data cube, the following picture emerges. The emission in the [OIII] is characterized by four components: (1) Bright filaments at LSR velocities between +10 and +100 km/s, having an FWHM velocity width around 25 km/s; (2) Diffuse emission linking the filaments and going up to velocities of +150 km/s, with a width around 20 km/s; (3) Emission of a very low intensity, very broad in velocity (FWHM about 70 km/s), centered around +85 km/s and present over almost the entire field; (4) Emission at v around +10 km/s with a width of about 15 km/s and little spatial structure. (In the sample crosscut of Figure 2 the latter two components are not very prominent.) Because the filaments and the diffuse emission between them are always connected in the position-velocity space, they cannot be uniquely separated.

In the Hα, components 1, 2 and 3 may be recognized as well. The distinction between the filaments and the diffuse inter-filament emission is less clear here, because the diffuse emission has a larger velocity width of about 30 km/s and is also relatively brighter (also in total intensity the filaments are less well defined in Hα than in [OIII]). There is also a spatially featureless component around v = +10 km/s (this is at geocentrical velocity 0 within the uncertainty), not unlike [OIII]-component 4; in contrast with that one, however, it shows no correlation (on arcmin-scales) with the other emission, so we interpret it as geocoronal Hα. Several times, two filaments occur at distinct velocities at the same position on the sky (as for example in Figure 2 around α = 311.455°). The velocity separation in such cases is typically about 90 km/s. This phenomenon occurs less frequently in [OIII].

In order to present an overview of the gas kinematics in two spatial dimensions, moments have been calculated for all individual spectra. Only intensities above a certain level have been taken into account, so that only emission from the filaments and the diffuse inter-filament component are included, while components 3 and 4 are discarded (as well as the atmospheric lines). The zeroth-moment map calculated in this way is the intensity map of Figure 1. The first-moment map, representing the radial velocity, is displayed in Figures 3 and 4 for Hα and [OIII]. For [OIII], there is a clear anti-correlation of intensity and radial velocity. For Hα, this anti-correlation is much less well-defined. There, the impression is rather that the filaments running from NW to SE at a radial velocity around +40 km/s lie superimposed on the diffuse emission around v = +120 km/s which can be seen uncontaminated in the NE. When the difference in radial velocity of Hα and [OIII] is computed for each point in the map, values fall largely between ±65 km/s. For the bright central filament, the radial velocity difference (Hα-[OIII]) is around -10 km/s in the section of overlap, and varies between ±10 km/s in the shifted part.

A map of the velocity width Δv of the spectra (from the second moment) is not presented here, but in [OIII] there is a clear correlation of Δv with intensity, which again in the Hα is much less well-defined.

A previous report on this work was presented at IAU Colloquium 101 (Greidanus & Strom 1988), dealing with the [OIII] emission of one of the other fields we observed in the Cygnus Loop. At the time of that presentation, we lacked an absolute velocity calibration, which has since been obtained. The result is that the velocity zero point defined there is within the uncertainty at heliocentric velocity 0, at LSR velocity +14 km/s; however, the direction of the radial velocity is actually the inverse of what was assumed. The characteristics of the [OIII] emission in that field are largely the same as for the field treated here.

3. Interpretation.

The kinematical behaviour of the [OIII] emitting gas is indicative of it lying in a rather homogeneous layer just behind a deformed shock front. In that case one would expect the product of radial velocity and intensity to be constant, as a purely geometrical effect: filaments appear at low radial velocities where the shock front is near tangency with respect to the line of sight, diffuse emission at higher radial velocities where the the shock front is more normally oriented. The velocity width would be highest in the filaments as a result of beam smearing. These effects are seen in the [OIII]. The intensity variations would not indicate density

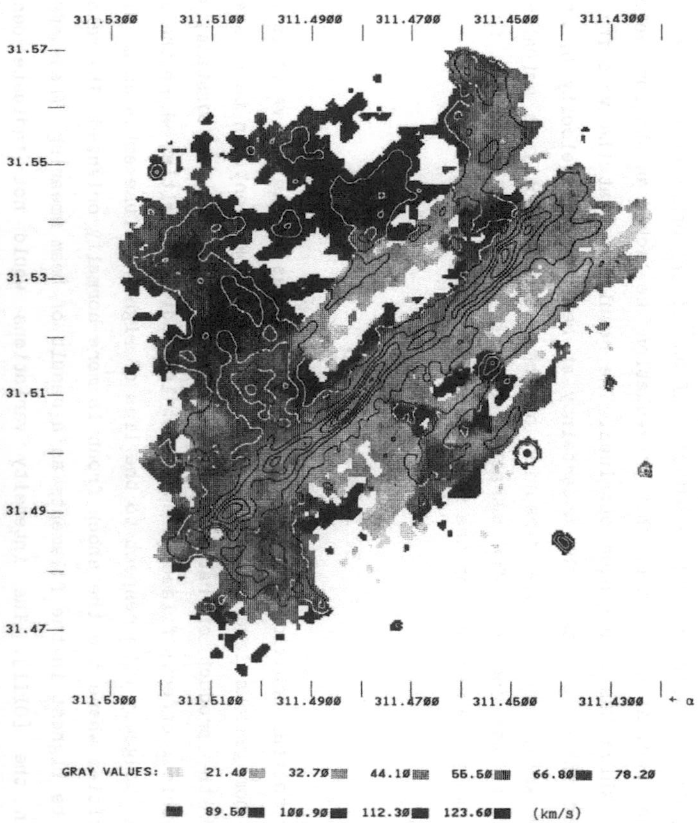

Figure 3. Radial velocity (LSR) of the Hα-emitting gas with Hα intensity contours overlaid.

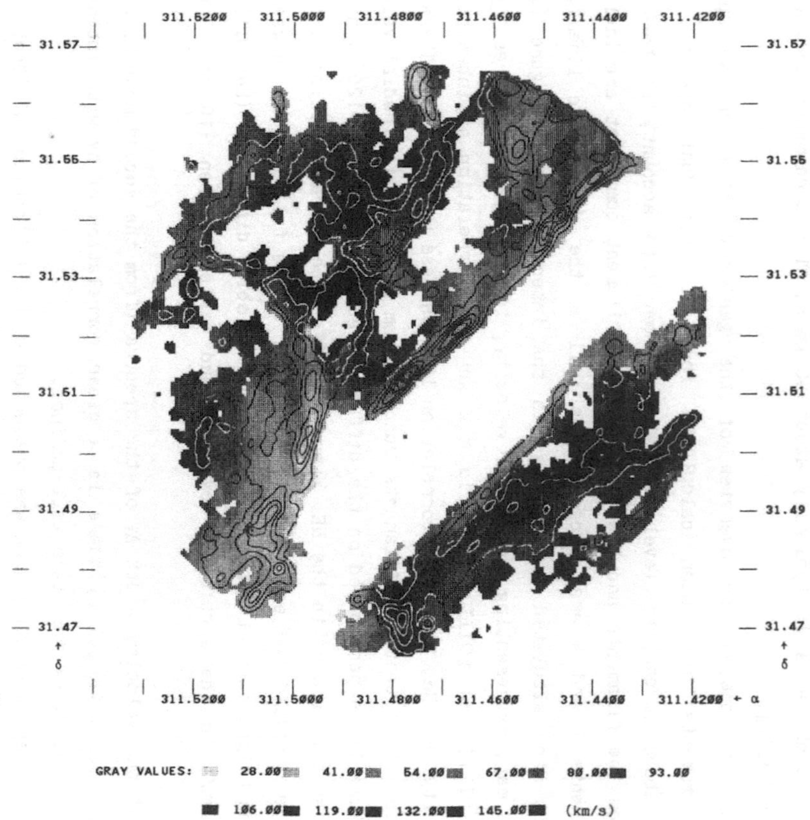

Figure 4. Radial velocity (LSR) of the [OIII]-emitting gas with [OIII] intensity contours overlaid.

variations in the medium the shock encounters (except that density fluctuations are needed to deform the shock front). This interpretation has recently received attention by Hester (1987).

In Hα, the sheet-like behaviour as found in the [OIII] is not nearly as obvious. If we want to explain the Hα emission also without assuming that the filaments are pre-shock density enhancements, then the indication is that by the time the gas radiates predominantly in Hα, its structure has become perturbed.

Acknowledgements.
We are gratefull for the use of the Isaac Newton Telescope, which is operated by the Royal Greenwich Observatory at the Observatorio del Roque de los Muchachos of the Instituto Astrofísica de Canarias, and for the use of computer facilities at the Royal Greenwich Observatory.

References.

Atherton, P.D. et al., 1982, Mon.Not.R.astr.Soc. 201, 661.

Greidanus, H., & Strom, R.G., 1987, in proc. IAU Coll. 101 "Supernova Remnants and the Interstellar Medium", eds. R.S. Roger & T.L. Landecker (Cambridge University Press), p. 443.

Hester, J.J., 1987, Astrophys.J. 314,187.

X-ray Observations of the Supernova Remnant G292.0+1.8

J.J. Claas

Laboratory for Space Research, P.O.Box 9504

2300 RA Leiden, The Netherlands

Summary

EXOSAT X-ray data of the supernova remnant G292.0+1.8 reveal a two temperature structure if the data are fitted with a collisional-ionization-equilibrium model: kT_{low}=0.5 keV and kT_{high}=2.0 keV. This two temperature structure indicates the possible existence of non-equilibrium-ionization effects. There is a small indication (2 % statistical significance level) for a low temperature difference between the bar and plateau component: $(kT_{low})_{bar}=0.36\pm0.06$ keV, $(kT_{low})_{plateau}=0.59\pm0.09$ keV. A possible explanation for this temperature difference is the overabundance by a factor 10 of oxygen for the bar component. The electron density of the central bar component, $(n_e)_{bar}$=(5-10) cm^{-3}, is found to be somewhat higher than the electron density of the plateau component, $(n_e)_{plateau}=2$ cm^{-3}.

1. Introduction

Early radio observations (Milne 1969, Shaver and Goss 1970) led to the discovery of the southern galactic SNR G292.0+1.8. Observations in various wavelenghtbands have been carried out since: radio (Braun *et al.* 1986, Lockhart *et al.* 1977), optical (Goss *et al.* 1979, Murdin and Clark 1979, Braun *et al.* 1983), infrared (Braun *et al.* 1986) and X-ray (Share *et al.* 1978, Agrawal and Riegler 1980, Clark *et al.* 1980, Tuohy *et al.* 1982). Combining these various observations one obtains the following preliminary picture. G292.0+1.8 is a young ($\approx$ 1100 yr) remnant consisting of two components: a faint plateau extending over 10 arcminutes on which is superimposed a bright elongated central region which is interpreted as a part of an expanding ring of oxygen rich ejecta. According to the model for a rotating precursor (Woosley and Weaver 1982) the expanding ring is probably caused by the equatorial expulsion of an oxygen rich envelope whereas the plateau component probably consists of ejected material from the outer envelope together with matter swept up from the ISM.

In this paper, we describe the analysis and interpretation of data obtained with the European X-ray satellite EXOSAT.

2. Instruments and observations

The present observations of G292.0+1.8 have been made by two experiments on board EXOSAT. Here we mention some characteristic features of these experiments.

The imaging telescope (de Korte *et al.* 1981), sensitive in the low energy (LE) range 0.05 - 2 keV, has a field of view of two degrees diameter and an on-axis spatial resolution of about 18 arcseconds. This LE telescope can be combined with several filters e.g. a 3000 Å Lexan (3Lx) and a Boron (Bo) filter. Differences in the spectral response of these filters provide the possibility of carrying out coarse spectrophotometry with this experiment.

The medium energy (ME), large area, proportional counter array (Turner *et al.* 1981) provides spectral information for the energy range 1.6 - 10 keV. Due to the small size of the remnant, the ME experiment does not provide any spatial information.

The observations used for the present analysis were obtained on January 19 and May 14, 1985 .

3. Data analysis

3.1 ME data analysis

Spectral parameters for the remnant were derived first by fitting the spectrum with a two component collisional ionization equilibrium (C.I.E.) model (Mewe, Gronenschild and van den Oord 1985). We used the interstellar absorption cross sections of Morrison and McCammon (1983) and adopted solar abundances. We fitted the two LE flux points, as determined by the two images i.e. 3Lx and Bo, simultaneously, which, in principle, enables a better determination of the spectral parameters and the interstellar neutral hydrogen column density. The spectral parameters are given in table 1. Since spectral fitting of the EXOSAT data leaves the N_H value undetermined, we further assume in our analysis that N_H equals the better determined N_H value found from the Einstein SSS data (Clark *et al.* 1980) i.e. $N_H \approx 3.8 \times 10^{21}$ cm^{-2}. The spectral parameters obtained with N_H fixed at 3.8×10^{21} cm^{-2} are also given in table 1.

Table 1

spectral fit parameters (two temperature C.I.E. model)

			N_H ($\times 10^{21}$ cm^{-2})	
			2^{+5}_{-2}	3.8 (fixed)
emission norm. cst.*	C_{low} (10^{50} cm^{-3}/pc^2)	=	30^{+20}_{-10}	43.5 ± 2.5
low temperature comp.	kT_{low} (keV)	=	$0.60^{+0.10}_{-0.29}$	0.50 ± 0.03
emission norm. cst.*	C_{high}(10^{50} cm^{-3}/pc^2)	=	$3.7^{+3.3}_{-1.7}$	$5.3^{+2.5}_{-1.5}$
high temperature comp.	kT_{high} (keV)	=	$2.4^{+1.6}_{-0.7}$	$2.0^{+0.5}_{-0.4}$
chi-squared	χ^2	=	21.7 (18 d.o.f.)	22.5 (19 d.o.f.)

* $Cd^2 = \int n_e n_H dV$

3.2 LE data analysis

Each of the LE images used in our analysis has been corrected for background and variations in the effective area across the detector and has been rebinned to 16 arcseconds pixels. Fig.1 shows a contour plot of the X-ray image for the Bo filter. This image reveals the following structure : The X-ray emission consists of two components: a faint plateau of rather ellipsoidal shape and a relatively bright central ridge bisecting the faint plateau in an East-West direction. The dimensions of the faint plateau are 8 by 6.5 arcminutes with the long axis coinciding with the central bar. The central bar has a width of about 2 arcminutes. Away from the central bar the faint plateau has a rather uniform surface brightness. An important feature which will be discussed later is the absence of limb-brightening at the boundary of the plateau.

In order to search for spectral morphology, especially differences between the central bar and the plateau, we overlayed the images with a grid of parallel strips. These strips, with a width of 16 arcseconds, i.e. the angular resolution of the images, were chosen parallel to the central bar feature, i.e. the position angle of the strips is 85 degrees. For each strip a distance was defined as the distance between the center of the strip and the center of the remnant along the vertical line with $\alpha = 11^h\ 22^m\ 16^s$. This grid was used to obtain intensity profiles for both images, respectively, by, for every strip, adding all counts contained in the strip. By adding the strips containing the central bar and plateau region, respectively, we find that the Bo/3Lx count rate ratio of the bar component is somewhat lower compared to the Bo/3Lx ratio of the plateau component (statistical significance level 2 %). Fig.2 shows the temperature profile obtained by converting the Bo/3Lx

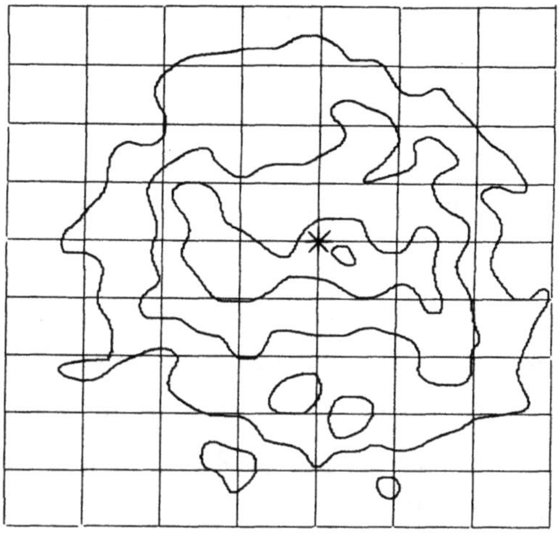

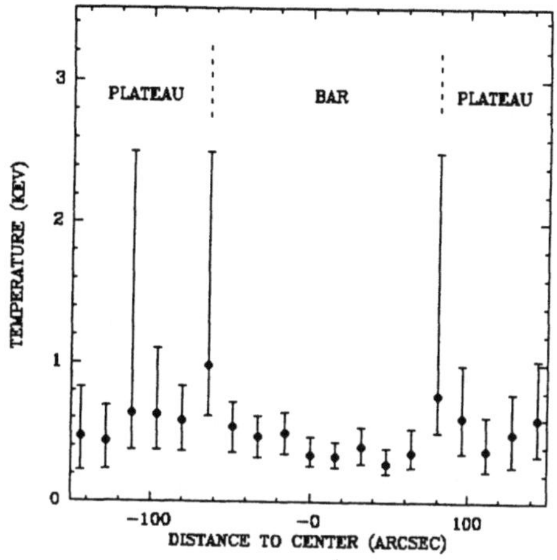

Fig.1 (*left*): The Bo image binned into 16 arcseconds pixels and smoothed slightly with a Gaussian of 12.8" FWHM. The observation time is 33886 seconds. Contour levels: 33, 103, 172 and 242 counts/arcmin.^{2} Gridspacing: $\Delta\alpha$=10 seconds, $\Delta\delta$=1 arcminute. The cross at the center denotes position α, δ = $11^h22^m15^s$, -58 ° 59′00″.

Fig.2 (*right*): Temperature profile derived by converting the Bo/3Lx count rate ratio into a temperature assuming N_H = 3.8 x 10^{21} cm^{-2}. The position of the bar and plateau component is indicated.

count rate ratio into a temperature, assuming solar abundances and N_H = 3.8 x 10^{21} cm^{-2}. By adding the strips containing the two components, respectively, we find kT_{bar} = 0.36 ± 0.06 keV and $kT_{plateau}$ = 0.59 ± 0.09 keV.

We also tried to calculate the temperature of the bar component by first subtracting the plateau component in both images. However, due to the large statistical uncertainty on the mean count rate of the plateau component, no accurate temperature determination was possible then.

From our LE/ME spectral fit we find that 95 % of the observed LE flux can be explained by the low temperature component only. Therefore we can use the 3Lx image to determine which fraction of the low temperature emission normalisation constant C_{low} is caused by the central bar and which fraction is caused by the plateau, including the temperature difference between these components. We find that $(C_{low})_{bar}$ ≈ 0.45 C_{low} and $(C_{low})_{plateau}$ ≈ 0.55 C_{low}. These emission normalisation constants are used to calculate the electron density n_e of both components. By assuming certain dimensions for both components, obtained from the Einstein HRI image (this image has a higher spatial resolution compared to the EXOSAT images), and adopting a distance of 3.6 ± 1.5 kpc (Braun *et al.* 1986) we find $(n_e)_{plateau}$ = (2.4 ± 0.7) cm^{-3} and $(n_e)_{bar}$ = (12 ± 6) cm^{-3}.

The relative young age of the remnant might be indicative of the existence of possible non-equilibrium-ionization (N.E.I.) effects. These N.E.I. effects may also be expected from the need for at least two temperature components (see section 3.1). We therefore also use a N.E.I. spectral model to fit the LE data. Since there is a great uncertainty with respect to the hydrodynamics of G292.0+1.8 (see section 4), we do not use a hydrodynamical model but only calculate the $n_e t$ value for both the central bar and plateau component in which n_e is the electron density (in cm^{-3}) of the shocked plasma and t the age (in seconds) of the shocked plasma. The calculations are carried out on the assumption that the electron temperature of the shocked plasma of both components equals the high temperature found when fitting the LE plus ME data with a C.I.E. spectral model (see section 3.1) i.e. 2.0 keV. Assuming the same structure for the bar and plateau component as used above, the results of these calculations show that, due to poor statistics, the ages of the shocked plasma of both components do not differ significantly. Only the electron density of the shocked plasma of both components differ. For a N_H value of 3.8 x 10^{21} cm^{-2} we find $(n_e)_{plateau}$ = (1.3 ± 0.5) cm^{-3} and $(n_e)_{bar}$ = (6.2 ± 3.6) cm^{-3}. These values are comparable with those found above when using the C.I.E. spectral model instead of the N.E.I. model used here. We conclude that the electron density of the shocked plasma is somewhat higher for the central bar than for the plateau.

4. Discussion

One of the main interesting questions about G292.0+1.8 is : what kind of hydrodynamics gives rise to the observed X-ray structure ? One of the implications of the interpretation of the X-ray structure of G292.0+1.8 as given by Tuohy *et al.* (1982) is the following :

The two spatial components i.e. the central bar and the plateau, should have distinctly different compositions; the bar component should be dominated by Oxygen, Neon, etc., as with the optical remnant, whereas the emission from the plateau should be characteristic of material with cosmic abundances. However, with both the LE and ME data it is not possible to determine this difference. Therefore we cannot exclude a possible overabundance in the bar for elements like O, Si and S as found by Clark *et al.* (1980) using the Einstein SSS spectrum. However, one of the results of our analysis is that there is an indication (2 % significance level) of a temperature difference between the central bar and the plateau component (see fig.2). The real temperature difference between these components may even be larger since, for deriving the temperature of the bar component, we did not correct for the count rate contribution from the plateau at the position of the bar component. This temperature difference may reflect the difference in density and/or chemical composition of the shocked material. We find that an overabundance of about a factor 10 in carbon, nitrogen and oxygen for the bar component can explain the temperature difference between the bar and plateau component. Although this does not prove the overabundance of oxygen, for example, in the bar component, it is consistent with the oxygen rich ejecta which have been identified at optical wavelengths. The expanding ring of these oxygen rich ejecta, when viewed edge-on, then forms the bar component seen in X-rays.

A possible explanation for the center-filled morphology is that of ejecta encountering a reverse shock (e.g. Braun *et al.* 1986). Therefore we tried to reproduce the observed ratio between the temperature kT_{low} of the shocked ejecta and the temperature kT_{high} of the shocked CSM/ISM by means of a reverse shock model (Chevalier 1982). From this model, however, we find a much larger temperature ratio than observed.

An important feature which directly influences the hydrodynamical modelling is the absence of any limb-brightening at the boundary of the plateau component. As already mentioned by Tuohy *et al.* (1982) this might be a time effect. G292.0+1.8 is of relatively young age ($\approx$1100 yr). If the remnant evolves further and more circumstellar/interstellar material is swept up, a limb-brightened shell will develop. Also the radial decrease in circumstellar density caused by the stellar wind of the progenitor star might explain the absence of limb-brightening.

Due to the relatively young age of G292.0+1.8, the existence of N.E.I. effects seems likely. The observed count rate ratio between the 3Lx and Bo image can be produced by a N.E.I. model. This, however, does not prove that these N.E.I. effects really exist since this count rate ratio can also be produced by just one C.I.E. temperature component which equals the low temperature component kT_{low} found in our C.I.E. spectral fit (95 % of the observed LE flux is caused by the low temperature component only). If, however, these N.E.I. effects do exist then the observed LE flux is, for at least a part, the enhanced low energy flux from shocked plasma with a temperature kT_{high}. This enhanced low energy flux mimics the C.I.E. low temperature component.

Since the electron densities derived for the plateau and bar component are about the same for both the C.I.E. and N.E.I. model, we cannot use the electron density to distinguish between these models. The electron density for the plateau component corresponds to a pre-shock hydrogen number density n_H of ≈ 0.4 cm^{-3} ($n_e/n_H = 1.23$ assumed) which is a normal value for the ISM density.

5. Conclusion

The following results have been found in our analysis of the X-ray data of G292.0+1.8: Two temperature components, $kT_{low} \approx 0.5$ keV and $kT_{high} \approx 2.0$ keV, are needed when we fit the data with a C.I.E. spectral model. This two temperature structure might indicate the existence of possible N.E.I. effects. The observed Bo/3Lx count rate ratio can be reproduced by means of a N.E.I. model, but it can also be explained by the C.I.E. model.

There is some indication (2 % statistical significance level) of a temperature difference between the central bar component, $kT_{bar} = 0.36 \pm 0.06$ keV, and the plateau component, $kT_{plateau} = 0.59 \pm 0.09$ keV. A possible explanation for this temperature difference is the overabundance by a factor 10 of oxygen for the bar component.

There are difficulties in explaining the low temperature component with a reverse shock since the predicted ratio between the low- and high temperature component is much larger than observed.

The electron density of the central bar component, $(n_e)_{plateau} \approx (5\text{-}10)$ cm^{-3}, is found to be somewhat higher then the electron density of the plateau component, $(n_e)_{plateau} \approx 1.8$ cm^{-3}.

Acknowledgements: I thank the 'Stiftung Volkswagenwerk' for financial support. The Laboratory for Space Research Leiden is financially supported by the Netherlands Organization for the Advancement of Scientific Research (NWO).

References:

Agrawal, P.C., Riegler, G.R. 1980, *Ap.J.(Letters)*, 237, L33
Braun, R., Goss, W.M., Danziger, I.J., Boksenberg, A. 1983, in *IAU Symposium 101, Supernova Remnants and their X-ray Emission*, ed. J.Danziger, P.Gorenstein (Reidel, Dordrecht), p.p.159
Braun, R., Goss, W.M., Caswell, J.L. Roger, R.S. 1986, *Astr.Ap.(Letters)*, 260, L65
Chevalier, R. A. 1982, *Ap.J.*, 258, 790
Goss, W.M., Shaver, P.A., Zealy, W.J., Murdin, P., Clark, D.H. 1979, *MNRAS*, 188, 357
de Korte, P.A.J., *et al.* 1981, *Space Sci. Rev.*, 30, 495
Lockhart, I.A., Goss, W.M., Caswell, J.L., McAdam, W.B. 1977, *MNRAS*, 179, 147
Mewe, R., Gronenschild, E.H.B.M., van den Oord, G.H.J. 1985, *Astr.Ap.*, 62, 197
Milne, D.K. 1969, *Austr.J.Phys.*, 22, 613
Morrison, R., McCammon, D. 1983, *Ap.J.*, 270, 119
Murdin, P., and Clark, D.H. 1979, *MNRAS*, 189, 501
Share, G., *et al.* : 1978, *IAU Circ.*, No. 3169
Shaver, P.A., Goss, W.M. 1970, *Austr.J.Phys.Astr.Suppl.*, 14, 133
Turner, M.J.L., Smith, A., Zimmerman, H.U. 1981, *Space Sci.Rev.*, 30, 513
Woosley, S.E., Weaver, T.A. 1982, in *Proc. NATO Conference on Supernovae*, ed. M.Rees, R.Stoneham, p.p.79

RECENT RESULTS FROM SUPERNOVA CALCULATIONS

James M. Lattimer
Dept. of Earth and Space Sciences, SUNY - Stony Brook
Stony Brook, NY 11794, USA

I. INTRODUCTION

On February 23, 1987, the brightest supernova in nearly 400 years, SN1987A, appeared in the Large Magellanic Cloud (LMC). The LMC is sufficiently close that for the first time, the neutrino burst, expected to occur as the core of a massive star collapsed to form a neutron star or a black hole, was detected. This was the Type II supernova theory's most dramatic prediction, and the neutrino detections alone will mark this supernova as an epochal event to be studied in textbooks for decades to come. This article will summarize the events surrounding the gravitational collapse and the birth of a neutron star. Discussion will be limited to Type II supernovae. The neutrino emission from a supernova is discussed and some comparisons with the particular case of SN1987A are made. It will be shown that the simple model of a cooling neutron star is in basic agreement with neutrino detections from this event.

Stellar evolution theory, coupled with the observed properties of supernovae, seems to unequivocably tie together massive stars, gravitational collapse, and Type II supernovae. On galactic timescales, massive stars are very short-lived; hence massive stars are only found near young, star forming regions of galaxies. Furthermore, the extended envelopes of massive stars must contain large amounts of hydrogen, except in some cases in which the envelope is depleted through a stellar wind or by mass transfer in a close binary system. Finally, the optical signature of a massive star explosion can be expected to vary depending on the mass of the star: the energy released and the amount of mass in the ejected envelope are certainly not invariants. The discrepancy between the energy expected to be released in a gravitational collapse event, $GM^2/R \simeq 2\text{–}3 \times 10^{53}$ ergs, and the observed envelope expansion energies, 10^{51} ergs, is, as we shall see, explained by the emission of neutrinos which carry away practically all the energy.

In many respects, the early evolution of a massive star parallels that of a moderate mass star like the Sun. However, while the Sun will end its life as a CO white dwarf, massive stars will undergo several subsequent burning stages and evolve cores that are too large to remain stable after nuclear energy generation ceases. The existence of the well-known Chandrasekhar mass, $\simeq 1.4 M_\odot$, which is the maximum mass that electron degeneracy pressure can support, leads to the fundamental divergence in the evolutions of low mass stars compared to those of high mass stars.

An important issue is whether or not the stages subsequent to He burning occur under degenerate conditions. If the pressure is dominated by electron degeneracy pressure, which is insensitive to temperature, the ignition of a nucelar fuel is accompanied by sometimes violent flashes and instabilities. Instabilities may lead to mass ejection, and prevent further evolution of the core. Stars less massive than $8M_\odot$ burn carbon degenerately. Stars more massive than $12M_\odot$ pass non-degenerately through all the remaining burning stages: carbon, neon, oxygen, and silicon. In between, the structural evolution is complex (e.g., Barkat, Reiss and Rakavy 1974, Nomoto 1984, Wilson et al. 1986).

The iron core mass increases with total stellar mass, primarily because heavier stars have larger temperatures and are less highly degenerate. For an equation of state that can be written in the polytropic form

$$P = K\rho^{4/3} , \tag{1}$$

the only mass that is stable in hydrostatic equilibrium is the Chandrasekhar mass

$$M_{ch} = 2.01824 \ (4\pi)(K/\pi G)^{3/2} . \tag{2}$$

Since the equation of state is dominated by relativistic, degenerate electrons, the core mass is, for all practical purposes, determined by the value of the Chandrasekhar mass. In this case K depends on the electron/baryon fraction Y_e ($K \propto Y_e^{4/3}$), and is also a slowly increasing function of the entropy, s. Approximately, $s \propto \ln M$; thus K and M_{ch} are larger for higher mass stars. It is currently thought that the iron core mass of a $12M_\odot$ star is $1.2M_\odot$, while stars greater than $25M_\odot$ may have iron cores approaching $2M_\odot$.

In the range 8–$10M_\odot$, high densities are reached following carbon burning as a degenerate neon-oxygen core develops. Before igniting neon, at these high densities electron capture reactions will dominate the evolution. In these reactions,

$$e^- + {}^{Z}A \rightarrow {}^{Z-1}A + \nu_e , \tag{3}$$

a proton is converted into a neutron, an electron is absorbed, and a neutrino emitted. Since the pressure is primarily due to electron degeneracy, and $K \propto Y_e^{4/3}$, pressure is lost, and a contraction of the core takes place. This, however, triggers further electron capture as is now explained. The electron chemical potential increases with density

$$\mu_e \simeq 11(\rho_{10} Y_e)^{1/3} \text{ MeV} , \tag{4}$$

where ρ_{10} is the density in units of $10^{10} gcm^{-3}$. The nuclear Q value, the difference of the neutron and proton chemical potentials μ_n-μ_p, is less sensitive to ρ. It is primarily determined by the proton fraction Z/A in nuclei and hence by the average proton fraction Y_e (ρ changes much more rapidly than Y_e). As long as $\mu_e - \mu_n + \mu_p > 0$, electron capture reactions will continue. Thus the cores of these stars undergo catastrophic collapse before an iron core develops, and while there is still unburned nuclear fuel in it.

In the range $>10M_\odot$, degenerate iron cores eventually develop. As in the lighter stars, core collapse is eventually triggered by electron captures. In the 10–$12M_\odot$ range, very few elements lighter than helium are thought to be ejected, while for stars $>12M_\odot$, progressively more heavy elements are ejected. Wilson et al. (1986) has calculated that the heavy element ejecta of a $15M_\odot$ star is $1.25M_\odot$, while that of a $50M_\odot$ star is $18M_\odot$.

II. GRAVITATIONAL COLLAPSE: GENERAL DEVELOPMENTS

Ever since the notion of neutron stars and their association with supernova was first proposed (Baade and Zwicky 1934), it has been realized that only a very weak coupling of the gravitational energy to the relatively loosely bound envelope would be necessary to eject it and explain the luminosity of Type II supernovae. For the last 20 years, it has been recognized that the vast majority of this energy would be lost to neutrinos, which couple extremely weakly to matter. Colgate and White (1966) first proposed that the coupling of neutrinos to the envelope matter might be sufficient to power an explosion. But subsequent calculations by Arnett (1966) indicated that neutrinos were too inefficient. Later, Wison (1971) showed that if the collapsing iron core is small enough ($<1.25M_\odot$), a purely hydrodynamical explosion, as initially proposed by Colgate and Johnson (1960) could result.

The discovery of neutral currents in the early 1970's complicated the picture. Freedman (1974) showed that these new interactions greatly increased the cross sections for scattering of neutrinos from heavy nuclei. While this initially seemed favorable for transporting energy to the envelope, Mazurek (1974) and Sato (1975) showed that, instead, the neutrinos become trapped within the core of the collapsed star and escape only relatively slowly. Thus, interest in the prompt, hydrodynamical mechanism increased.

By the early years of this decade, the physics of gravitational collapse clarified. Improvements in our understanding of weak interactions (especially concerning electron capture rates), the equation of state, and new computer calculations which couple hydrodynamics and neutrino transport, have all contributed. Before discussing some of the relevant physics in detail, it is useful to list the major developments.

1. Electron capture during the collapse is inefficient because neutrino trapping occurs shortly after the collapse begins. The number of leptons eventually trapped, $Y_L=Y_e+Y_\nu \simeq 0.35$–$0.38$ is only slightly smaller than the number in the precollapse iron core, $Y_e \simeq 0.41$–0.42.

2. The collapse is essentially adiabatic, because of the small change of lepton number due to neutrino trapping. The adiabat has a rather low entropy, $S/k \simeq 1$ per nucleon, because of neutrino cooling of the precollapse iron core (Bethe et al. 1979).

3. Nuclei are not dissociated during the collapse (Lamb et al. 1978). The density of nuclear excited states is so large, the proton fraction is so high, and the entropy of the

collapse path so low, that most of the matter remains in nuclei all the way to nuclear densities, $\rho_s \simeq 2.7\times10^{14}$ g cm^{-3}.

4. When the central density approaches ρ_s, the nuclei merge into arelatively incompressible nucleon fluid. The collapse is halted only above this density.

5. Baryonic pressure is not significant until $\rho>\rho_s$, because most baryons remain in nuclei. Therefore, the equation of state during collapse is essentially polytropic, $P=K\rho^{\gamma}$, with $\gamma\simeq1.30$. $\gamma<4/3$ because of residual attractive nuclear effects, and because electrons are slowly converted into neutrinos as the density is increased. K, which depends on Y_L and s, is nearly constant because neutrinos are trapped, and because the collapse is adiabatic.

6. Collapsing matter obeying a polytropic equation of state can be described by a self-similar solution (Yahil 1983), which shows that a collapsing core separates into an inner, homologous ($v\propto r$) part, and an outer, supersonically infalling part that is left behind. The inner core mass turns out to be about 10% greater than the Chandrasekhar mass for the same K, and is thus proportional to $Y_L^{\,2}$.

7. When the collapse is halted, a shock wave is produced at the boundary between the inner and outer parts of the core, because outside of this point, sound travels more slowly than the matter is moving. Due to energy conservation, the initial energy of the shock is well approximated by the binding energy of the inner core, and is proportional to $GM^{5/3} \propto Y_L^{\,10/3}$. The inner core is bound because its mass exceeds its Chandrasekhar mass.

8. A successful shock must quickly propagate through the outer core, and eject the envelope before the infalling matter pushes it inwards. The shock is primarily dissipated in dissociating the nuclei in its path, which takes about 9 MeV/nucleon, or 18×10^{51} ergs/$M_{\odot}$. The mass it must dissociate, that of the outer core, is the initial iron core minus the inner core mass. Since the inner core mass $\propto Y_L^{\,2}$, and the initial shock energy $\propto Y_L^{\,10/3}$, whether the shock is successful or not depends strongly on Y_L and the initial iron core mass.

9. In the case of an unsuccessful shock, some calculations (Wilson 1984) have indicated that neutrino heating (Bethe and Wilson 1985, Lattimer and Burrows 1984) of matter behind the shock will eventually resuscitate it. Otherwise, the shock will be forced back by the ram pressure of the infalling material, and a black hole will eventually form.

III. THE EQUATION OF STATE

The major features of the equation of state at subnuclear densities have been delineated by Lamb et al. (1978) and Lattimer et al. (1985), who developed a finite-temperature compressible liquid drop model for nuclei based on the pioneering study of Baym, Bethe and Pethick (1971). In this model, the nuclei are essentially viewed as a dense liquid ($\rho\simeq\rho_s$) in equilibrium with a vapor of nucleons. Once a theory of nucleon-nucleon interactions is adopted, the free energy of both phases is minimized with respect to their densities and proton fractions, the size of the nuclear droplets, and to the relative volume of each phase.

However, by neglecting the finite size of the nuclei, the bulk approximation, some of the more general results can be observed.

Let both nucleon fluids be describable by a free energy per baryon, $f(\rho,x)$, where ρ is the nucleon number density and x is the proton fraction. To examine the densities and temperatures at which nuclei may coexist with a vapor phase, first expand the free energy about its value, -16 MeV per nucleon, at T=0, x=1/2 and $\rho=\rho_s$:

$$f \simeq -16 \text{ MeV} + S_v(1-2x)^2 + (K_s/18)(1-\rho/\rho_s)^2 - (2m/\hbar^2)(\pi/12\rho)^{2/3}T^2 . \tag{5}$$

Here m is the nucleon mass. The expansion parameters are the incompressibility K_s, whose value is uncertain but probably between 100 and 240 MeV, and the symmetry coefficient $S_v \simeq 30$ MeV. Equilibrium of two phases (N=nuclei, o=vapor) is possible when the total free energy per unit volume

$$F(\rho,Y_e) = u\rho_N f_N(\rho_N,x_N) + (1-u)\rho_o f_o(\rho_o,x_o) \tag{6}$$

is minimized. Here $u = (\rho-\rho_o)/(\rho_N-\rho_o)$ is the volume fraction occupied by the denser phase, N. Also, we have $\rho Y_e = \rho_N x_N u + \rho_o x_o(1-u)$. One finds the conditions

$$\mu_{no} = \mu_{nN}; \ \mu_{po} = \mu_{pN}; \ P_o=P_N . \tag{7}$$

For $Y_e=1/2$, the fluid pressure is

$$P = \rho^2\partial f/\partial\rho = (\rho^2 K_s/9\rho_s)(\rho/\rho_s - 1) + (4\pi^2\rho/243)^{1/3}(m/\hbar^2)T^2 , \tag{8}$$

where $\rho = \rho_N$ or ρ_o. Therefore, for $Y_e=1/2$ and T=0, equilibrium is only possible between two phases with unequal densities when their pressures vanish. One easily sees this implies $\rho_N=\rho_s$ and $\rho_o=0$. That is, at T=0 nuclei may exist up to $\rho=\rho_s$, at which point u=1 and they occupy all of space. At still higher densities, a single phase, a uniform fluid of nucleons, exists.

At moderate temperatures equilibrium is also possible. Figure 1 illustrates the coexistence boundary as a function of density and temperature for $Y_e=1/2$. The maximum temperature, T_c in the case $Y_e=1/2$, at which coexistence is possible, is determined by the conditions

$$\partial P/\partial\rho = \partial^2 P/\partial\rho^2 = 0 . \tag{9}$$

The temperature, density and entropy per baryon of the critical point are thus:

$$T_c = (5\hbar^2 K_s/16m)^{1/2}(5\rho_s/8\pi)^{1/3} \simeq 13\text{–}18 \text{ MeV}$$

$$\rho_c = (5/12)\rho_s \tag{10}$$

$$s_c = (8\pi/5\rho_s)^{1/3}(5mK_s/16\hbar^2)^{1/2} \simeq 3\text{–}4 \ .$$

This result demonstrates that nuclei persist to rather large temperatures, and do not dissociate in stellar collapse, during which $s \simeq 1$.

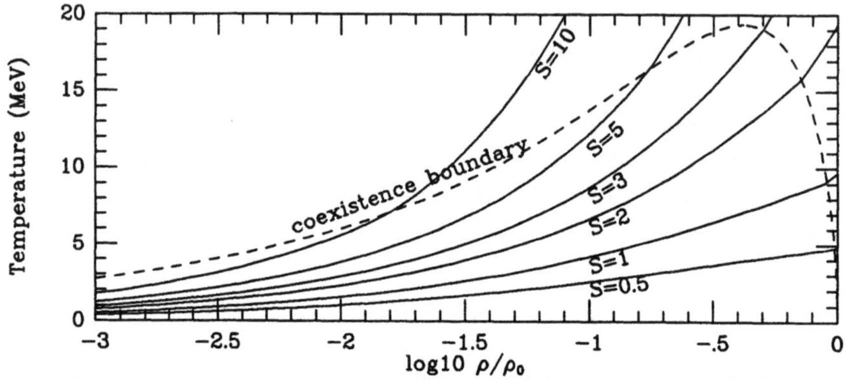

Figure 1. The region for coexistence of two nuclear phases in equilibrium (Y_e=1/2). Within the dashed line, the coexistence boundary, a dense nuclear state coexists with a less dense vapor state. Outside the dashed region, no such equilibrium is possible, and a uniform sea of nucleons exists. Adiabats, labeled by their entropy per baryon, are shown.

Now consider the equilibrium at finite, but small temperatures ($s \lesssim 1$), for which the vapor density ρ_o is no longer zero, but still very small. Taking $P_o \simeq 0$, one finds

$$\rho_N/\rho_s \simeq 1 - (2\pi/3\rho_s)^{2/3}(3m/K_s\hbar^2)T^2 \simeq 1 - (T/30 \text{ MeV})^2 \tag{11}$$

which shows that the density inside nuclei remains less than, but close to, ρ_s. We may thus ignore the last two terms in eq. (5). When Y_e=1/2, it is easy to see that $\mu_n \simeq \mu_p \simeq$-16 MeV. In the asymmetric case one finds

$$\mu_p \simeq -16 - S_v(1-2x_N)(3-2x_N) << 0. \tag{12}$$

The abundance of the rare, non-degenerate, protons outside nuclei is then

$$\rho_p = 2(mT/2\pi\hbar^2)^{3/2}\exp(\mu_p/T) \simeq 5\times10^{-4}\ T^{3/2}\exp(\mu_p/T)\ \text{fm}^{-3}. \tag{13}$$

For $x_N \simeq 0.4$, T=2 MeV and $\rho=10^{11}$ gcm^{-3}, we find $\mu_p \simeq$-29 MeV, and $\rho_p/\rho \simeq 2\times10^{-5}$. Generally, the free proton abundance is small, but is sensitive to both Y_e and T. As discussed earlier, the electron capture rate during collapse depends mostly on the proton abundance and is therefore relatively small. As x_N falls due to electron captures, μ_p becomes more negative, which mitigates the decrease in x_N and Y_e (Fig. 2).

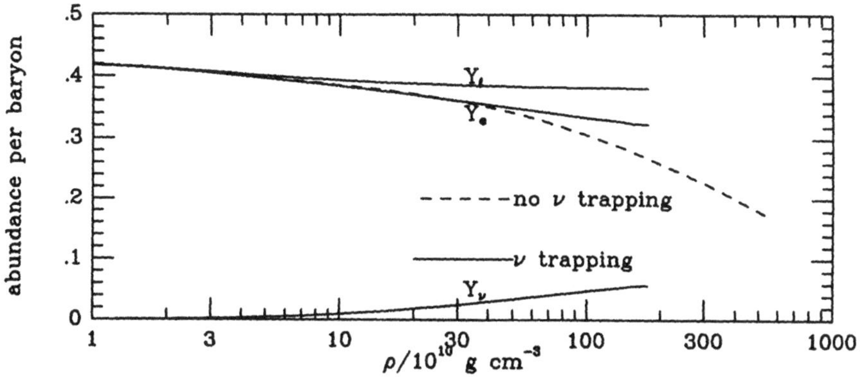

Figure 2. Typical lepton abundances during stellar collapse are shown by solid lines. Y_L is the total lepton/baryon fraction, and Y_e and Y_ν are the electron and neutrino fractions, respectively. The dashed curve shows the Y_L ($= Y_e$) fraction in the case that neutrinos freely stream out of the collapsing star.

Including finite-size nuclear effects does not alter this. The surface and Coulomb energies depend on x_N and these contributions together decrease the effective nuclear symmetry energy. A rough approximation would be to replace S_v in eqs. (5) and (12) by $S_v^{eff} \simeq 2S_v/3$. However, the relative contributions to the nuclear symmetry energy and the value of S_v itself remain somewhat uncertain. The greater S_v^{eff} is, the less efficient electron capture will be, and the larger Y_e will be during the collapse.

IV. NEUTRINO TRAPPING

If neutrinos did not become trapped, electron capture would continue as higher densities are encountered with the result that Y_e would be small, $\lesssim 0.1$, at densities exceeding ρ_s (see Fig. 2). With knowledge of the largest relevant neutrino scattering cross sections, that of elastic, coherent, scattering off nuclei (Freedman 1974), one may estimate the trapping density. The mean free path of neutrinos is approximately

$$\lambda_\nu \simeq 60\rho_{12}^{-1}(6X_n+5X_p + A(1-x_N)^2X_A)^{-1}(10\ \mathrm{MeV}/\epsilon_\nu)^2\ \mathrm{km}, \qquad (14)$$

where ρ_{12} is the density in units of 10^{12} gcm^{-3}, ϵ_ν is the mean neutrino energy, X_n, X_p, and X_h are the mass fractions of vapor neutrons, vapor protons, and nuclei respectively, and A is the mean atomic weight ($\simeq$70–90). Therefore, the mean free path becomes comparable to the size of the collapsing star, $R \simeq (3M/4\pi\rho)^{1/3}$, when $\rho \simeq 3\times10^{10}$ gcm^{-3}. Thus, the flow of neutrinos is impeded rather early in the collapse. The actual density of trapping is somewhat higher because neutrinos are still able to diffuse out. Only when the diffusion is less rapid than the collapse will the lepton fraction be frozen for the duration of the collapse. The standard expression for the diffusion and free-fall timescales give

$$\tau_d = 3R^2/c\lambda_\nu \simeq .3\rho_{12} \text{ s}; \quad \tau_f = (3/8\pi G\rho)^{1/2} \simeq 1.2\times10^{-3} \rho_{12}^{-1/2} \text{ s}. \tag{15}$$

The latter assumes pressureless matter. A more realistic value can be derived from self-similar models (Yahil 1983): $\tau_c \simeq \sqrt{33}\,\tau_f$. This implies $\tau_d = \tau_c$ when $\rho \simeq 9\times10^{10}$ gcm^{-3}. This result is insensitive to the electron capture, but does depend on the density profile of the collapsing star. This trapping density is low enough that relatively little neutronization takes place during infall. The latest calculations imply that the number of trapped leptons, $Y_L = Y_e + Y_\nu$, is in the range 0.36–0.39. The precise value of this parameter is extremely important in determining the outcome of collapse calculations.

It has already been shown that the bulk nuclear pressure is small, i.e., $P_N = P_o \simeq 0$. The pressure is therefore dominated by the lepton degeneracy pressure

$$P_L = \rho(\mu_e Y_e + \mu_\nu Y_\nu)/4 \simeq 6.4\times10^{-3}(\rho_{12} Y_L)^{4/3} \text{ MeV fm}^{-3}, \tag{16}$$

where the approximate result that $4Y_\nu \simeq Y_e$ is used to form the final equality. There is, however, a small, but significant correction due to the nuclear Coulomb lattice energy:

$$P_{lat} = -(9/40\pi)(Z^2e^2/r_N{}^4)u^{4/3}(1-u^{2/3}). \tag{17}$$

Here r_N is the mean nuclear radius and $Z = x_N A$ is the mean nuclear charge. Employing the equilibrium condition for the optimal nuclear size (Baym, Bethe and Pethick 1971), which states that the nuclear surface energy is twice the Coulomb energy, and assuming $x_N \simeq Y_L \simeq 0.4$ for definiteness, one finds

$$P_{lat}/P_L \simeq -0.05\ (1+u^{1/3})/[(1-u^{1/3})(1+u^{1/3}/2)^2]^{1/3}\ . \tag{18}$$

At low densities, $u \rightarrow 0$, this ratio varies little with u, and the overall equation of state is polytropic with an adiabatic index of 4/3. But at 10^{12} gcm^{-3} and 10^{14} gcm^{-3}, the magnitude of this ratio increases by factors of 1.16 and 2.15, respectively. An effective adiabatic index with a value slightly less than 4/3, about 1.3 (Lattimer et al. 1985), will approximate the equation of state.

V. GRAVITATIONAL COLLAPSE AND SHOCK FORMATION

The self-similar solutions of Yahil (1983) show that the core separates into inner and outer parts. When it bounces, the entire inner core bounces as a unit. Upon rebounding, the large velocity differences between the inner and outer cores form a shock that quickly accelerates outwards. The inner core is left behind as a quasi-static, marginally bound, remnant (a baby neutron star!) since its rebound and oscillations are effectively damped by the shock. From energy conservation, the shock's energy is equal to the binding energy of the newly formed neutron star, and will scale as $GM^2/R \propto M^{5/3} \propto K^{5/2} \propto Y_L^{10/3}$. Had the inner core's mass been equal to the Chandrasekhar mass, the binding energy would have been zero. But, in fact, the total mass exceeds M_{ch} as the self-similar solutions show. Since the energy obtained from adding a unit mass to an object in hydrostatic equilibrium is just $\partial E/\partial M = -GM/R$, one expects the total energy to be

$$E \simeq -GM_{ch}(0.1\ M_{ch})/R \simeq 8\times10^{51}\ (Y_L/0.38)^{10/3}\ \text{ergs} \tag{19}$$

where the factor 0.1 refers to the excess of the inner core's mass over M_{ch}.

Can the shock succeed? This is a complicated numerical question, but the answer is directly related to the amount of mass through which the shock must pass before leaving the iron core. Beyond the iron core, the density rapidly decreases and shocks accelerate. Most of the work done by the shock is to dissociate nuclei, at a cost of 9 MeV/nucleon or 18×10^{51} ergs/$M_\odot$. The mass in this region, and the energy needed to dissociate it, are proportional to $(Y_{Fe}^2 - Y_L^2)$, where $Y_{Fe} \simeq 0.41$–0.43 is the effective lepton fraction in the precollapse iron core. It is quite clear that the parameter Y_L is of paramount importance (Fig. 3). At present, it appears that even in lower mass stars, which have the smallest iron cores, if $Y_L \lesssim 0.38$, the shock alone will fail to eject the mantle and envelope of the star (Burrows and Lattimer 1985, Baron et al. 1985).

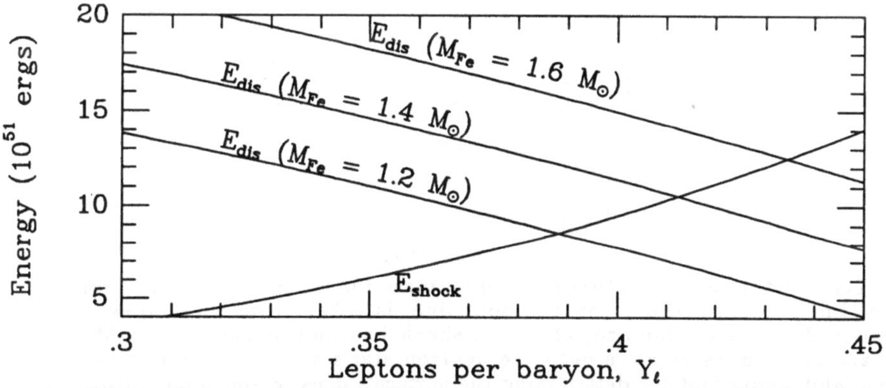

Figure 3. Schematic representation of the energetics of gravitational collapse supernovae. E_{shock} is the binding energy of a newly formed neutron star core. E_{dis} is the amount of energy required to completely dissociate the shell of mass between the neutron star core and the outer edge of the collapsing star's iron core, whose mass is M_{Fe}. Even in the case $M_{Fe} = 1.2\ M_\odot$, a Y_L smaller than 0.39 may not produce a successful explosion, since the energy required to dissociate the overlying matter exceeds the initial energy of the supernova shock.

It has alternatively been proposed that in the event of an unsuccessful shock, neutrino heating behind the shock eventually reenergizes it (Wilson 1985). A great deal of effort is presently going into confirming this mechanism. Needless to say, there is a great deal of controversy surrounding the supernova mechanism. Nevertheless, several things are certain: supernovae of Type II do explode, they leave collapsed remnants, and they emit neutrinos.

VI. NEUTRINO EMISSION FROM SUPERNOVAE

The binding energy of a newly formed neutron star is less than 10^{52} ergs, and most of this energy is initially in the shock. The total neutrino emission up to the bounce is only about 10^{51} ergs. Eventually, a cold, catalyzed neutron star will form with binding energy

$$E \simeq 3GM^2/5R \simeq 1.3\times10^{53}\ (M/M_\odot)^{5/3}\ \text{ergs}. \tag{20}$$

The observed energy of a supernova in photon radiation is about 10^{49} ergs, and the kinetic energy of the expanding gases is about 10^{51} ergs. Neutrinos must carry off the vast majority of a supernova's energy, and they do this primarily in the time shortly after the neutron star's birth. Applying equation (14), one sees that $\lambda_\nu \simeq 2$ cm, if a density 3 times nuclear and a lepton fraction of 0.4 are used ($X_H=0$). In fact, degeneracy corrections modify this result, so that $\lambda_\nu \simeq 10$ cm. Therefore, the diffusion time is about 3–10 s. A peak neutrino luminosity of nearly 10^{53} erg/s is expected. It is interesting that this is equivalent to the combined luminosity of all the stars in the observable universe!

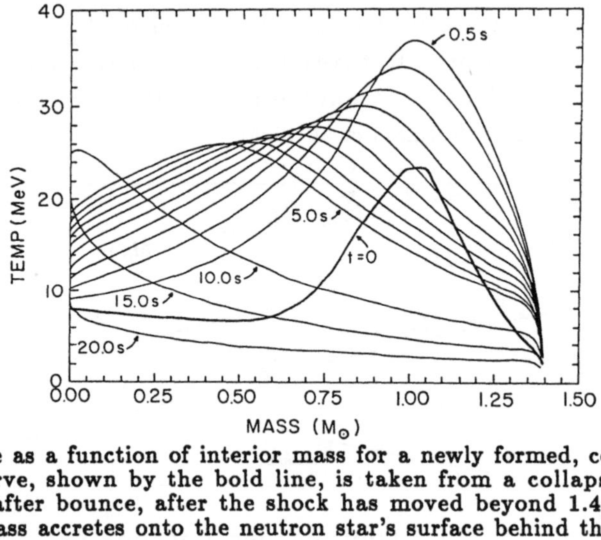

Figure 4. Temperature as a function of interior mass for a newly formed, cooling neutron star. The time=0 curve, shown by the bold line, is taken from a collapse calculation approximately 20 ms after bounce, after the shock has moved beyond 1.4 $M_\odot$. In this model, no additional mass accretes onto the neutron star's surface behind the shock. This assumption, while important for determining the surface's density and temperature, does not affect the details of the interior cooling.

A newly formed neutron star is lepton rich (compared to a cold star with $Y_e \simeq 0.05$) and <u>hot</u>. Somewhat suprisingly, a neutron star heats up during the initial stages of its cooling, because its outer layers compress and because the degeneracy energy of the neutrinos is converted into heat as they diffuse out (Burrows, Mazurek and Lattimer 1981, Burrows and Lattimer 1986) (Fig. 4). Near the center, neutrinos have energies $\epsilon_\nu \simeq \mu_\nu \simeq 240\ (\rho/\rho_s)^{1/3}$ MeV, but by the time they emerge from the star (at $\rho \simeq 10^{10}$–10^{11} gcm^{-3}), their energies have been degraded to about 10–20 MeV. Near the center, most of the neutrinos are of the electron type and are very degenerate, ($\mu_\nu/T \simeq 10$), so that there are virtually no anti-electron neutrinos. However, the neutrino flux emerging from the star is divided almost equally into thermal neutrinos of all types (e, μ, and τ). Because the interactions of $\bar{\nu}_e$, ν_μ and ν_τ with matter in the star are somewhat smaller than those of ν_e, their "neutrinospheres" (i.e., the surface from which they emerge from the star) are slightly further in and therefore hotter than that of the ν_e's. Calculations (Mayle 1985) show that the emerging ν_e energies are 10–12 MeV, the $\bar{\nu}_e$ energies are 15–20 MeV, and the ν_μ and ν_τ energies are about 20–30 MeV.

It is truly amazing that the unequivocable signatures of a cooling neutron star, namely $E \simeq 2$–4×10^{53} ergs, $\epsilon_\nu \simeq 10$–20 MeV, and emission timescale $\simeq$ 10 s, were observed by the two detectors, Kamiokande II (Japan) and IMB (USA), from SN1987A (Hirata et al. 1987, Bionta et al. 1987). A simultaneous signal with similar properties was seen by Baksan (USSR; Alekseev et al. 1987). (The signal observed at Mt. Blanc, recorded some 4 hours earlier, indicates a total energy of about 10^{54} ergs and ϵ_ν about 4 MeV. These data are also inconsistent with the optical observations.) Fig. 5 displays how well current theory was able to accommodate the actual accumulation of signal from SN1987A. The model (Burrows and Lattimer 1986) was calculated before the supernova was observed. We have since increased $\bar{\nu}_e$ energies by 40% to correct an approximation used in the original calculation, that the $\bar{\nu}$ opacity is equal to the ν opacity. In fact, the former opacity is smaller, due to the deficiency of protons in the outer layers of the neutron star. This change affects only the emerging neutrino energies, and not the generic nature of the neutron star cooling (Fig. 4).

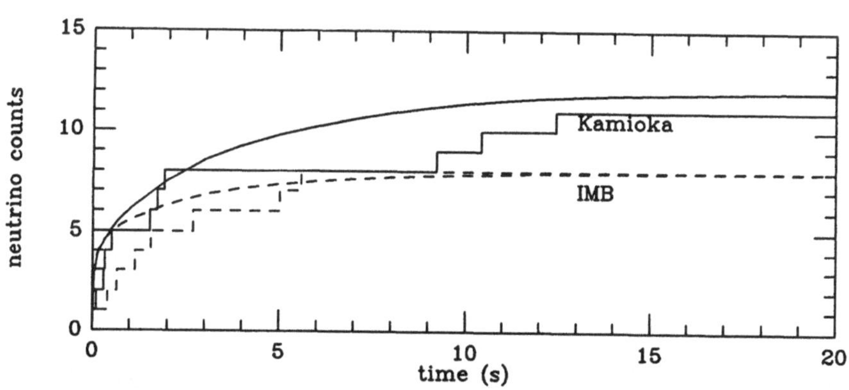

Figure 5. Theoretical and observed accumulations of SN1987A events for the Kamioka and IMB neutrino detectors. The zeros of time for the two detectors are assumed equal. The theoretical curves are obtained with the neutron star cooling models of Burrows and Lattimer (1986).

VII. CONCLUSIONS

SN1987A has presented scientists with both reassurances and challenges. The basic agreement of the neutrino flux and spectrum with those of cooling neutron stars strongly suggests that gravitational collapse resulted in the formation of a neutron star. The multisecond timescale over which the neutrinos were recorded, and their moderate individual energies, demonstrate the diffusion of neutrinos out of a dense object. The magnitude of the total inferred energy points to the formation of a neutron star rather than a black hole. The standard interpretation of a Type II supernova has been upheld.

Nevertheless, there are important questions left unanswered. What is the actual mechanism by which the star exploded? Is the shock generated by the core's bounce powerful enough to blow off the star's envelope? Or do neutrinos provide a secondary energy source that will effectively reenergize a stalled shock, after a "pause that refreshes"? Although the neutrino signal from SN1987A was too meagre to answer these questions, those from a galactic supernova, 5–50 times closer, very well might. It was fortunate that SN1987A was observed since neither the IMB nor the Kamioka detectors had been primarily designed to detect supernovae and had become sensitive enough only about a year earlier. With the same kind of luck that permitted the detection of SN1987A we may soon know.

This research was supported in part by the U.S. Dept. of Energy under grant DE-FG02-87ER40317.A000 at the State University of New York at Stony Brook.

REFERENCES

Alekseev, E.N., Alekseeva, L.N., Volchenko, V.I., and Krivosheina, I.V. 1987 JETP Letters, **45**, 589.
Arnett, W.D. 1966, Can. J. Phys., **44**, 2553.
Baade, W. and Zwicky, F. 1934, Phys. Rev., **45**, 138.
Barkat,, Z., Reiss, Y., and Rakavy, G. 1974, Astrophys. J. Letters, **196**, 633.
Baron, E., Cooperstein, J. and Kahana, S. 1985, Phys. Rev. Lett., **55**, 126.
Baym, G., Bethe, H.A., and Pethick, C.J. 1971, Nucl. Phys. A, **175**, 225.
Bethe, H.A., Brown, G.E., Applegate, J., and Lattimer, J.M. 1979, Nucl. Phys. A, **324**, 487.
Bethe, H.A. and Wilson, J.R. 1985, Astrophys. J., **295**, 14.
Bionta, R.M. et al. 1987, Phys. Rev. Letters, **58**, 1494.
Burrows, A. and Lattimer, J.M. 1985, Astrophys. J. Letters, **299**, L19.
Burrows, A. and Lattimer, J.M. 1986, Astrophys. J., **307**, 178.
Burrows, A., Mazurek, T.J., and Lattimer, J.M. 1981, Astrophys. J., **251**, 325.
Colgate, S.A. and Johnson, H.J. 1960, Phys. Rev. Letters, **5**, 235.
Colgate, S.A. and White, R.H. 1966, Astrophys. J., **143**, 626.
Freedman, D.. 1974, Phys. Rev. D., **9**, 1389.
Hirata, K. et al. 1987, Phys. Rev. Letters, **58**, 1490.
Lamb, D.Q., Lattimer, J.M., Pethick, C.J., and Ravenhall, D.G. 1978, Phys. Rev. Lett., **41**, 1623.
Lattimer, J.M. and Burrows, A. 1984, in Problems of Collapse and Numerical Relativity, ed. D. Bancel and M. Signore (D. Reidel: Dordrecht) p. 147.
Lattimer, J.M., Pethick, C.J., Ravenhall, D.G., and Lamb, D.Q. 1985, Nucl. Phys. A, **432**, 646.
Mayle, R. 1985, Ph.D. thesis, Univ. Cal. Berkeley.
Mazurek, T.J. 1974, Nature, **252**, 287.
Nomoto, K. 1984, Astrophys. J., **277**, 780.
Sato, K. 1975, Prog. Theor. Phys., **53**, 595 and **54**, 1325.
Wilson, J.R. 1971, Astrophys. J., **163**, 209.
Wilson, J.R. 1985, in Numerical Astrophysics, ed. J. Centrella, J. LeBlanc and R. Bowers (Boston: Jones and Bartlett), p. 422.
Wilson, J.R., Mayle, R., Woosley, S.E., and Weaver, T.A. 1986, Proc. of the 12th Texas Symp. on Rel. Astrophys., Ann. N.Y. Acad. Sci., **470**, 267.
Woosley, S.E. and Weaver, T.A. 1986, Ann. Rev. Astron. and Astrophys., **24**, 205.
Yahil, A. 1983, Astrophys. J., **265**, 1047.

SUPERNOVA STRUCTURE AND LIGHT CURVES

W. Kundt
Institut für Astrophysik der Universität Bonn, Auf dem Hügel 71,
D 53 Bonn, FRG

Abstract: It is concluded from simple assumptions that:

1) The SN piston has a relativistic sound speed; a likely realisation of the piston is a magnetic torque that transfers the collapsing core's angular momentum and subsequently converts into relativistic pair plasma.

2) In pressure contact with the pair plasma, a SN shell tears into small-filling-factor filaments during the ejection event.

3) The exponential tails of SN lightcurves can be explained by photon diffusion: Emission-line photons are trapped in the expanding filamentary shell; they take ≈ years to leak out.

For ejected shell masses ≥ 1 $M_\odot$, these conclusions are not very sensitive to the mass and nuclear chemistry. They are therefore thought to apply to all types of SN, whereby blue-giant progenitors yield smaller (bolometric, time-integrated) luminosities than red-giant progenitors.

1. SN Dynamics

We know from the historical SNSs and from SN spectra that the bulk of SN ejecta escape at velocities $v \gtrsim 10^{3.8}$ Km s^{-1}. When one assumes that the ejection is effected by the pressure of a hot gas, or plasma, this plasma must have a sound speed v_s which largely exceeds v. I.e.

$$T >> 10^{9.5} \text{ K } (m/m_p) \qquad (1)$$

must hold for the temperature of the SN piston throughout the ejection event, where m/m_p (≥ 1) is the average ionic mass in proton units.

It is not clear to me whether exploding white dwarfs can reach temperatures $\geq 10^{10}$ K for time intervals of ≥ 1 sec and whether their nuclear explosion energy can be transferred at a high efficiency (of $\geq$ 0.1) to an ejected shell. But Oemler and Tinsley (1979) have given convincing arguments that SN events of type I have massive progenitors. White dwarfs are thought to lose more mass on average during their binary evolution than they gain. And Blaauw (1985) has demonstrated that there are at least as many pulsars near the Sun as progenitor stars of mass $\gtrsim 5$ $M_\odot$; a comparable number of binary neutron stars would sharpen this reasoning (Blair & Candy, 1985). We are therefore in need, statistically, of producing neutron stars in the distinct majority of SN explosions (cf. Kundt, 1985). Independently, it is not clear to me whether SN shells can have masses much smaller than 1 $M_\odot$.

In what follows, I restrict considerations to SN explosions with progenitor star masses distinctly larger than Chandrasekhar's limit of 1.4 $M_\odot$ (of a white dwarf). Stellar evolution calculations for such massive giant stars predict mass density profiles $\rho(r) \sim r^{-3}$ sandwiched between a dense core, of mass ≈ 1 $M_\odot$, and an extended low-density envelope so that the mass M(r) inside of radius r grows logarithmically with r (cf. Woosley & Weaver, 1986). All I need here is the estimate that a significant fraction of the star's mass is spread out beyond a radius $r_s = 10^{12\pm1}$ cm. r_s is orders of magnitude larger than the core radius r_c inside of which most of the collapse energy is liberated. The runway of the SN piston, therefore, reaches from r_c out to r_s and beyond.

Inequality (1) for the piston's temperature T must hold throughout the effective runway, of reach r_e ($\leq r_s$), during which T(t) drops as r^{-2} (for an adiabatically expanding non-relativistic plasma, degenerate or not). At the onset of the ejection, we therefore get the necessary condition:

$$T(r_c) \gg 10^{9.5}\ \mathrm{K}\ (m/m_p)\ (r_e/\epsilon r_c)^2\ , \qquad (2)$$

where $\epsilon^{-1}\ v$ is the ejection velocity at radius r_e, $\epsilon \leq 1$.

Two cases should be distinguished. (i) The softest possible ejection is of sweepup character: it takes place throughout the complete runway, at velocities $\leq v$. In this case, r_e equals r_s, and $\epsilon = 1$. The acceleration time t is of order $r_s/v \gtrsim 10^{3\pm1}$ s, and the initial temperature must (formally) exceed $10^{19.5}$ K, according to condition (2). Such a high (relativistic) temperature is, however, ruled out (already) by energy estimates: core collapse does not reach temperatures much beyond 10^{11} K. Besides, neutrino cooling has a cooling time of order $10^{0.5}$ s $r_6^{-2}\ T_{11}^{-4}$ for core temperatures T_c above 10^{10} K (estimated with the blackbody law at radius r, which is an upper limit to Fermion radiation). Piston temperatures above $10^{10.5}$K are therefore unrealistic on timescales exceeding some 10s (cf. Burrows & Lattimer, 1987; Spergel et al, 1987).

(ii) Remains the (complementary) case that the necessary radial momentum is transferred, by the piston, to a small fraction $\epsilon < 1$ of the ejected shell mass ΔM. This fraction must then acquire a velocity $\epsilon^{-1}\ v\ (> v)$, in order for the whole shell to end up with a radial velocity of v after momentum sharing. The energy ΔE stored intermittently in $\epsilon \Delta M$ scales as $\Delta E \sim \epsilon^{-1}$, i.e. is larger than the asymptotic energy (and will be dissipated); ϵ can therefore not be too small: $\epsilon \gtrsim 10^{-2}$. ϵ must even be larger than $10^{-1.5}$ if relativistic velocities are to be avoided, which would cause excessive losses. Even for $\epsilon \approx 10^{-1}$, the temperature of the shocked shell material would reach $10^{-1}\ m_p\ v^2/k = 10^{11}$ K v_{10}^2 and give rise to excessive neutrino cooling. Independently of these additional problems, inequ. (2) still requires relativistic temperatures $T(r_c) \gg 10^{13}$ K for the starting temperature of the piston - which largely exceed the feasible (and observed, for SN 1987 A) core temperatures. In short: adiabatic cooling losses prevent a non-relativistic piston from transferring enough radial momentum to the overlying SN shell, because of its large spatial extent; cf. fig.1.

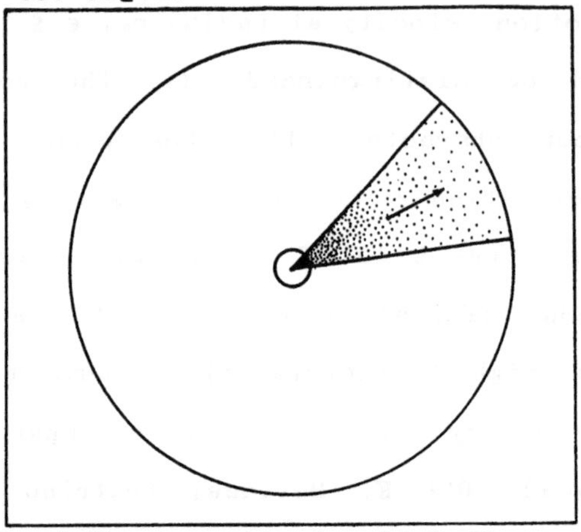

Figure 1: Sketch of the geometry to which the momentum balance of a SN explosion is applied. For symmetry reasons, the radial momentum acquired by an ejected spherical sector must equal the time-integrated force exerted across its inner (spherical) boundary.

This conclusion turns invalid when one drops the restriction to a gaseous piston with a non-relativistic sound speed. On the one hand, adiabatic cooling losses scale as r^{-1} in the relativistic case, i.e. are much weaker. More importantly, a relativistic piston can have negligible radiative losses as it does not quickly share its energy with neutrinos - because of its low number density (for given pressure) - whereas electromagnetic radiation is trapped inside the shell.

How to generate a relativistic piston? In 1976 I proposed a transfer of the contracting core's angular momentum to the surrounding shell via toroidal magnetic fields, an idea which has been independently pursued by Kardashev (1970), Bisnovatyi-Kogan et al (1976) and Srinivasan (1985). The fields grow strong by differential winding and transfer the core's angular momentum to an adjacent shell of comparable moment of inertia. Thereafter, the fields reconnect in the thinning medium and convert to highly relativistic charges which in turn convert to relativistic $e^{\pm}$-pairs on collision with photons (whenever their Lorentz factor γ exceeds MeV/hν). In this way, the core's maximal rotation energy, of order $10^{52.5}$ erg, is transformed into corotational motion of some inner shell plus a stuffing of its interior with pair plasma. Both together are thought to drive the rest of the shell to the observed velocities.

This scenario of a magnetically driven SN explosion appears to be consistent with the following facts and interpretations:

1) All well-mapped SNSs have filamentary morphologies, reminiscent of magnetic Rayleigh-Taylor instabilities during their acceleration. The filaments are the likely cause of refractive interstellar scintillations (e.g. Wolszczan & Cordes, 1987).

2) Neutron stars, the stellar remnants of SN explosions, appear to have uniform masses (of $(1.4 \pm 0.2)\ M_{\odot}$). The uniformity asks for an efficient, reproduceable piston.

3) The angular velocities of new-born pulsars tend to be much less than critical, with periods between 10^{-2} s and 1 s, but range almost up to a neutron star's centrifugal stability limit of $\lesssim$ 1 ms. (High spin frequencies at birth should be interpreted as due to inefficient coupling in the explosion; they correlate with weak fields).

4) If the relativistic pair plasma manages to pierce the ejected shell, it can give rise to a radio and X-ray flare, as observed in particular in type Ib explosions. The mystery spot of SN 1987 A and the non-circular radio sources in M82 (Bartel et al, 1987) may similarly find explanations as bipolar flows driven by pair plasma jets (cf. Kundt, 1987a).

5) Jim Lattimer's contribution to this workshop has given independent reasoning against neutrino-driven SN explosions.

2. Young SN shells

Above, it has been concluded that the SN piston cannot be a non-relativistic plasma but that a magnetic piston may be feasible. I now want to further explore the magnetic piston once it has decayed into relativistic pair plasma.

Some 10^{52} erg =: E of pair plasma inside a sphere of radius $r = 10^9$ cm exert a pressure

$$p_{rel} = E/4\pi\, r^3 \approx 10^{24} \text{ dyn cm}^{-2}\ r_9{}^{-3} . \tag{3}$$

The average number density $\langle n_{therm} \rangle$ of a solar mass of hydrogen, say, distributed inside a similar volume, is

$$\langle n_{therm} \rangle = 3N/4\pi r^3 \approx 10^{29.5} \text{ cm}^{-3}\ M_{(o)}\ r_9{}^{-3} . \tag{4}$$

This thermal plasma, assumed homogeneous for the moment, can (temporarily) ram-pressure confine the pair plasma at an (expansion) velocity v of

$$v \approx (2\ p_{rel}/\varrho_{therm})^{1/2} = 10^{9.3} \text{ cm s}^{-1} . \tag{5}$$

During expansion, $p_{rel}(r)$ decreases adiabatically as r^{-4}, with:

$$\int r^2\ p_{rel}\ dt \approx \int r^2\ p_{rel}\ dr/v \approx (p_{rel}\ r^3/\ v)_{initial} \approx 10^{42} \text{ dyn s}. \tag{6}$$

This momentum, when transferred to a surrounding shell mass ΔM, gives it a (radial) velocity v of

$$v = 4\pi \int r^2\ p\ dt/\Delta M = 10^{9.2} \text{ cm s}^{-1}/\Delta M_{(0.5)} , \tag{7}$$

where $\Delta M_{(0.5)} := \Delta M/10^{0.5}\ M_\odot$ is the shell mass in units of 3 $M_\odot$; cf.

fig. 1. This velocity agrees, order-of-magnitude wise, with observed shell velocities at birth. It shows - via momentum conservation - that 10^{52} erg of relativistic plasma can boost a surrounding shell of 3 $M_\odot$ to some 10^9 cm s^{-1} . According to eq. (5), the inner parts of the shell may (or may not) be pushed to somewhat higher than ultimate velocity. If they are, the kinetic energy $\Delta M\ v^2/2$ of the ejected shell is smaller than that of the driving pair plasma: the boosted shell is heated.

Shocked hydrogen at a speed v heats up to $T_{shock} \approx 0.1\ m_p\ v^2/k \approx 10^9\ K\ v_9{}^2$. Part of this heat energy is reconverted into radial motion, via adiabatic expansion, but a large fraction thereof is radiated, in the form of neutrino and electromagnetic radiation. The temperature T of the ejected shell will thus be non-uniform and decreasing with time. Near the shell's surface, radiation pressure post-accelerates a boundary layer to large velocities, upto and beyond 3 v, as is observed in the form of the broad lines. As a result, the shell's temperature is expected to fall quickly from 10^9 K to 10^5 K and further to 10^4 K. Can the cooling matter stay in pressure balance with the expanding piston?

In order to answer this question, let us divide the thermal pressure $p_{therm} \approx 2\ n_{therm}\ kT$ by the relativistic pressure p_{rel}, allowing for adiabatic relaxation at least out to the edge of the progenitor star's shell. This yields the volume-filling factor

$$f := p_{therm}/p_{rel} = \xi\ T_{9.5}\ \ \Delta M_{(0.5)} << 1\ , \qquad (8)$$

where the correction factor ξ takes care of adiabatic expansions: p_{rel} drops as r^{-4} as long as the pair plasma is confined, p_{therm} drops as r^{-5} from the pickup radius out to the radius r_s where the whole shell has reached terminal velocity. Near r_s, ξ therefore tends to be < 1, and together with $T_{9.5}$ << 1 I find f << 1. This means that in pressure equilibrium with the piston, an ejected shell cannot be volume filling. It will be torn into bits and pieces, as the result of Rayleigh-Taylor instabilities during pressure buildup. SN shells - if driven by relativistic plasma - cannot be continuous.

Once an ejected shell has taken shrapnel morphology, it can be overtaken by the pair plasma for not too dense a circumstellar medium (=CSM). Red giants have slow, high-density winds, with $\dot{M}$ of order $10^{-5}\ M_\odot/yr \approx 10^{21}$ g/s, whence

$$n_{CSM} = \dot{M}/4\pi r^2\ v_{wind}\ m \approx 10^{12}\ cm^{-3}\ \dot{M}_{(-5)}/v_6\ r_{13}^2\ . \tag{9}$$

For a pressure of $p_{rel} = 10^8$ dyn $cm^{-2}\ r_{13}^{-3}$, cf. equ. (3), the pair plasma is strong enough to sweep the CSM at ≥ one third the speed of light (for $n_{12} \leq 1$). More likely, however, it will form jets (c.f. Kundt, 1987 a).

The shell fragments, on the other hand, will be ram-pressure confined by the CSM, with

$$p_{ram} = \varrho_{CSM}\ v^2/2 \approx 10^{5.5}\ dyn\ cm^{-2}\ \dot{M}_{(-5)}/v_6\ r_{13}^2, \tag{10}$$

at a density

$$n_{therm} = p_{ram}/2\ k\ T = 10^{17}\ cm^{-3}\ \dot{M}_{(-5)}/v_6\ r_{13}^2\ T_4\ . \tag{11}$$

This density drops below the often reported $\lesssim 10^{10}\ cm^{-3}$ at radial distances $r \geq 10^{16.5}$ cm, corresponding to several months after the explosion. It drops much earlier for thinner windzones, viz. for $\dot{M}_{(-5)} \ll v_6$.

Figure 2 is a symbolical sketch of a young SN shell. The drop in mass density $\varrho(r)$ towards large radii is known to resemble a high power law, $\varrho(r) \sim r^{-p}$ with $p \approx 11$, e.g. for SN 1987 A (Dopita, 1988). When the shell is filamentary - as is indicated by the dots - a SN is seen as (at least) three different ring-like areas for every frequency: an inner circle which is optically thick, a surrounding ring which is optically thin but can be treated as though it was homogeneous, (for linear processes), and an outer ring whose density is so low that there is either zero or one filament along every line of sight. The latter ring tends to contribute symmetrical emission lines (only) to the spectrum.

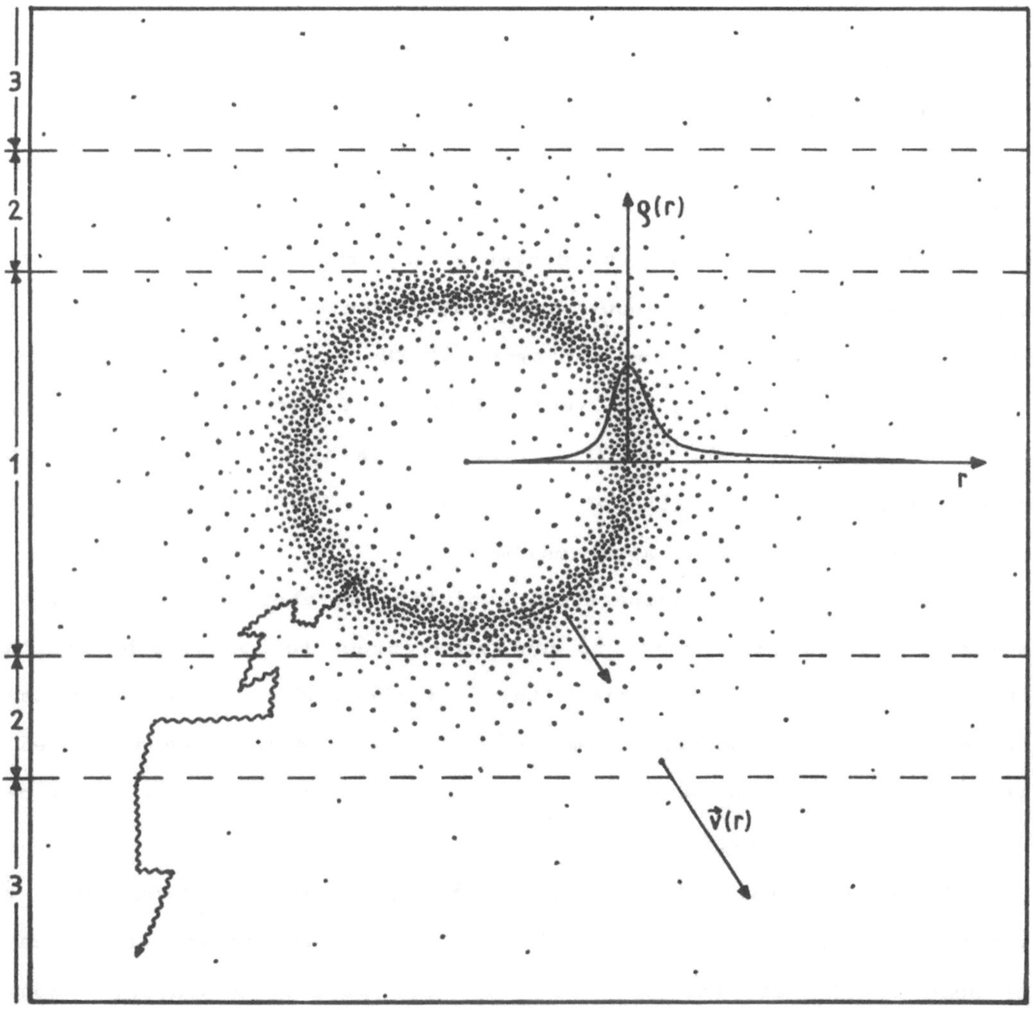

Figure 2:
Expected cross section through a young SN shell. The dots symbolize filaments of small volume-filling factor; their density $\varrho(r)$ decreases to both sides from the peak, as steep powers of radial distance (not drawn quantitatively). The velocities $\vec{v}(\vec{r})$ of individual splinters are strictly proportional to $\vec{r}$. Line-photons bounce back and forth between filaments ; they can be stored for years in the expanding shell without being reabsorbed.

3. SN lightcurves

The most impressive property of a SN is its brightness: during its $\lesssim$ month of maximum luminosity, a SN radiates some 10^{50} erg, significantly at visible frequencies. Even the faint SN 1987 A emitted 10^{49} erg. This energy is, however, only some 10% of the kinetic energy of the ejec-ted shell, and only some $10^{-3.2}$ of the total liberated energy, assumed via neutron star formation (e.g. Kundt, 1985). A SN is not an efficient lamp.

During its rapid expansion, the light of a SN increases initially because of a faster growth in area ($\sim t^2$) than evolution in temperature. After maximum light, the decrease in temperature dominates over the growth in size for about one month; thereafter, the effective temperature stabilizes - for all types of SNe - at some value between 5000 K and 7000 K, for $\lesssim$ two years (Kirshner & Kwan, 1975; Kirshner, 1977; Bouchet et al, 1987). This remarkable result has been corroborated by Della Valle et al (1988) as B-V $\simeq$ 0 for both type I and type II, up to 820 d after maximum.

The change in the temperature evolution at about one month after maximum, from rapid decrease to constancy, seems to coincide both with the beginning of an exponential 'tail' of the lightcurve and with the transition of the spectrum from continuum to emission-line-dominated. Kirshner (1977) stresses this very similar evolution of all types of SNe. How can it be explained?

The exponential tail of type I SN lightcurves, lasting for $\gtrsim$ 2 years, is commonly explained by heating due to the radioactive decay of some 0.1 $M_\odot$ of ^{56}Ni to ^{56}Fe via ^{56}Co (Woosley & Weaver, 1986). ^{56}CO-decay has been equally invoked to explain the late lightcurve of the type II SN 1987 A (Schaeffer et al, 1987; Pinto & Woosley, 1988; Kirshner, 1988; Woosley & Phillips, 1988). This explanation meets, however, with the following difficulties:

1) The exponential tails of type II SNe tend to have half the e^{-1}-folding timescale of those of type I SNe.

2) Even within a particular class of SNe, the e^{-1}-folding timescale can differ by a factor of 3: SN 1986 G (of type Ia) fell by 3 mag/10^2d, S And 1885 (of type I) evolved even faster, whereas typically, $\dot{m}(I)$ = 1.5 mag/10^2d; for heterogeneities see also Branch (1987: type Ia), Schaefer (1987: type II) and Wheeler et al (1987: type Ib). See also fig. 3.

3) In many cases, including SN 1987 A, the amounts of ^{56}Co and ^{56}Fe inferred from the optical or γ-ray spectra have fallen a long way short of the required amount; explanations were presented to bridge discrepancies by a factor of $\lesssim 10^2$ (cf. Gehrels et al, 1987).

4) SN lightcurves have been reported to rise again after several years: SN 1006 was seen again in 1016 (Minkowski, 1966; Wang, 1980), SN 1572 (Tycho) in 1604 and SN 1604 (Kepler) in 1664 (refs. as above); SN 1961 V had another small peak in 1964 (Utrobin, 1984). Also, SN shells can radiate $10^{37.5}$ erg s^{-1} after $\gtrsim$ 300 yr.

5) For reasonable shell masses, the cooling times are initially of order a day so that a constant (observed) T_{eff} asks for a remarkable thermostat. Moreover, a short cooling time is in conflict with a high optical depth in most of the emission lines.

6) The constant effective temperature has been explained by a photosphere whose temperature is controlled by recombination. An exponential light curve implies, in this model, an exponentially shrinking radius of the photosphere. Only an exponentially declining fraction of the radioactive heat would thus be radiated in the continuum, i.e. the continuum would have to disappear quickly, and T_{eff} would have to change.

From what has been said, I understand that the postulated radioactive stove is in conflict with the facts. I can even think of three other potential stoves whose energies are comparable to the 10^{49} erg from ^{56}Co-decay:

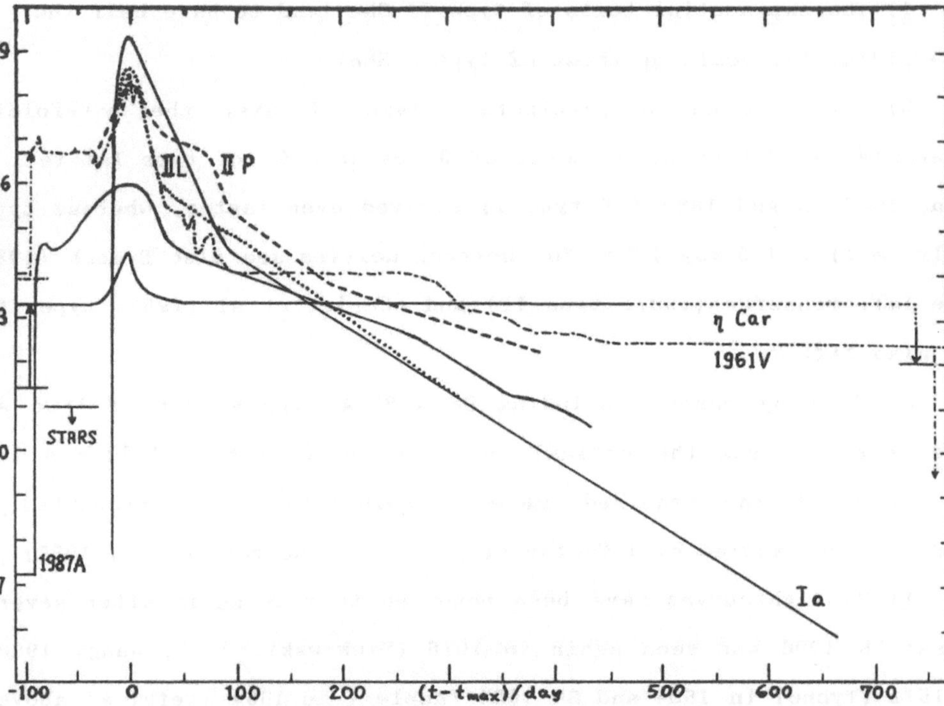

Figure 3: A representative sample of SN lightcurves, taken from (Kundt, 1987b). Plotted is the logarithmic (intrinsic) luminosity, in magnitudes (left-hand scale) or erg s^{-1} (right-hand scale), versus time after maximum (in days) for SNe of various types. For all the curves, the integrated bolometric luminosity $\int Ldt$ is near 10^{50} erg except for that of SN 1987A (with $\int Ldt = 10^{49}$ erg), and the luminosity exceeds that of the brightest stars by factors between 10 and 10^3. Note the different risetimes and slopes of the exponential tails. The role played by η Carinae is not clear.

1) 10^{49} erg are just 1% of the kinetic shell energy; ram pressure heating of the ejecta is a potential stove.

2) A newly formed neutron star, of internal temperature $T \approx 10^{10}$K, stores some 10^{49} erg T_{10}^2 thermally. No (mature) neutron star with a surface temperature above $10^{5.8}$ K has been detected (Brinkmann & Ögelman, 1987); apparently, neutron star cooling is faster than predicted (cf. Nomoto & Tsuruta, 1983; Romani, 1987; Van Riper, 1988). If young neutron stars cool via volcanos, they may have average surface temperatures in the vicinity of 10^8 K throughout a few years.

3) If a SN explosion is driven by some 10^{52} erg of relativistic pair plasma, as concluded above, a certain percentage of that energy will still be present after takeoff. This pair plasma will radiate synchrotron and/or inverse Compton radiation, in interaction with the local (wave) fields.

Potentially, therefore, there is no lack of energy to explain the tail of a SN light curve. It is more problematic to understand why at the onset of the tail, the effective temperature evolution changes from rapid decline to a constant, and why at the same time, the spectrum develops emission-line structure. This behaviour looks like a radiation transfer phenomenon: At a time when the shell gets transparent in the continuum, it still remains opaque in most of the lines. An exponential tail results as the solution of a diffusion problem ; (cf. Arnett, 1982, who speaks of a "fascinating problem in radiative transfer").

4. Leaky-box model for SNSs

A few quantitative estimates are in order. An ejected shell of mass ΔM, composed of ions of average mass m, has a radial column number density N (at radius r) of

$$N = \Delta M/4\pi r^2 m = 10^{24.5}\ \mathrm{cm}^{-2}\ \Delta M_{(0.5)}\ (m/m_p)\ r_{16}^{-2}\ . \qquad (12)$$

A hydrogen shell of 3 $M_\odot$ therefore becomes transparent in the continuum for $r \gtrsim 10^{16}$ cm, corresponding to an optical depth of $\tau_{cont}\,(\gtrsim \tau_T) \lesssim 1$. At a speed v of $\lesssim 10^9$ cm s^{-1}, r = 10^{16} cm is reached in $t_{tr} \gtrsim 10^7 s \approx 3$ months. In other words: a SN shell gets transparent in the continuum after a time

$$t_{tr} = 10^7\ s\ (\Delta M_{(0.5)}\ m_p/m)^{1/2}\ v_9^{-1}\ . \qquad (13)$$

This time corresponds to the onset of the exponential tail. Clearly, at $t = t_{tr}$ the shell has still a huge opacity in most of the emission lines whose resonance cross section

$$\sigma_{line} = gf\,\pi\, e^2\lambda/\beta\, m_e\, c^2 = 10^{-12}\ \mathrm{cm}^2\ (gf\ \lambda_{-4.5}/\beta_{-4.5}) \tag{14}$$

can be some 10^{12} times the continuum cross section ($\gtrsim \sigma_T$) for an oscillator strength $gf \approx 1$, wavelength $\lambda \approx 10^{-4.5}$ cm, and for a Doppler offset $\beta = v/c$ corresponding to thermal velocities at $T = 10^4$ K. In some of the lines, therefore, a SN shell stays opaque for centuries, depending on excitation conditions.

How can one understand the constant radiation temperature of a transparent shell? Exponential tails have luminosities $L = 10^{40\pm2}$ erg s^{-1}, cf. fig. 3. At such luminosities, the cooling time t_{cool} is of order

$$t_{cool} \approx \Delta M\, kT/m\, L = 0.6\ \mathrm{day}\, \Delta M_{(0.5)}\, T_4/L_{41}\ , \tag{15}$$

i.e. short compared with the evolution time scale. A constant shell temperature would therefore ask for a stable, permanent heat input. Alternatively, the constant radiation temperature can simply evidence a frozen energy distribution of the photons trapped in the expanding shell: The shell is a (line) photon bag.

Why don't the trapped photons suffer adiabatic expansion losses? The reason is that for resonance scattering, an ion or atom re-emits an absorbed photon at a frequency which is independent of the photon's history; there is no average degrading of photon energy.

According to eq. (14), a photon in an expanding shell is absorbed inside a 'trapping sphere'of size s_{trap} whose absorbers have the resonant Doppler velocity.

$$s_{trap} \approx v_{th}/|\nabla v| \approx r\, v_{th}/v_{SN} = 10^{12}\ \mathrm{cm}\ t_7\ T_2^{1/2}/v_9 \tag{16}$$

for a thermal velocity $v_{th} \approx 10^5$ cm s^{-1} (corresponding to a temperature of $T = 10^2$ K), at a radius $r = 10^{16}$ cm corresponding to $t \approx 10^7$ s. This trapping length is large compared with the mean free path λ_γ of a photon

$$\lambda_\gamma = 1/n\, \sigma_\gamma = 10^2 \text{ cm } n_{10}\, T_2^{1/2} \qquad (17)$$

of (resonance) cross section σ_γ given in eq. (14), inside a gas or plasma of density $n = 10^{10}$ cm^{-3} corresponding to the average number density of a shell of 3 $M_\odot$ at $r = 10^{15.5}$ cm, cf. equ. (12). If the shell is filamentary, n inside filaments is higher than average and determined by ram pressure balance. From equs. (16) and (17) we find that s_{trap}/λ_γ is large, of order 10^{10}, and can be even larger if the large-scale velocity dispersion of the ejected shell exceeds the thermal dispersion at a temperature of 10^2 K. Line photons cannot easily leak away from their birth sites: an 'on-the-spot' approximation is legitimate.

Line photons can, however, be reconverted to heat if their scattering event overlaps with the near passage of an electron. This absorption probability W is the ratio of the photon-emission timescale $t_{em} = A_{21}^{-1} = (\Delta\omega)^{-1}$ and the electron encounter time scale $t_{enc} = 1/n_e \cdot \langle \sigma_{enc}\, v_e \rangle$,

$$W = t_{em}/t_{enc} = 10^{-8}\, n_{10}\, T_2^{1/2}\, A_9^{-1} \quad . \qquad (18)$$

It means that a line photon is thermalized, at an electron density n_e near 10^{10} cm^{-3}, after some 10^8 scattering events, corresponding to an optical depth (in the line) of $\tau_{line} \gtrsim 10^4$. From a homogeneous shell, therefore, few line photons would escape even for a low degree of ionization.

When, however, the shell is filamentary, a photon spends most of its time in the pair-plasma quasi-vacuum between filaments. It occasionally hits the edge of a filament and is generally released after only a small number of scatterings. In this way, the escape time from the shell is lengthened by τ_{fil}^2 over the straight-line escape time r/c:

$$t_{esc} = \tau_{fil}^2 \; r/c \approx \tau_1^2 \; yr \; r_{16} \quad , \tag{19}$$

where τ_{fil} is the filament-encounter opacity of the shell, assumed of order 10. Only some 10 encounters with filaments are therefore needed by a line photon in order to delay its leakage by one year.

It remains to be shown that photon leakage declines exponentially with time. Inside the expanding shell, loss-free (line) photon diffusion satisfies

$$0 = r^2 \, dn_\gamma/dt + \partial_r \, (r^2 \; D \, \partial_r \; n_\gamma) \quad , \tag{20}$$

where d/dt is the comoving time derivative, and $D = \lambda_\gamma c/3$ is the photon diffusion coefficient. The separation ansatz $n_\gamma = m(x)\, n(t)$, with $x := r/r(t)$, ($r(t)$ = comoving shell radius), leads to

$$\partial_t \; \ln n(t) = - r^{-2} \; \partial_x \; (x^2 \; D \, \partial_x \; \ln m(x)) \approx \text{const} \quad , \tag{21}$$

so that the luminosity $L(t) \approx 4 \pi \; r_{phot}^2(t) \; n_\gamma \; c \, h\nu$ evolves according to

$$\partial_t \; L = L \, \partial_t \; \ln \, (r_{phot}^2 \, (t) n_\gamma \,) =: -L/\tau \quad , \tag{22}$$

i.e. according to an exponential decrease if and only if τ = const. By definition, $\tau = [-2(v/r)_{phot} + a]^{-1}$ where $v_{phot} = \partial_t \; r_{phot}$, the photospheric velocity of the emission lines, is of unknown sign but expected to be small (in magnitude) compared with $\partial_t \; r(t)$ and where a is the separation constant of eq. (21). The sum of the two is of order

$$\tau \approx \Delta r/v = 10^7 \text{ s } (\Delta r)_{16}/v_9 \quad , \tag{23}$$

in agreement with the observations, and will hardly decrease with time. Note that a strictly constant τ would be inconsistent with the observations, in particular for lightcurves of type II (cf. fig. 3). Note also that τ tends to grow with the onset time ($\sim \Delta r$) of the tail, again in agreement. In other words: equs. (21), (22) support a photon-leakage interpretation of SN lightcurve tails.

5. How much is understood?

A number of questions have been asked repeatedly concerning SNe: What makes them go off? What do they leave behind? What is the structure of their ejected shells? What are their progenitor stars? As explained above, my preferred answers read respectively: a magnetic torque - a neutron star - filaments of small filling factor - $M \gtrsim 5\ M_{\odot}$.

More in detail, I cannot see a clear dividing line between type I and type II events other than that the progenitors of type I lack hydrogen. All SN shells are filamentary. The different shell morphologies can be understood as due to different shell decompositions and different densities and temperatures of the CSM.

Hydrogen shells insulate better than helium shells do: for them, maximum light occurs after $\lesssim 10^2$ days instead of after $\lesssim 15$ days. Moreover, a generally larger shell mass (of type II) implies a smaller bulk velocity for the same ejection energy, again lengthening the evolution.

If the evolution times of types Ia and Ib are the same, their shell masses ought to be the same. Yet type Ib events correlate better with H II regions than type Ia events do. Are the progenitors of type Ia a little older than those of type Ib, or have they received a larger kick velocity during the first SN explosion in the system? A smaller integrated luminosity is expected for an exploding blue (super-) giant than for a red giant, because of larger adiabatic expansion losses of the overpressure photon gas. Given red and blue giants with and without hydrogen, we have already four different types of SN. Do we need more? It would surprise me if there was more than one ejection mechanism, leading to a very similar velocity distribution and temperature evolution of the ejecta and to a filamentary morphology.

Concerning the appearance of an evolved SN shell, there are the further possibilities that the newly formed neutron star can be single or binary, i.e. a pulsar (Crab), a dormant binary, or an X-ray binary (SS 433). The expected variety - and rate - of neutron star formation are so large that I cannot see any demand for exploding white dwarfs, forming black holes, or cosmic strings.

Even so, there are a handful of objects in the sky with an integrated light output equal to that of a SN but with very different lightcurves. Among them are SN 1961V and η Car, cf. Kundt (1987b). Their precursors have luminosities largely in excess of those of the brightest reported stars in the Local Group. Another object is SN 1984 E whose blue progenitor ejected $\geq 0.4\ M_{\odot}$ in a discrete event, between 2.2 and 23 years before maximum light (Gaskell & Keel, 1987). Are these progenitors enhancing their luminosities via contraction?

Acknowledgements: My interest in SNe was strengthened by Mehmet Özel in 1985 and furthered by Radha Banhatti in 1986. More recently, valuable input has been received from Reinhold Schaaf, Axel Jessner, and Nino Panagia.

References

Arnett, W.D., 1982: Astrophys.J. 253,785

Bisnovatyi-Kogan, G.S., Popov, Yu.P. & Samochin, A.A., 1976: Astrophys.Sp.Sci. 41, 287

Blaauw, A., 1985, in: Birth and Evolution of Massive Stars and Stellar Groups, eds. Boland & van Woerden, Reidel, p. 211.

Blair, D.C. & Candy, B.N., Mon.Not.R.Astron.Soc. 212, 219

Bouchet, P., Stanga, R., Moneti, A., Le Bertre, Th., Manfroid, J., Silvestro, G. & Slezak, E., 1987, in: SN 1987A, ed. I.J. Danziger, ESO Garching, p. 79

Branch, D., 1987: Astrophys.J. 316, L81

Brinkmann, W. & Ögelman, H., 1987: Astron.Astrophys. 182, 71

Burrows, A. & Lattimer, J.M., 1987: Astrophys. J. 318, L63

Della Valle, M., Cappellaro, E., Ortolani, S. & Turatto, M., 1988: The Messenger 52, 16

Dopita, M.A., 1988: Nature 331, 506

Gaskell, C.M. & Keel, W.C., 1987: Sky & Telescope 74, 234; also: preprint 1988

Gehrels, N., MacCallum, C.J. & Leventhal, M., 1987: Astrophys.J. 320, L19

Kardashev, N., 1970: Sovj.Astron. 14, 375

Kirshner, R.P., 1977: Ann.N.Y.Acad.Sci. 302, 81

Kirshner, R.P., 1988: National Geographic 173, 640 (May)

Kirshner, R.P. & Kwan, J., 1975: Astrophys.J. 197, 415

Kundt, W., 1976: Nature 26, 673

Kundt, W., 1985: Bull.Astron.Soc. India 13, 12

Kundt, W., 1987a: Astrophysical Jets and their Engines, NATO ASI C 208,Reidel, 1

Kundt, W., 1987b, in: ESO workshop on the SN 1987A, ed. Danziger, ESO Garching, p. 633

Minkowski, R., 1966: Astron.J. 71, 371

Leising, M.D., 1988: Nature 332, 516

Nomoto, K. & Tsuruta, S., 1983, in: SNRs & their X-Ray Emission, eds. Danziger & Gorenstein, Reidel, p. 510

Oemler, A. & Tinsley, B.M., 1979: Astron.J. 84, 985

Pinto, P.A. & Woosley, S.E., 1988: Nature 333, 534

Romani, R.W., 1987: Astrophys.J. 313, 718

Schaefer, B.E., 1987: Astrophys.J. 323, L51

Schaeffer, R., Cassé, M., Mochkovitch, R. & Cohen, S., 1987: Astron. Astrophys. 184, L1

Spergel, D.N., Piran, T., Loeb, A., Goodman, J. & Bahcall, J.N., 1987: Science 237, 1471

Srinivasan, G., 1985: invited lecture at XIXth General Assembly of IAU, New Delhi

Utrobin, V.P., 1984: Astrophys.Sp.Sci. 98, 115

Van Riper, K.A., 1988: Astrophys. J. 329, 339

Wang, J.-M., 1980: Chinese Astronomy 4, 407

Wheeler, J.C., Harkness, R.P., Barker, E.S., Cochran, A.L. & Wills, D., 1987: Astrophys.J. 313, L69

Wolszczan, A. & Cordes, J.M., 1987: Astrophys.J. 320, L35

Woosley, S.E., & Phillips, M.M., 1988: Science 240, 750

Woosley, S.E. & Weaver, T.A., 1986: Ann.Rev.Astron.Astrophys. 24, 205

WHAT ARE THE MASSES OF SNIb PROGENITORS?

Sidney van den Bergh
Dominion Astrophysical Observatory,
National Research Council of Canada

ABSTRACT

The spectra, frequency, and spatial distribution of supernovae are all consistent with the hypothesis that SNIb have precursor masses $\mathcal{M} > 15\ \mathcal{M}_\odot$ and that SNII are produced by stars that have main-sequence masses in the range $8 < \mathcal{M} < 15\ \mathcal{M}_\odot$.

I. INTRODUCTION

Five years ago (van den Bergh 1983) astronomers first began to realise that there was a major discrepancy between the fact that only two kinds of supernovae (SNI and SNII) were observed whereas there exist three distinct types of young supernova remnants, of which Cassiopeia A, the Crab nebula and the remnant of Tycho's supernova of 1572 are the prototypes. The first indication of a solution to this dilemma came when it was realized (e.g. Panagia 1985, Wheeler and Levereault 1985, Uomoto and Kirshner 1985) that SNI, which are characterised by a complete absence of hydrogen lines, consist of a mixture of two spectroscopically distinguishable classes of objects. Speculations suggesting that there are actually two distinct types of SNI had previously been published by Bertola and Sussi (1965) and by Shklovskii (1984).

II. SUPERNOVA PROGENITORS

It is now almost universally accepted that SNIa, which exhibit an absorption feature near 6150Å, are produced by the deflagration of a CO white dwarf (Iben and Tutukov 1984). This conclusion is consistent with the observation that SNIa are the only type of supernovae that have ever been observed in elliptical galaxies.

The fact that SNIb and SNII have their highest intrinsic frequency in spiral galaxies of type Sc strongly suggests that both of these kinds of supernovae have massive progenitors. For supernovae of type II this speculation is supported by the observation (Maza and van den Bergh 1976) that these objects are strongly concentrated in spiral arms and therefore have massive young progenitors.

Of the 12 supernovae that have been classified spectroscopially as belonging to type Ib [Wheeler, Harkness and Cappellaro 1987, Panagia (this conference)] 6 appear to be associated with HII regions. This figure is likely to be an underestimate because only giant HII regions can be observed in distant spirals. Some of the SNIb, that appear not to be associated with HII regions, might therefore still be located in intrinsically faint HII regions such as the Orion nebula.

III. THE MASSES OF THE PROGENITORS OF SNIb.

The fact that <u>at least</u> half of all SNIb are associated with HII regions demonstrates that they must have massive O-type progenitors. From a study of northern O stars in the Bright Star Catalogue (Hoffleit 1982) van den Bergh (1988) finds the following distribution of O-type stars:

Spectral type	05 + 06	07 + 08	09 + 09.5
In bright HII region	12	22	5
In faint HII region	4	8	9
Not in HII region	8	18	28

The data in the table show that the frequency with which SNIb occur in HII regions is very similar to that for stars of type O, which all have $\mathcal{M} > 15\,\mathcal{M}_\odot$. This observations strongly suggests that SNIb also have progenitors with masses in excess of $15\,\mathcal{M}_\odot$. This conclusion agrees with spectroscopic observations of SNIb (Begelman and Sarazin 1986) which all indicate that such objects have compositions consistent with the assumption that they had massive WR type progenitors which had lost their hydrogen envelopes before they experienced core collapse.

IV. SUPERNOVAE AND SUPERNOVA REMNANTS

Of the 7 young oxygen-rich supernova remnants similar to Cas A, that are presently known, 5 appear to be associated with HII regions (van den Bergh 1988). This observation agrees with evolutionary model calculations (Woosley and Weaver 1986, Chiosi and Maeder 1986) which show that oxygen-rich SNR's are produced by supernovae that have massive progenitors. The idea that SNIb produce oxygen-rich remnants is strongly supported by observations of SN1983N (Gaskell <u>et al</u>. 1986) and 1985F (Filippenko and Sargent 1985) which show that oxygen is almost certainly over-abundant in these objects.

Davidson <u>et al</u>. (1982), Nomoto (1983) and Henry (1986) argue that the observed abundance pattern in the Crab nebula is consistent with that expected from the explosion of a star with a main-sequence progenitor of 8-10 $\mathcal{M}_\odot$. This suggests that plerionic remnants, such as the Crab, are produced by stars at the lower end of the mass spectrum of young stars that experience core collapse. This conclusion is entirely consistent with the observation that plerionic SNR's are not strongly associated with HII regions. The fact that the envelope of the Crab nebula contains hydrogen gas shows that it must, by definition, have been produced by a supernova of type II.

A mass spectrum of star formation of the form $\psi \propto \mathfrak{M}^{-2.5}$, the SNIb and SNII rates of van den Bergh, McClure and Evans (1987), and a Hubble parameter $H = 75$ km s^{-1} Mpc^{-1} are consistent with the assumption that SNIb have main-sequence progenitors with $\mathfrak{M} > 15\,\mathfrak{M}_{\odot}$ and that SNII have progenitors with $8 < \mathfrak{M} < 15\,\mathfrak{M}_{\odot}$.

By a process of elimination this leaves SNIa as the progenitors of collisionless shock-type SNR's, such as Tycho and Lupus (= SN1006). The fact that no abundance anomalies are observed in the optical spectra of these objects militates in favour of the idea that they were produced by low-mass supernovae that did not eject massive, and hence easily observable, shells consisting entirely of elements heavier than hydrogen.

REFERENCES

Begelman, M.C. and Sarazin, C.L. 1986, Astrophys.J.(Letters) 302, L59.
Bertola, F. and Sussi, M.G. 1965, Contr. Obs. Astrofis. Asiago No. 176.
Chiosi, C. and Maeder, A. 1986, Annu. Rev. Astron. Astrophys. 24, 329.
Davidson, K. et al. 1982, Astrophys.J. 253, 696.
Filippenko, A.V. and Sargent, W.L.W. 1985, Nature 316, 407.
Gaskell, C.M., Cappellaro, E., Dinerstein, H.L., Garnett, D.R., Harkness, R.P. and Wheeler, J.C. 1986, Astrophys.J.(Letters) 306, L77.
Henry, R.B.C. 1986, Publ. Astr. Soc. Pacific 98, 1044.
Hoffleit, D. 1982, The Bright Star Catalogue (New Haven: Yale University Observatory).
Iben, I. and Tutukov, A.V. 1984, Astrophys.J.Suppl. 54, 335.
Maza, J. and van den Bergh, S. 1976, Astrophys.J. 204, 519.
Nomoto, K. 1983, in IAU Symposium 101, Supernova Remnants and Their X-Ray Emission ed. J. Danziger and P. Gorenstein (Dordrecht: Reidel), p.139.
Panagia, N. 1985, in Supernovae as Distance Indicators, ed. N. Bartel (Berlin: Springer), p.14.
Shklovskii, I.S. 1984, Sov. Astr. Letters 9, 250.
Uomoto, A. and Kirshner, R.P. 1985, Astron. Astrophys. 149, L7.
van den Bergh, S. 1983, Astrophys.J. 268, 129.
van den Bergh, S. 1988, Astrophys.J. 327, in press.
van den Bergh, S., McClure, R.D. and Evans, R. 1987, Astrophys.J. 323, 44.
Wheeler, J.C. and Levereault, R. 1985, Astrophys.J.(Letters) 294, 17.
Wheeler, J.C., Harkness, R.P. and Cappellero, E. 1987, preprint.
Woosley, S.E. and Weaver, T.A. 1986, Annu. Rev. Astron. Astrophys. 24, 205.

TYPE Ib SUPERNOVAE: WHAT THEY MAY BE AND WHAT THEY ARE NOT

Nino Panagia[1]
Space Telescope Science Institute, Baltimore
and
University of Catania

Victoria G. Laidler[2]
Astronomy Programs, Computer Sciences Corporation

ABSTRACT

We show that the progenitors of Type Ib supernovae are likely to be moderately massive stars ($M \sim 7\ M_\odot$) in binary systems and we argue that the hypothesis that they originate from very massive stars ($M > 20\ M_\odot$) is untenable both on an observational and a statistical basis.

The realization that there is a separate subclass of Type I supernovae (SNe) to be denoted as Type Ib came after the detailed study of the SN 1983N in M83 (Panagia et al 1988; see also Panagia 1985, Wheeler and Levreault 1985, Uomoto and Kirshner 1985). It showed beyond any doubt that SN 1983N is distinctly different from the "classical variety" of Type I SNe (obviously denoted as Type Ia SNe) in a number of important aspects. Since then about a dozen SNe have been classified as Type Ib SNe, some newly discovered and others found just by re-examining old spectra or paying due attention to the comments that the observers gave at the time that the original obervations were made (e.g. Bertola 1964). In addition to the defining condition of Type I SNe, i.e. the absence of hydrogen lines from the

---

[1]Affiliated with the Astrophysics Division, Space Sciences Department of European Space Agency.

[2]Staff member, Space Telescope Science Institute.

spectrum, the distinctive characteristics of Type Ib SNe can be summarized as follows (cf. Panagia et al 1986, Weiler and Sramek 1988):

- The 6150 Å feature is absent from the spectrum.
- The overall spectral distribution is redder (Δ(B-V)$\sim$ 0.5) and fainter ($\sim$ 1.5 magnitudes) than for Type Ia SNe.
- The optical light curve is essentially "normal", i.e. quite similar to that of Type Ia SNe.
- the IR light curve is single-peaked, the maximum occurring a few days after the optical maximum.
- They are, or may be, strong radio emitters with a steep spectrum and a quick temporal decline.
- They are found only in spiral galaxies.
- They are located in spiral arms.
- They are possibly "associated" with (i.e. projected on, or near to) an HII region.

All these properties can be "read" in a rather simple manner and the picture that emerges is that of the explosion of stars which are "compact" (hence *not* red supergiants) and have a small envelope similar to the case of Type Ia SNe (because of the similarity of the light curves), but have a chemical composition different from that of Type Ia progenitors (absence of the 6150 Å band) and a lower amount of ^{56}Ni synthesized in the explosion (redder and fainter emission).

Their radio emission requires the presence of a circumstellar envelope created by mass loss corresponding to $\dot{M}/v_{exp} \sim 3\ 10^{-7}$ [$M_\odot$ yr^{-1}]/ [km s^{-1}] (Weiler et al 1986). Such a value is orders of magnitude too high for any early type star (which excludes the WR hypothesis for Type Ib SNe progenitors) and is possible only for a red supergiant: this implies the presence of a relatively massive companion (i.e. several solar masses; Sramek et al 1984).

The lifetime of their progenitors must be shorter than 3 10^8 years, and therefore their original mass larger than 5.5 $M_\odot$, in order to satisfy the condition posed by their being located in spiral arms. In fact, *if* Type Ib SNe are intrinsically associated with HII regions, their progenitors should be quite short-lived and, therefore, much more massive. Since this is the only argument which may favor the idea that Type Ib SNe progenitors are very massive stars, we have considered this point in quite some detail.

First of all we have checked the validity of the "association" of Type Ib SNe with HII regions which is claimed for about 50% of them (Wheeler et al 1987). By overlaying the best positions of Type Ib SNe [among those reported in the Barbon et al Catalog of SNe (1984) and those astrometrically determined either in the optical or in the radio] on galaxy images (taken from Schmidt survey plates) with the use of the GASP[3] software, we find that 6 ± 1 out of 12 SNe appear to fall within 5" of the image of a knot (i.e. presumably an HII region). On the other hand, only in one case (SN 1981I in NGC 4051) does the SN seem to fall on top of an HII region. Therefore, the "association" with an HII region actually means *close proximity.* And since 5" at a distance even as short as 4 Mpc corresponds to 100 pc, such a *proximity* may in fact be just fortuitous. For example, had the LMC been at 4 Mpc instead of 54 kpc, SN 1987A would appear to be *associated* with the 30 Doradus nebula while it is *not.*

A possible way to clarify this issue is to compare these results with similar statistics made for the case of Type II SNe (Huang 1987). In that case, out of 29 SNe for which there were good position measurements and good galaxy images with which to make the overlays, 25 objects were found to fall within an average distance of 5" from an HII region . This immediately indicates that, independently of whether *any* such association is real in *any* case, the "association" of Type Ib SNe is *much* looser than for Type II SNe. Assuming that this difference is entirely due to "evaporation" of the stars from their birth place, the difference implies lifetimes of the SN Ib progenitors considerably longer (3-10 times) than those of SN II progenitors and, consequently, original masses considerably lower (2-3 times lower for SN Ib than for SN II). Since Type II SNe are believed to originate from progenitors with masses in the range 8-20 $M_{\odot}$ (e.g. Maeder 1987; see also van den Bergh, this Conference), the progenitors of Type Ib must have had masses within the possible range 5-10 $M_{\odot}$.

Such a range of progenitor masses can be further narrowed down by con-

[3]GASP is the Guide Star Astrometric Support Program available at the Space Telescope Science Institute.

sidering that the frequency of Type Ib explosions in spiral galaxies is about 1/3 that of Type II SNe (Branch 1986, van den Bergh, McClure and Evans 1987). Therefore, assuming that stars more massive than 8 $M_\odot$ make Type II SNe, and adopting an initial mass function proportional to $M^{-2.35}$, the possible mass range for SN Ib progenitors turns out to be about 6.5-8 $M_\odot$. This agrees well with the direct estimate of $M > 6.5\ M_\odot$ made by Sramek et al (1984) for SN 1983N on the basis of its radio emission. Also, a relatively "modest" mass for the progenitor can naturally explain why the mass ejected in the explosion (as implied by the "normal" optical light curve) is a few solar masses at most (Branch 1988).

We conclude that the only viable scenario to account for Type Ib events is that of a star with original mass around 7 $M_\odot$, which is a member of a binary system in which the companion is slightly less massive (say, $\sim 5\ M_\odot$). The primary follows its evolution to the end becoming a rather massive degenerate star, which explodes when the secondary has reached the stage of red supergiant. The alternative hypothesis of a very massive progenitor (i.e. $M > 20\text{-}30\ M_\odot$) is ruled out on the basis of the mass loss characteristics required to account for the radio emission, the "light" envelope implied by the behaviour of the optical light curve, and the "association" with HII regions which is much looser than for Type II SNe. A full account of this study will be presented in a forthcoming paper.

REFERENCES

Barbon, R., Cappellaro, E., Ciatti, F., Turatto, M., Kowal, C.T., 1984, *Astron. Astrophys. Suppl.*, **58**, 735.

Bertola, F., 1964, *Ann. Astrophys.*, **27**, 319.

Branch, D., 1986, *Astrophys. J. (Letters)*, **300**, L51.

Branch, D., 1988, *Proc. IAU Colloquium No. 108*, in press.

Huang, Y.-L., 1987, *Pub. Astron. Soc. Pac.*, **99**, 461.

Maeder, A., 1987, *ESO Workshop "SN 1987A"*, ed. I.J. Danziger, p. 251.

Panagia, N., 1985, in *"Supernovae as Distance Indicators"*, ed. N. Bartel (Berlin: Springer), p. 14.

Panagia, N., Sramek, R.A., Weiler, K.W., 1986, *Astrophys. J. (Letters)*, **300**, L55.

Panagia, N. et al, 1988, in preparation.

Sramek, R.A., Panagia, N., Weiler, K.W., 1984, *Astrophys. J. (Letters)*, **285**, L59.

Uomoto, A., Kirshner, R.P., 1985, *Astron. Astrophys.*, **149**, L7.

van den Bergh, S., McClure, R.D., Evans, R., 1987, *Astrophys. J.*, **323**, 44.

Weiler, K.W., Sramek, R.A., Panagia, N., van der Hulst, J.M., Salvati, M., 1986, *Astrophys. J.*, **301**, 790.

Weiler, K.W., Sramek, R.A., 1988, *Ann. Rev. Astron. Astrophys.*, **26**, in press.

Wheeler, J.C., Levreault, R., 1985, *Astrophys. J. (Letters)*, **294**, 17.

Wheeler, J.C., Harkness, R.P., Cappellaro, E., 1987, *Proc. 13th Texas Symposium on Relativistic Astrophysics*, in press.

ULTRAVIOLET OBSERVATIONS OF SN 1987A

Nino Panagia[1]
Space Telescope Science Institute, Baltimore
and
University of Catania

ABSTRACT

The results of IUE observations of SN 1987A are presented and discussed. One of the most important conclusions is that the direct progenitor of the supernova was a blue supergiant (Sk -69 202) that, however, had gone through a regular phase of red supergiant before turning back to the left in the HR diagram.

1. INTRODUCTION

SN 1987A was first observed with IUE from the GSFC station (Kirshner et al 1987a) just a few hours after the discovery announcement. At VILSPA the observations, carried out as part of the ESA-SERC Target-of-Opportunity Program for observing bright supernovae, were started in the early morning of the 25th of February, on the first available European shift (Wamsteker et al 1987a). Since then, the close collaboration of the two observing teams, as well as the generous time allocation by the corresponding Agencies, have assured that the coverage of such an important event was quite complete. Some of the IUE results have already been discussed in papers published in scientific journals (Wamsteker et al 1987a, Cassatella et al 1987, Panagia et al 1987a, Fransson et al 1987, de Boer et al 1987, Gilmozzi et al 1987, Dupree et al 1987, Kirshner et al 1987, Sonneborn et al 1987) and more will be presented in a number of papers which are either in press (Fransson et

[1]Affiliated with the Astrophysics Division, Space Sciences Department of European Space Agency.

al 1988, Blades et al 1988) or in preparation. The main results obtained so far can be summarized as follows:

1. Initially the UV flux was rather high indicating high photospheric temperatures (T > 14000 K). The fast UV flux drop (orders of magnitude in a few days; cf. e.g Panagia et al 1987a) implies a small initial radius, say, of the order of few tens solar radii, and, therefore, excludes that the star was a red supergiant when it exploded.

2. The B3 I star Sk -69 202 has been ascertained to be the SN progenitor: this was first suggested in Panagia et al (1987a) and subsequently confirmed by the more detailed analyses presented by Gilmozzi et al (1987) and Sonneborn et al (1987).

3. The strong UV flux of the early epochs has offered a unique opportunity to obtain high dispersion spectra to study the interstellar medium in our Galaxy and in the LMC toward SN 1987. A large number of components for both highly ionized and low ionized species have been detected (de Boer et al 1987, Dupree et al 1987, Blades et al 1988) indicating a rather complex structure for the intervening matter.

4. The presence of narrow emission lines of highly ionized species, detected in the short wavelength spectrum since late May (Wamsteker et al. 1987b), has provided evidence for the progenitor having been a red supergiant before coming back toward the blue side of the HR diagram. Since this a new and quite fascinating topic I will concentrate the rest of my talk on the discussion of these emission lines and their implications.

2. THE UV EMISSION LINES

At the early epochs the supernova short wavelength spectrum was decaying steeply with time, almost a factor of ten per day for the first few days after explosion (e.g Panagia et al 1987a; cf. Fig. 1). Such a precipitous drop of the SW spectrum was stopped by the presence of the two neighbor stars, creatively named star 2 and star 3, which already ten days after the explosion contributed more than 80% to the observed flux shortward of 1700 Å (Panagia et al 1987a, Panagia 1987). By mid March the SW spectrum

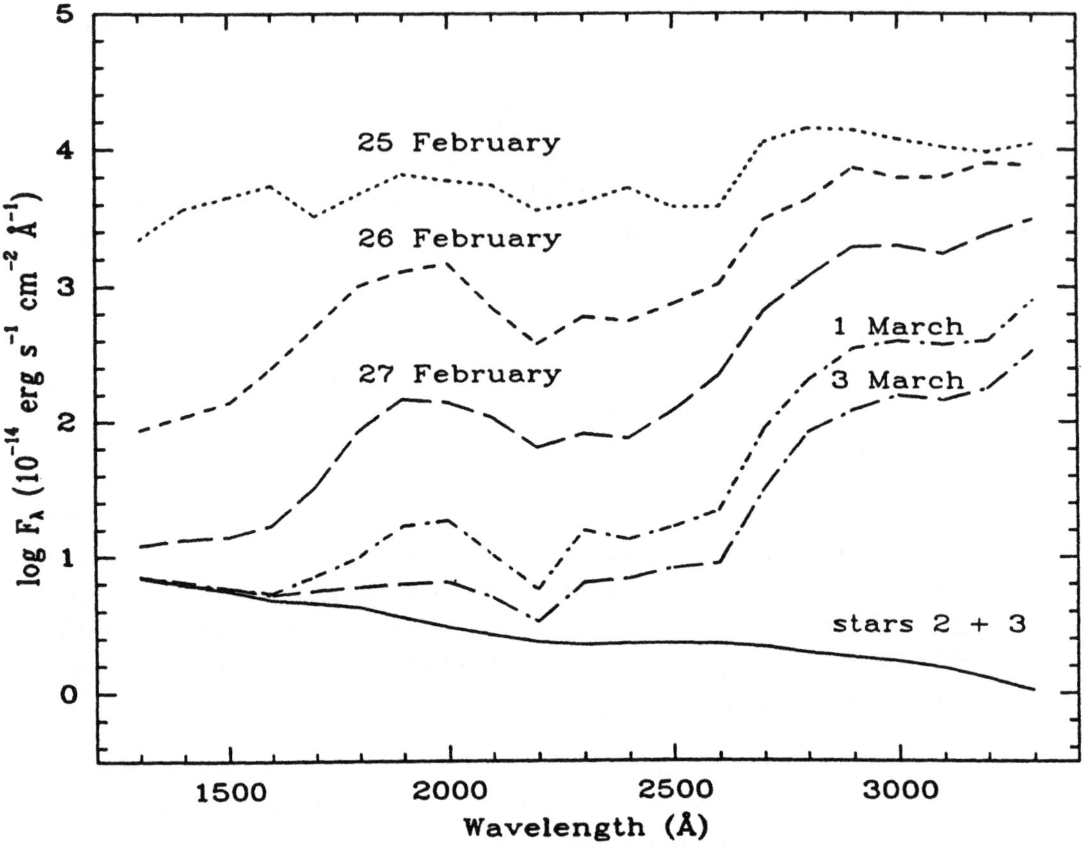

Figure 1 . The IUE spectra, rebinned into 100 Å bands, are shown for the first observing week, February 25 through March 3. For comparison, the total spectrum of stars 2+3 is also shown as a solid line.

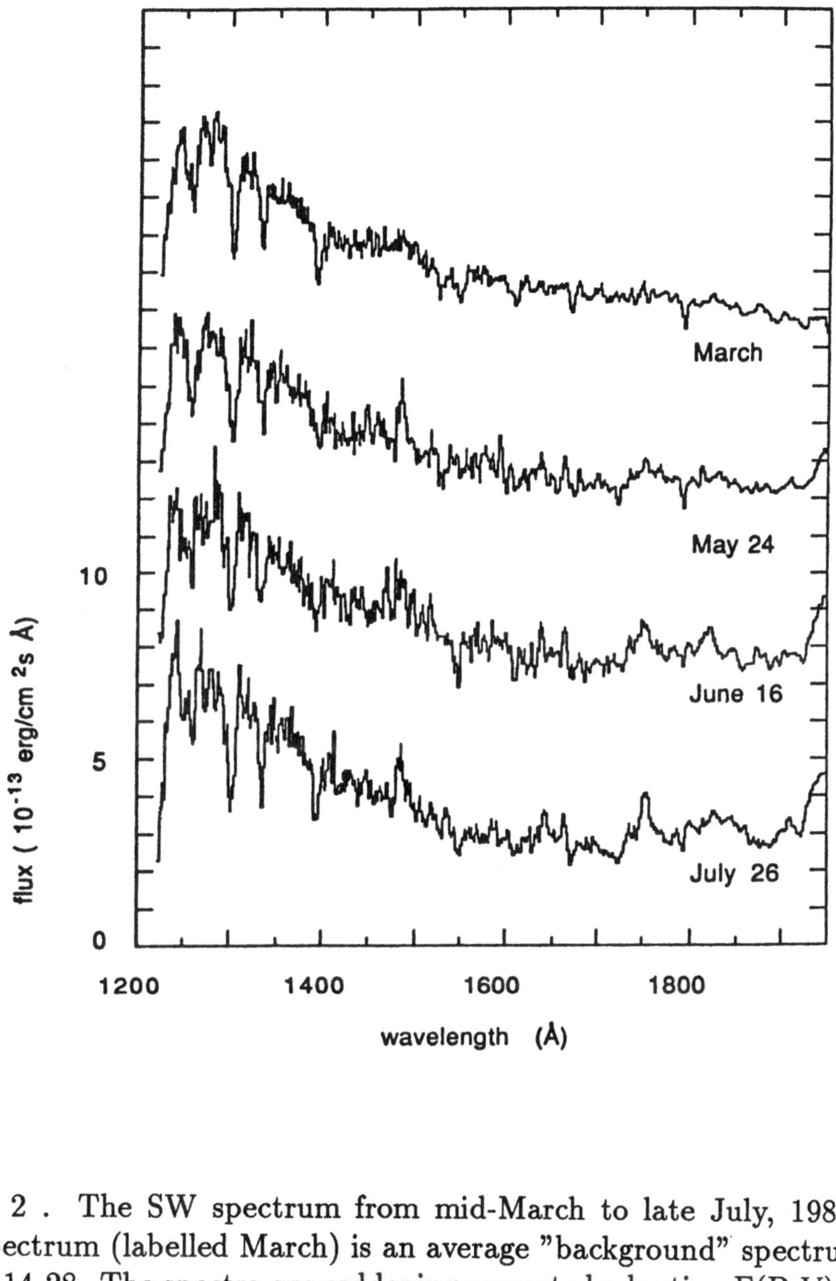

Figure 2 . The SW spectrum from mid-March to late July, 1987. The first spectrum (labelled March) is an average "background" spectrum from March 14-28. The spectra are reddening corrected adopting E(B-V) = 0.20. Each spectrum is displaced by 5 10^{-13} erg cm^{-2} s^{-1} $Å^{-1}$.

was entirely dominated by the stellar contamination and, therefore, the observed flux was not varying appreciably anymore.

However, at the end of May something new happened: the continuum started rising again, but only longward of about 1850 Å and, more importantly, the presence of narrow emission lines, i.e. NV 1240 Å, NIV] 1483 Å, HeII 1640 Å, OIII] 1663 Å, NIII] 1750 Å, CIII] 1909 Å (cf. Fig. 2) was noticed (Wamsteker et al 1987b). They were unresolved in the low resolution spectra taken with the SWP camera, thus implying widths narrower than 1500 km s^{-1}. Also, they were initially rather weak but they increased steadily with time (Fig. 3) following an essentially linear behaviour (see e.g. Fig. 4). The narrow width and the long persistence of the emission lines indicate that they originate in a medium with small velocity dispersion and a suitably low electron density (say, less than few 10^4 cm^{-3}) so as to make the recombination times longer than about a year. A relatively low value of the density is confirmed by the peak wavelength of the 1483 Å blend of NIV (Cassatella 1987).

More detailed information was obtained at later times when the increased intensities made it feasible to observe the short wavelength spectrum with high resolution. A first exposure was obtained on November 25 and the lines were found to be still unresolved, thus implying a velocity dispersion in the emitting region of less than 30 km s^{-1} (Panagia et al 1987b). The emission line peaks appeared to be displaced by 284 $\pm$ 6 km s^{-1} relative to the laboratory rest wavelengths. That radial velocity virtually coincides with the strongest LMC component observed in the interstellar line spectrum of SN 1987A both in the ultraviolet (de Boer et al 1987, Blades et al 1988) and in the optical (Vidal Madjar et al 1987). Therefore, not only there is a small velocity dispersion in the emitting gas but also that the systemic velocity of the emitting region relative to the SN is rather low, say, less than 10-20 km s^{-1}. Moreover, an explicit estimate of the electron density can be obtained from the intensity ratio of the CIII] lines 1906.68 Å and 1908.73 Å which indicates a value of n_e = 2-3 10^4 cm^{-3}.

Two subsequent high dispersion observations were made, on February 25 and April 7, 1988: they not only confirm the previous results but also indicate that some evolution may have taken place in the meanwhile, in that the emission peaks seem to be red-shifted by several km s^{-1} relative to the November 25, 1987, spectrum. As an illustration, Figure 5 displays the spectral region around the CIII] doublet for the two epochs of November 25, 1987, and February 25, 1988.

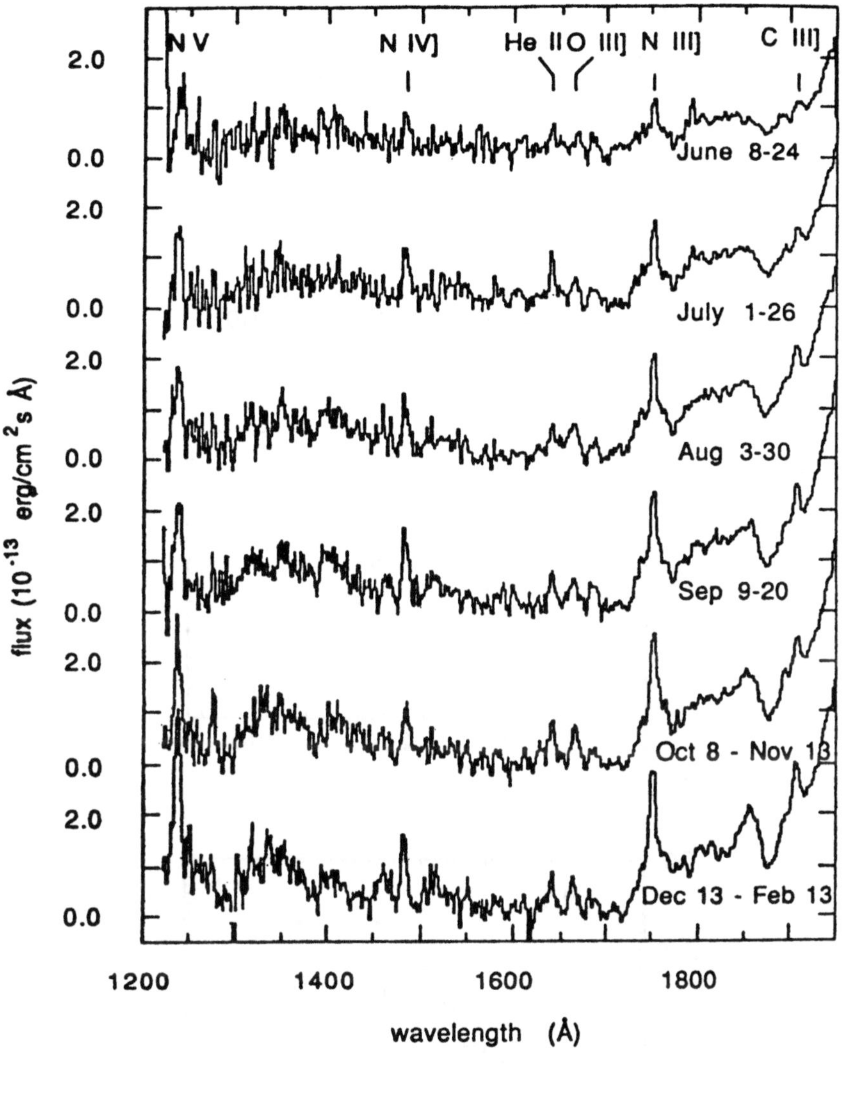

Figure 3. The averages of the difference spectra (i.e. subtracted of the mean March spectrum) are presented for a number of epochs. Note the steady increase of the NIII] 1750 Å and NV 1240 Å lines.

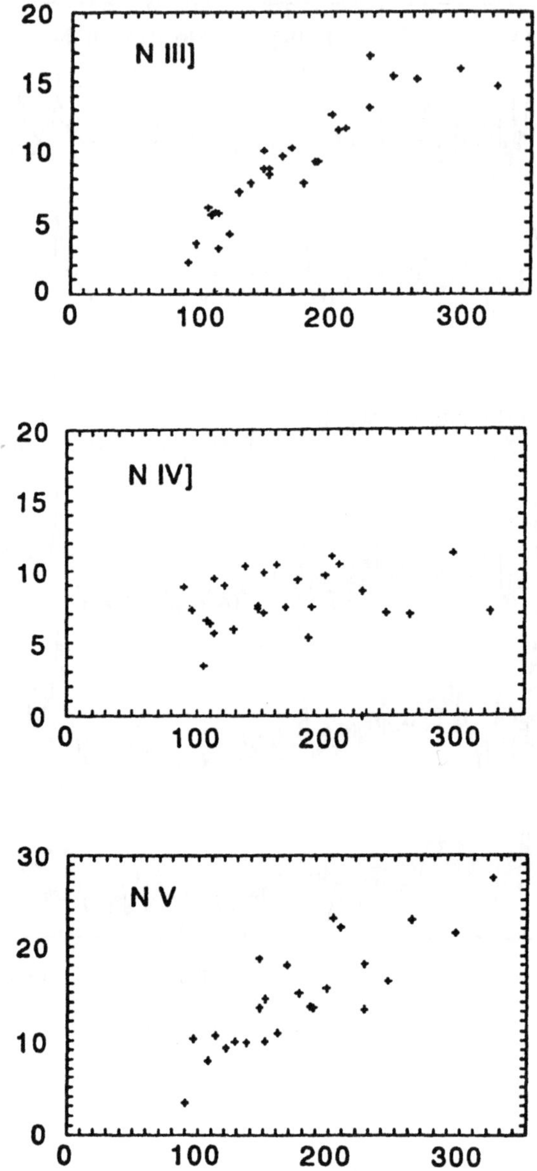

Figure 4. The time evolution of the nitrogen lines. The intensities are in 10^{-13} erg cm^{-2} s^{-1} and the time is measured in days since the explosion.

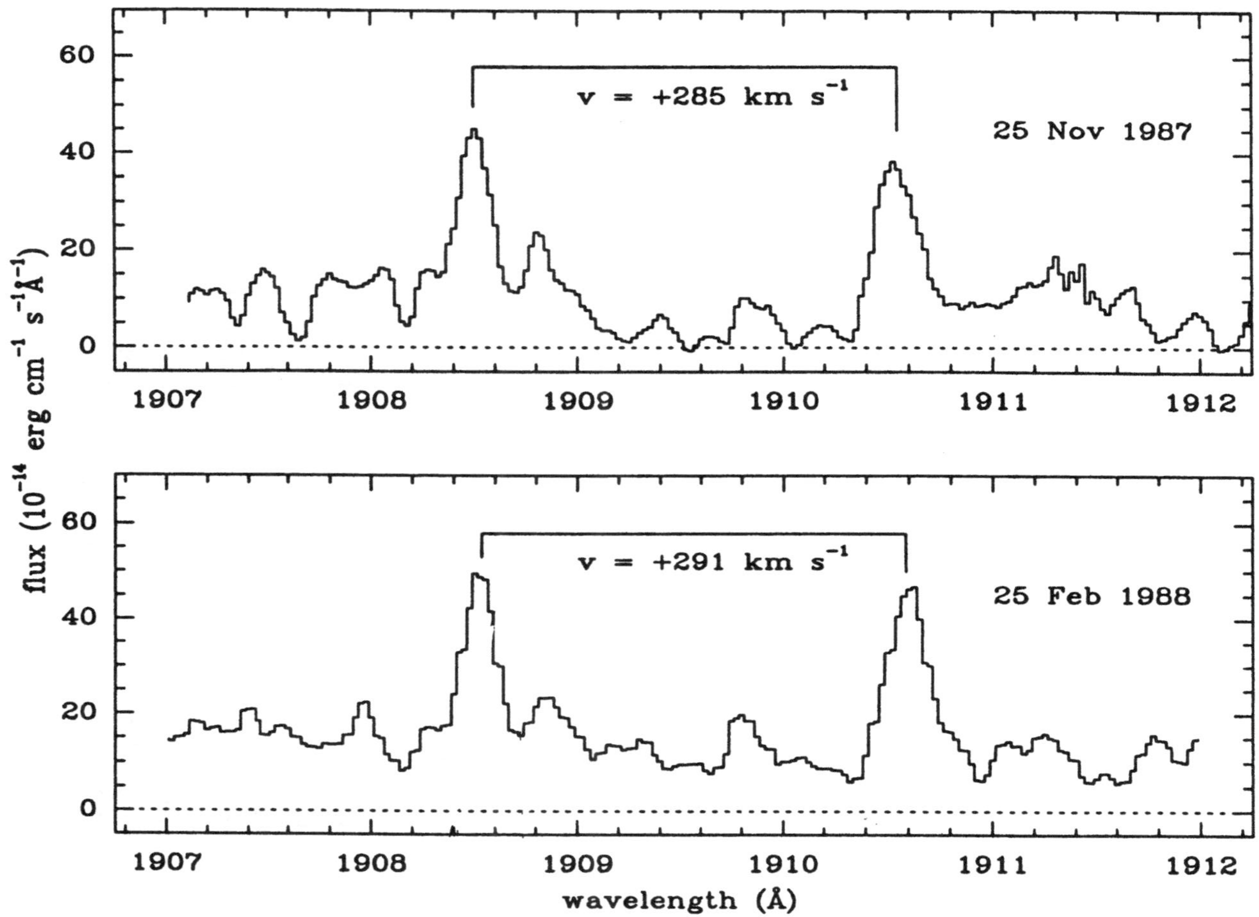

Figure 5. The high resolution spectrum of the CIII] region on November 25, 1987 (upper panel) and February 25, 1988 (lower panel).

Meanwhile, narrow emission lines of HI, HeII and [OIII] were also detected in the optical (Wampler 1988) which confirmed the UV results and added a definite value for the line width on late November of HPFW = 18 km s^{-1} and direct evidence from the [OIII] I(4363)/I(4959+5007) ratio for electron temperatures in excess of 30000 K.

The detailed analysis of the IUE data done by Fransson et al (1988) leads to the following conclusions:

i) Nitrogen is highly overabundant relative to both carbon and oxygen: in particular, N/C = 7.8 ± 3.9 and N/O = 1.6 ± 0.8, which are 37 and 12 times higher than the corresponding solar ratios.

ii) The electron density is about 2.5 10^4 cm^{-3} and the electron temperature is higher than 30000 K.

iii) The total size of the emitting region is at least one light year: this follows from the consideration that the line intensities are still increasing a year after the explosion, while the ionization is likely to have occurred at the initial UV burst of the SN explosion. The apparent evolution of the emission line intensities is due to light travel effects for which we see larger and larger fractions of the emitting region as time proceeds. Eventually the intensity rise will level off when radiation from the whole region will be received.

iv) The emitting mass as of November 1987 is at least 0.08 $M_\odot$ (but probably not much higher than twice that value). Combining this result with the estimates of the density and the diameter, the width of the emitting region turns out to be very small, definitely much smaller than its radius.

v) The expansion velocity of the emitting gas is less than 30 km s^{-1} and, possibly as low as 10 km s^{-1}: this follows from the narrow, unresolved profiles of the UV lines and the observations of the optical line profile as reported by Wampler (1988).

This observational picture of the emitting region is suggestive of a circumstellar shell of CNO processed material, such as expected in the wind of a moderately massive red supergiant. In fact both the abundances and the implied mass loss rate (i.e. roughly $\dot{M}$ ~ (shell mass)/[(shell radius)/(expansion velocity) ~ 1.5 10^{-6} $M_\odot$ yr^{-1}) are perfectly consistent with that hypothesis. The red supergiant wind, however, must have been compressed quite a bit in order to reach the observed density: this is the effect of the pressure exercized by the much faster stellar wind ejected by the SN progenitor in the subsequent phase when it had become a blue supergiant.

Although one can anticipate that refinements in both the overall picture and the physical parameters will be obtained when the observations will extend so as to cover the leveling-off phase of the line intensities, the basic interpretation should not change appreciably.

On the other hand, there is still a lot of excitment to expect with this supernova. In particular, the interaction of the SN ejecta, which are moving at speeds as high as 30000 km s^{-1} in the outermost layers, with the UV line emitting shell will produce another phase of intense emission, possibly both in the UV and in the X-ray domains. Thus, the fireworks will start cracking in about ten years from now: be there...

REFERENCES

Blades, J.C., Wheatley, J.M., Panagia, N., Grewing, M., Pettini, M., and Wamsteker, W., 1988, *Astrophys. J.*, in press.

Cassatella, A., 1987, *ESO Workshop "SN 1987A"*, ed. I.J. Danziger, p.101.

Cassatella, A., Fransson, C., van Santvoort, J., Gry, C., Talavera, A., Wamsteker, W. and N. Panagia, N., 1987, *Astron. Astrophys.*, **177**, L29.

de Boer, K., Grewing, M., Richtler, T., Wamsteker, W., Gry, C., and Panagia, N., 1987, *Astron. Astrophys.*, **177**, L37.

Dupree, A.K., Kirshner, R.P., Nassiopoulos, G.E., Raymond, J.C., and Sonneborn, G., 1987, *Astrophys. J.*, **320**, 597.

Fransson, C., Grewing, M., Cassatella, A., Panagia, N., and Wamsteker, W., 1987, *Astron. Astrophys.*, **177**, L33.

Fransson, C., Cassatella, A., Gilmozzi, R., Kirshner, R.P., Panagia, N., Sonneborn, G., and Wamsteker, W., 1988, *Astrophys. J.*, in press.

Gilmozzi, R., Cassatella, A., Clavel, J., Fransson, C., Gonzalez, R., Gry, C., Panagia, N., Talavera, A. and Wamsteker, W., 1987, *Nature*, **328**, 318.

Kirshner, R.P., Sonneborn, G., Crenshaw, D.M., and Nassiopoulos, G.E., 1987, *Astrophys. J.*, **320**, 602.

Panagia, N., 1987, *ESO Workshop "SN 1987A"*, ed. I.J. Danziger, p. 55.

Panagia, N., Gilmozzi, R., Clavel, J., Barylak, M., Gonzalez Riestra, R., Lloyd, C., Sanz Fernandez de Cordoba, L. and Wamsteker, W., 1987a, *Astron. Astrophys.*, **177**, L25.

Panagia, N., Gilmozzi, R., Cassatella, A., Wamsteker, W., Kirshner, R.P., and Sonneborn, G., 1987b, *I.A.U. Circ.* No. 4514.

Sonneborn, G., Altner, B., and Kirshner, R.P., 1987, *Astrophys. J. (Letters)*, **323**, L35.

Vidal-Madjar, A., Andreani, P., Cristiani, S., Ferlet, R., Lanz, T., and Vladilo, G., 1987, *Astron. Astrophys.*, **177**, L17.

Wampler, J., 1988, *I.A.U. Circ.* No. 4541, and private communication.

Wamsteker, W., N. Panagia, Barylak, M., Cassatella, A., Clavel, J., Gilmozzi, R., Gry, C., Lloyd, M., van Santvoort, J., and Talavera, A., 1987a, *Astron. Astrophys.*, **177**, L21.

Wamsteker, W., Gilmozzi, R., Cassatella, A. and Panagia, N., 1987b, *I.A.U. Circ.* No. 4410.

X RAYS FROM SN 1987A: A PARTIALLY OBSCURED PLERION.

Rino Bandiera, Franco Pacini and Marco Salvati
Osservatorio Astrofisico di Arcetri
Largo E. Fermi 5, I-50125 FIRENZE (Italy)

Introduction

Soon after the discovery of SN 1987A McCray *et al.* (1987) predicted the appearance of X-ray emission. According to their model, γ rays are produced at the centre of the SN, either due to a central pulsar or to decay of radioactive nuclei. While diffusing outwards the γ rays are degraded into X rays by Compton scattering. For some time after the explosion all photons are down-scattered below $\approx 10\,\mathrm{keV}$, and cannot get over the photoelectric barrier; but, as the envelope expands and its Thomson depth decreases, a time will be reached at which the first photons marginally escape: from that time the SN can be detected in X rays. With the further decrease of the Thomson opacity, the average energy of the observed photons increases, and eventually γ-ray photons can freely escape. Taking the density stratification and the mass of Ni^{56} from Woosley *et al.* (1987), Xu *et al.* (1988) produced a quantitative model for the X-ray emission.

X rays from SN 1987A have been actually detected (Dotani *et al.* 1987, Sunyaev *et al.* 1987), but observations contradict most of the expectations of the model. The first X rays were detected just 4 months after the explosion, while an 8-month delay was predicted; the emission extends below the theoretical cut-off at 10 keV; the X-ray light curve has not yet shown any sign of decline, whereas the γ-ray lines are already decreasing (Matz 1988); finally, the theory cannot explain the observed short time-scale fluctuations.

Revisions of the original model have been proposed, in order to solve some of the above inconsistencies. The presence of a soft X-ray component is attributed to the interaction of the ejecta with circumstellar matter (Masai *et al.* 1987); then the coincidence of the rise epoch for the soft and hard component would remain unexplained; furthermore, there is no evidence of the radio emission which should accompany the shock. In order to simulate the prompt appearance of X rays and their flat light curve, Itoh *et al.* (1987) assume that the radioactive material is partially mixed into the outer layers; the persistence of a substantial X-ray emission after more than one year, however, is hard to reproduce even if the chemical distribution is very finely tuned. Finally, none of these revisions can explain the short term variability, nor the behaviour of the X and γ rays, inverted with respect to the expectations.

The synchrotron model

In an alternative model the emission originates from a synchrotron nebula, fed by a newly born pulsar (Bandiera *et al.* 1988). We assume standard values for the pulsar magnetic field ($4 \cdot 10^{12}$ G) and the expansion velocity of the shell ($5 \cdot 10^{8}\,\mathrm{cm\,s^{-1}}$); the X-ray emission above 40 keV (Sunyaev *et al.* 1987) can then be fitted with a power law index $\gamma = 1$ for particle injection, and a pulsar period $P_0 = 18\,\mathrm{ms}$. The plerion radiates $3 \cdot 10^{39}\,\mathrm{erg\,s^{-1}}$; therefore radioactive decay is still needed to explain the bolometric light curve. The key advantage of this model over the radioactive decay is that no decrease in the X-ray emission should occur: on the contrary, for $\gamma = 1$, the plerion emission must increase as $t^{1/2}$, in (at least qualitative) agreement with the observations.

A basic requirement of the synchrotron model is that the ejecta surrounding the plerion must break down into clouds or filaments at very early times, leaving an inter-cloud medium which is transparent also to low-energy X rays. Both observation and theory are already providing independent evidence for massive deviations from spherical symmetry.

The central source is partially obscured by the fragments. Photoelectric opacity depends strongly on the photon energy, hence, if the fragments have an internal structure, with a dense core and a gradually thinner envelope, their effective area will also depend on the photon energy: more specifically, they will look bigger at lower energies. Furthermore, if the blobs move with non-radial velocities, the fraction of the source which is obscured varies with time: this effect can naturally account for the short time variations in the flux.

The effect of obscuration on the observed flux can be described analytically, in the limit where the blobs are small compared with the source ($d \ll D$). If n is the average number of clouds covering the source, one can define $\tau = nd^2/D^2$ as the average number of clouds along any given path. Then the observed luminosity (L) in terms of the unobscured one (L_0) is $L = e^{-\tau} L_0$; moreover, a typical fluctuation of the observed luminosity due to random cloud motion is $\Delta L/L = (d/D)\sqrt{e^\tau - 1}$: large fluctuations can be obtained only if the blobs have a size comparable to the source, or if the opacity is large.

In order to refine the above estimates, and to explore the case of large clouds, we have developed a Monte Carlo code. The left panel of Fig. 1 shows a fit to the spectrum observed by GINGA on September, corresponding to an obscuration by an average number of clouds $n = 10$; each cloud has been chosen with an effective size $d = 0.4D$ below 16 keV, and decreasing as $E^{-0.5}$ above 16 keV. These parameters fit the overall X-ray spectrum with the additional condition that typical fluctuations are of about a factor 2. In the right panel, instead, the same theoretical spectrum is compared with that observed last January, when X-ray emission increased considerably for a few weeks. The spectral change observed during this phase is consistent with a temporary decrease in the number of clouds covering the source; instead, the near equality of the flux levels before and after the brightening is not easily understandable within this framework.

Conclusions

There are in the data details which our model in its present version cannot address; most important of all is the sudden increase observed in January, which appears to be a flare rather than an opacity drop. However, the clear physical connection between low and high energy X rays, and the global shape of the light curve point to a heavy fragmentation and to a steady internal source. A crucial expectation is a progressive thinning out of the cloud swarm, until the plerion emission becomes completely visible.

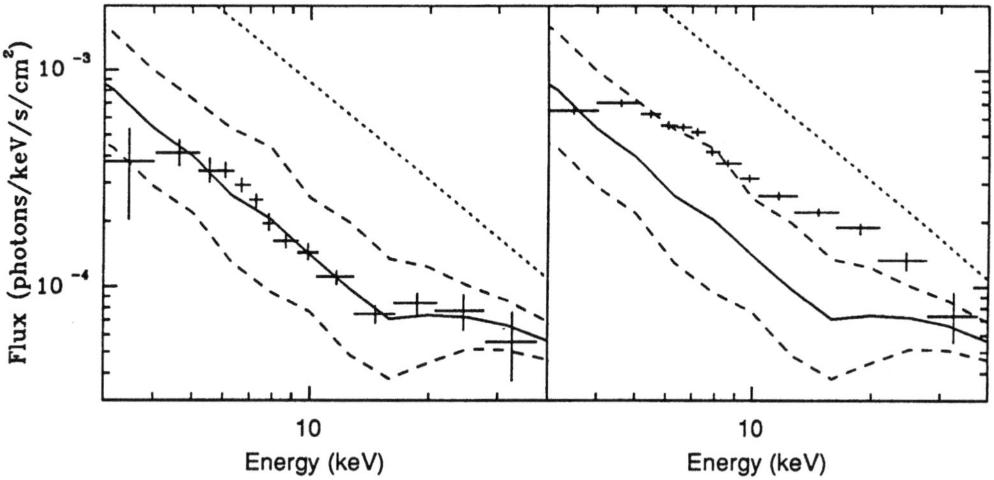

Figure 1: on the left panel our model is fitted to the X-ray spectrum of SN 1987A taken on September 3 from the GINGA satellite; the dotted line is the unobscured flux from the source; the solid line is the fit to the data, accounting for obscuration; the area included in between the two dashed lines represents the typical range for variability. On the right panel the same model is compared with the X-ray spectrum based on observations from January 19 to January 22.

References.

Bandiera, R., Pacini, F., Salvati, M. 1988, *Nature*, **332**, 514.

Dotani, T. *et al.* 1987, *Nature*, **330**, 230.

Itoh, M., Kumagai, S., Shigeyama, T., Nomoto, K., and Nishimura, J. 1987, *Nature*, **330**, 233.

Matz S. S. 1988, *IAU Circ.* No. 4568.

McCray, R., Shull, J.M., and Sutherland, P. 1987, *Astrophys. J.*, **317**, L73.

Sunyaev, R. *et al.* 1987, *Nature*, **330**, 227.

Woosley, S.E., Pinto, P.A., and Ensman, L. 1987, *Astrophys. J.*, **324**, 466.

Xu, Y., Sutherland, P., McCray, R., and Ross, R.R. 1988, *Astrophys. J. (Letters)*, in the press.

SUPERNOVA VLBI

Norbert Bartel
Harvard–Smithsonian Center for Astrophysics
60 Garden Street, Cambridge, MA 02138 USA

ABSTRACT. Over the last five years, six supernovae or young supernova remnants have been observed with Very-Long-Baseline-Interferometry (VLBI). I will review the observations and results for each of the supernovae, and I will indicate how supernova VLBI can uniquely contribute to the investigations of the evolution of the expanding shockfront, of a possible pulsar nebula in its center and of the circumstellar medium, and to the determination of the extragalactic distance scale.

1. WHY SN–VLBI?

Investigations of supernovae (SNe) and young supernova remnants (SNRs) with VLBI provide unique information to a number of fields of interest that complement SN and SNR research with other techniques. Most notable are the possibilities and opportunities to:

1. make a movie of an exploding star,
2. investigate the evolution of SNe and young SNRs over a large fraction of their lifetimes,
3. derive properties of the circumstellar medium of the supernova's progenitor,
4. determine, even if the date of the explosion is unknown, the degree of deceleration or acceleration and thereby, in the latter case, infer the possible existence of a pulsar, and
5. estimate the host galaxies' distances and the value of the Hubble constant.

Over the last five years, six SNe or young SNRs have been observed with VLBI at different times in their evolution after the date of explosion. One of them, SN1987A in the Large Magellanic Cloud, was observed only 5.2 d after the neutrino burst, the event that best defines the epoch of core collapse and consequently of the explosion. However, we have obtained so far only a lower, albeit very important, bound on the angular radius and angular expansion velocity of the supernova's radiosphere.

SN1979C in M100 in the Virgo cluster was observed 3.7 years after the explosion and at three later epochs. By virtue of its having been monitored for a large fraction of its lifetime, and its relatively large distance from us, SN1979C has been, scientifically, the most rewarding supernova of the ones observed with VLBI. Since for each epoch an angular radius was determined, the observations allowed estimates of the supernova's angular expansion velocity and of bounds on any decleration or acceleration of it. Combined with a determination of the radial expansion velocity of the line-emitting and -absorbing gas, the distance to the Virgo cluster and Hubble's constant could be estimated.

SN$\sim$1955 (SNR41.9+58) in the nearby galaxy M82 is, with an age between 20 and 50 yr, among the oldest SNe or SNRs that have been observed with VLBI. It is also so far the only one for which an image has been obtained.

Incidentally, the promptness of VLBI observations with respect to a SN explosion and consequently the potential percentage of the coverage with VLBI observations of the SN's evolution has been, for our observations, inversely related to the information obtained of the morphology of a SN/SNR. For instance, SN1987A was observed with VLBI as early as five days after the explosion but was not even detected. In contrast, SN~1955 was observed with VLBI relatively late in its evolution, namely a few decades after its explosion, but it provided the most detailed information on the morphology of a SN or young (extragalactic) SNR yet. However, anticipated future observations may provide the sensitivity and u-v coverage to image some of the other six SNe/SNRs with a quality approaching that of the image of SN~1955 and afford us the opportunity to make a movie of an exploding star.

In the remainder, I will review the observations and their results for the six SNe and SNRs investigated with VLBI, in the order of increasing delay, t, between the epoch of explosion and the start of VLBI monitoring.

2. SNe AND SNRs OBSERVED WITH VLBI

a) SN1987A: $t = 0.01$ yr

With an unprecedented promptness, the radio emission of SN1987A reached its maximum of ~ 140 mJy at 1.4 GHz (Turtle *et al.* 1987) only three days after the neutrino burst (Aglietta *et al.* 1987; Bionta *et al.* 1987; Hirata *et al.* 1987). The only antennas in the southern hemisphere that were conceivably capable of resolving the expanding shell of SN1987A were one of the antennas in Tidbinbilla, Australia, and the 26-m diameter in Hartebeesthoek, South Africa. After frantic organization, Mark III VLBI observations were made at 2.3 GHz at $t = 5.2$, 6.2, and 7.2 d with the latter antenna and NASA's 34-m diameter DSS42 antenna in Tidbinbilla (Shapiro *et al.* 1988). Unfortunately, at the time of the observations, the SN was already too weak and too extended to be detected with the above interferometer. No fringes were found from the supernova's radio emission on any of those three observing days, although fringes were obtained for the calibrator sources on all three days with amplitudes agreeing (to within the corresponding combined standard errors) with those obtained from VLBI measurements made at the same resolution, but five years earlier (Preston *et al.* 1985; G. Nicolson 1987, priv. communication).

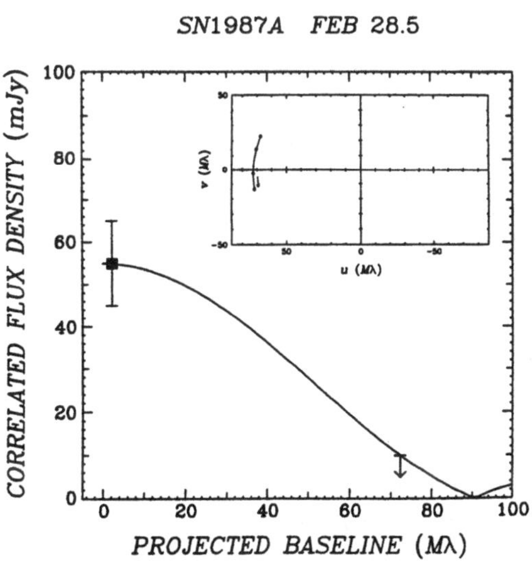

Figure 1. The total flux density and an upper bound on the correlated flux density of SN1987A on ~ 28.5 Feb. 1987, together with the prediction of a model of an optically thin uniform sphere with a radius of 1.6 mas. The prediction is similar to that from a model of an optically thin shell with an outer angular radius of 1.25 mas. The u-v track for the VLBI observations is plotted in the inset.

The upper bound on the correlated flux density of SN1987A obtained on the first day of the VLBI observations was compared with the corresponding total flux density (Figure 1) to derive a lower bound on the angular radius of the radiosphere (Bartel *et al.* 1988; Jauncey *et al.* 1988). For an optically thin shell model with a shell thickness of $\sim 15\%$ of the shell's outer radius, the lower bound on the radius is $\theta_{radio} > 1.25 \pm 0.07$ mas. Here and hereafter, the quoted errors are meant to be standard errors (σ), with statistical and systematic contributions combined, unless otherwise stated. The use of any other physically plausible model would result in an up to $\sim 30\%$ larger lower bound (see, e.g., Marscher 1985). Given a distance to SN1987A of 50 ± 5 kpc (Feast and Walker 1987), the lower bound on the angular radius corresponds to a lower bound (with the combined 1σ error subtracted) on the linear radius, R_{radio}, of $R_{\text{radio}} > 8.3 \times 10^{14}$ cm $= 12 \times 10^3\ R_\odot = 55$ AU, at $t = 5.2\ d$. If one assumes that the radiosphere expanded linearly from zero size at $t = 0$, the lower bound on the radius corresponds to a lower bound on the expansion velocity, v_{radio}, of the radiosphere: $v_{\text{radio}} > 19 \times 10^3$ km s^{-1}.

Since the distance to the Large Magellanic Cloud is known to within about 10%, SN1987A allowed for the first time a comparison of the linear radii and the expansion velocities of a supernova's photosphere and radiosphere with those of the supernova's line-emitting and -absorbing regions. In Figure 2, the lower bounds on the radius and expansion velocity obtained from the VLBI radio data are compared with the corresponding radii and velocities obtained from optical photometric (Menzies *et al.* 1987) and spectroscopic (Hanuschik and Dachs 1987; Blanco *et al.* 1987) data.

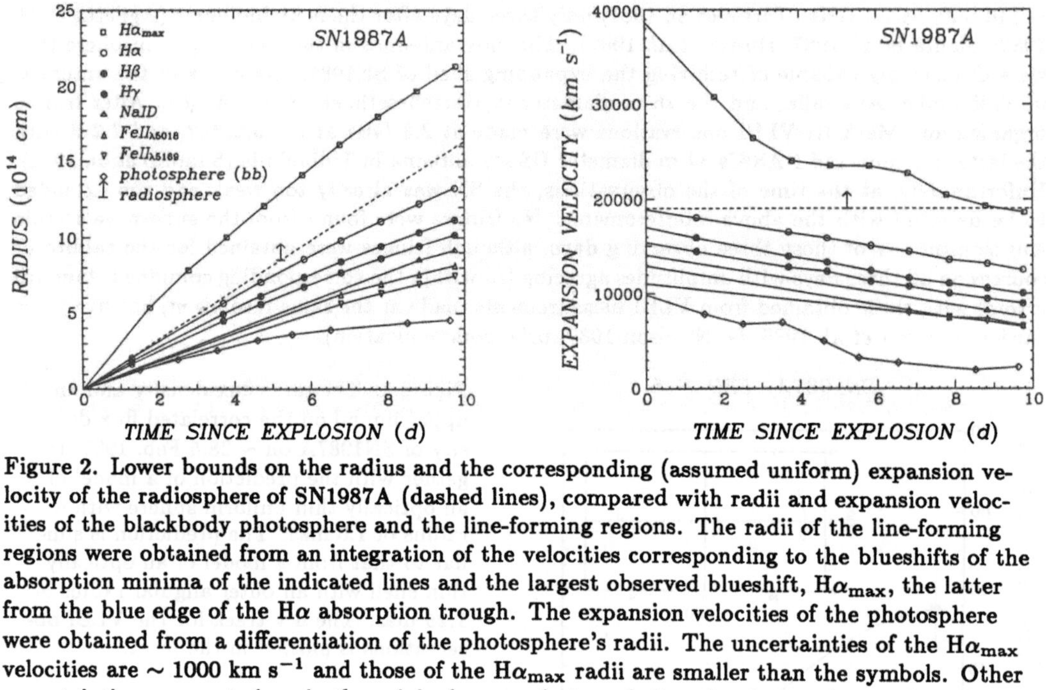

Figure 2. Lower bounds on the radius and the corresponding (assumed uniform) expansion velocity of the radiosphere of SN1987A (dashed lines), compared with radii and expansion velocities of the blackbody photosphere and the line-forming regions. The radii of the line-forming regions were obtained from an integration of the velocities corresponding to the blueshifts of the absorption minima of the indicated lines and the largest observed blueshift, Hα_{max}, the latter from the blue edge of the Hα absorption trough. The expansion velocities of the photosphere were obtained from a differentiation of the photosphere's radii. The uncertainties of the Hα_{max} velocities are ~ 1000 km s^{-1} and those of the Hα_{max} radii are smaller than the symbols. Other uncertainties were not given in the original papers but are believed to be not larger than those of the Hα_{max} velocities and radii, respectively.

At the time of the VLBI observations, the radius and the expansion velocity of the radiosphere were both at least a factor 2.5 larger than those of the blackbody photosphere and, respectively, at least 10% and 25% larger than the radius and velocity inferred from the Hα-line absorption minimum. The results not only add to our knowledge of supernovae but are also important in limiting the uncertainties accompanying the use of supernovae as distance indicators (see, e.g., Bartel 1985, 1986; Bartel *et al.* 1985), if the physical processes responsible for the radio emission from SN1987A are typical for SNe in general (see section 3 of this paper).

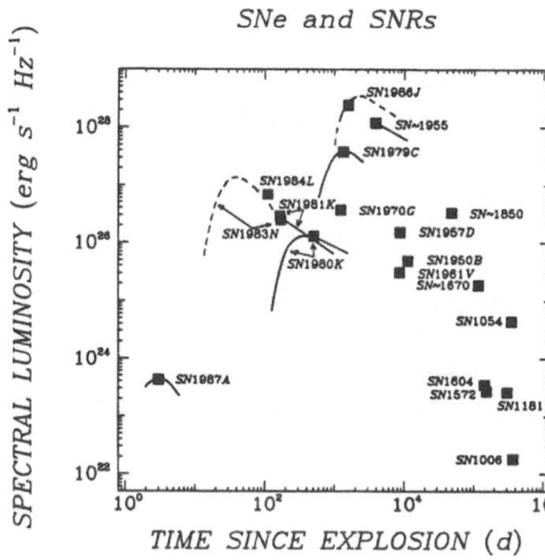

Figure 3. The spectral luminosity at 1.5 GHz and, for SN1987A only, at 1.4 GHz, vs. the age of radio supernovae and young supernova remnants. The squares indicate the largest measured spectral luminosities of the corresponding sources, the solid curves represent all measured spectral luminosities, and the dashed curve predicts spectral luminosities according to Chevalier (1984*a*). The time at which the flux density of a SN reaches maximum is frequency dependent; typically threefold shorter at 8 than at 1.5 GHz. Note that SN~1955 (SNR41.9+58) is the brightest source among more than 20 hot spots in M82, which are most likely also SNe or SNRs. Their spectral luminosities are a few times to hundredfold smaller than the spectral luminosity of SN~1955, and their ages range from a few years to a few hundred years (Unger *et al.* 1984; Kronberg *et al.* 1985; Kronberg and Sramek 1985; Bartel *et al.* 1987). The data for SN1987A are from Turtle *et al.* (1987), for SN1986J from Rupen *et al.* (1987) and Bartel *et al.* (1988), for SN1961V from Cowan *et al.* (1988), for SN~1955 in M82 from Kronberg *et al.* (1985) and Bartel *et al.* (1987), and for SN~1850 in NGC4449 from de Bruyn (1983). The data from all other sources are from Weiler *et al.* (1986) and references therein.

Are these processes typical for SNe in general? In Figure 3, I plot the spectral luminosity at 1.5 GHz of SNe and young SNRs versus the time since their explosions. Since the spectral luminosity of SN1987A is two to five orders of magnitude smaller than the spectral luminosities of other recent SNe, and since its peak was reached one to three orders of magnitude faster than for the other SNe, it was speculated that the prompt radio burst was perhaps a precursor of, and quite distinct from, a more prominent outburst. Such a precursor would have been undetectable for more distant supernovae and, moreover, because of its likely occurrence well before maximum light, unobservable from other SNe discovered well after the epoch of explosion.

However, more than a year has passed without SN1987A's flaring up again at radio frequencies (0.8–8 GHz). Also, the circumstellar interaction model was quite successful in fitting the evolution of the prompt burst's spectrum and light curve (Storey and Manchester 1987; Chevalier and Fransson 1987) and suggested a density of the circumstellar medium two to three orders of magnitude lower than the corresponding densities for SN1979C and SN1980K. This lower density is consistent with the progenitor of SN1987A having been a blue supergiant (wind velocity ~ 500 km s^{-1}), in contrast to the progenitors of SN1979C and SN1980K, believed to have been red supergiants (wind velocities ~ 10 km s^{-1}) (Chevalier and Fransson 1987).

Thus, it appears now that the relatively low radio luminosity and the early turn-on of radio emission were caused by the relatively low density of the circumstellar medium encountered by the SN1987A shockfront. Further, it appears that the physical processes responsible for the radio emission were indeed typical for SNe in general.

b) SN1980K: $t = 2.5$ yr

The supernova SN1980K in the galaxy NGC6946 reached its maximum flux density of ~ 2.5 mJy at 1.4 GHz ~ 0.4 yr after the explosion that occurred on, or near, 1980 Oct. 17 (Weiler *et al.* 1986). VLBI observations were made at 2.3 GHz on 1983 May 7 with the sensitive NASA 64-m antennas at Goldstone, CA and Madrid, Spain. The supernova was detected in two out of three adjacent 13-min scans. The visibility amplitudes determined in the two scans, and an upper bound determined in the third scan, are shown in Figure 4. Since the visibility amplitudes of the three segments are expected to be approximately Gaussianly distributed, we can compute their mean value and standard error and infer that SN1980K was unresolved at the epoch of our VLBI observations, 2.54 yr after the assumed date of 1980 Oct. 17 for the explosion. The value for the angular radius of a shell model is: $\Theta \lesssim 0.5 \pm 0.5$ mas, equivalent to $\Theta < 1$ mas (Bartel 1988). The prediction from this model is also shown in Figure 4.

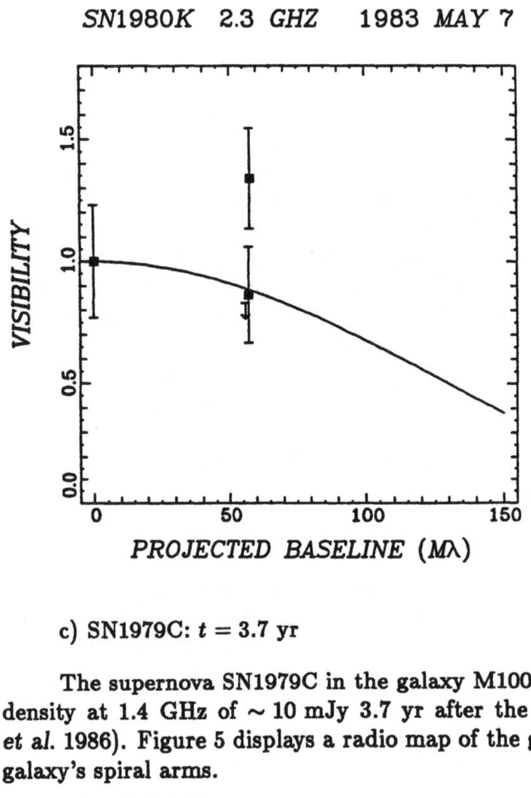

Figure 4. Visibility amplitudes and the prediction from a shell model. The visibility amplitude at zero spacing was obtained from interpolating between several measurements made with the VLA by Weiler *et al.* (1986) at 1.4 and 5 GHz only two days to seven weeks prior to our VLBI observations.

c) SN1979C: $t = 3.7$ yr

The supernova SN1979C in the galaxy M100 in the Virgo cluster reached its maximum flux density at 1.4 GHz of ~ 10 mJy 3.7 yr after the explosion on, or around, 1979 Apr. 1 (Weiler *et al.* 1986). Figure 5 displays a radio map of the galaxy with the supernova located in one of the galaxy's spiral arms.

VLBI observations commenced on 1982 Dec. 8 and have resulted in angular radius determinations at 5 GHz at four consecutive epochs. The measured visibility amplitudes and the predictions from a model for the supernova's brightness distribution are shown in Figure 6. The angular radius determinations, Θ, are plotted, as a function of the time since explosion, t, in Figure 7. A weighted least-squares fit of the form $\Theta \propto t^m$ gives $\Theta = 0.42 \pm 0.03$ mas for the time of our first VLBI observations at $t = 3.69$ yr, and $m = 1.03 \pm 0.15$, consistent with uniform expansion.

These data provide the first direct measurement of the expansion of a supernova. The expansion is consistent with being uniform, as predicted on the basis of a fit of the circumstellar interaction model (Chevalier 1982) to the radio light curves at two frequencies (Chevalier 1984*b*; Weiler *et al.* 1986).

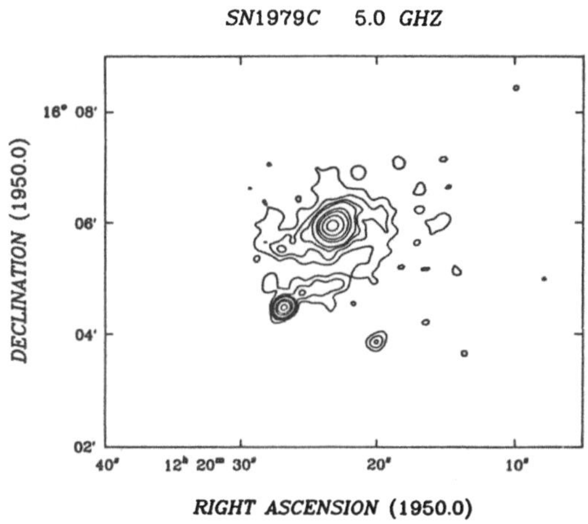

Figure 5. Radio map of SN1979C, located at the southern edge of a spiral arm of the galaxy M100. The map was made with the VLA on 1982 Dec. 8.

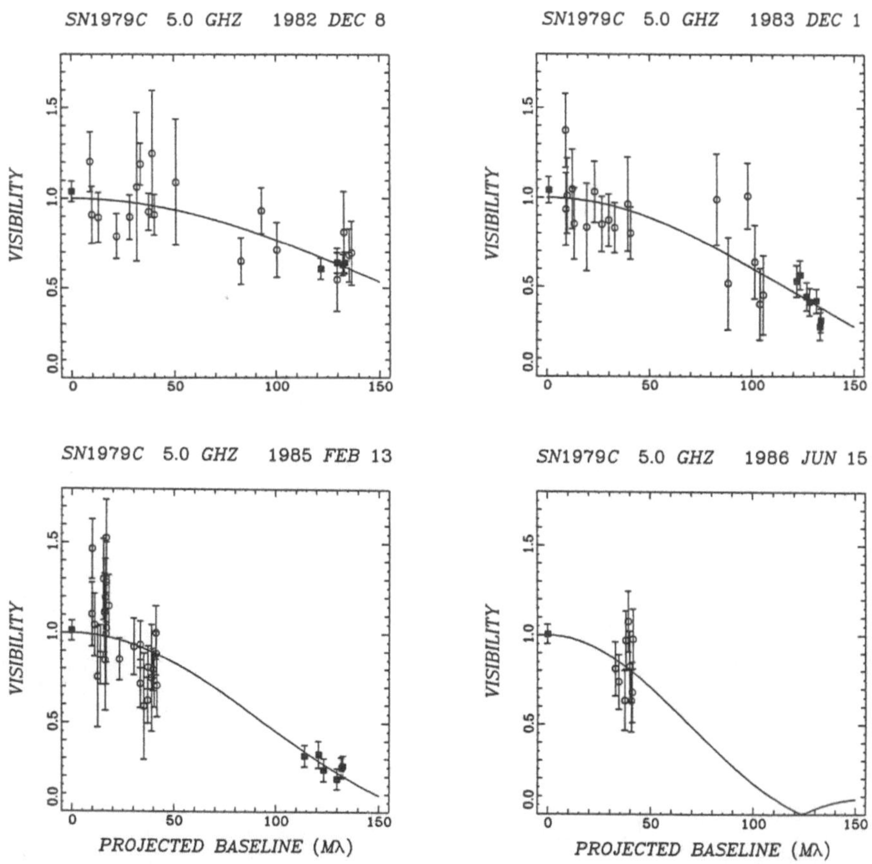

Figure 6. Measured visibility amplitudes at 5.0 GHz and predictions from the fit of a uniform sphere model. The filled squares show the most significant data points, obtained with the VLA alone and with the Bonn-VLA interferometer. At epoch 1986 June 15, the antenna at Bonn was malfunctioning and therefore prevented the recording of data from SN1979C with the transatlantic interferometers.

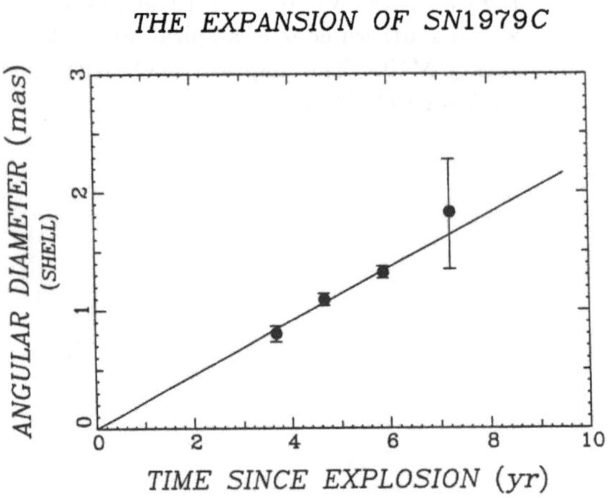

Figure 7. The angular diameter determinations for a shell model. For the extreme models of a ring and a uniform sphere, the ordinate scale has to be multiplied by 0.8 and 1.3, respectively. The solid line represents uniform expansion ($m = 1$), which is consistent with our weighted least-squares solution for m.

d) SN1986J: $t = 3$ to 12 yr

The supernova SN1986J in the galaxy NGC891 (Figure 8), at 0.56 the distance to the center of the Virgo cluster (Aaronson *et al.* 1982), reached its maximum flux density of ~ 120 mJy in mid 1987 (Weiler 1988, priv. communication). This epoch occurred several years after the (unobserved) explosion, which, on the basis of the supernova's light curve (Chevalier 1987) and its last nondetection, occurred sometime between ~ 1975 and ~ 1983.

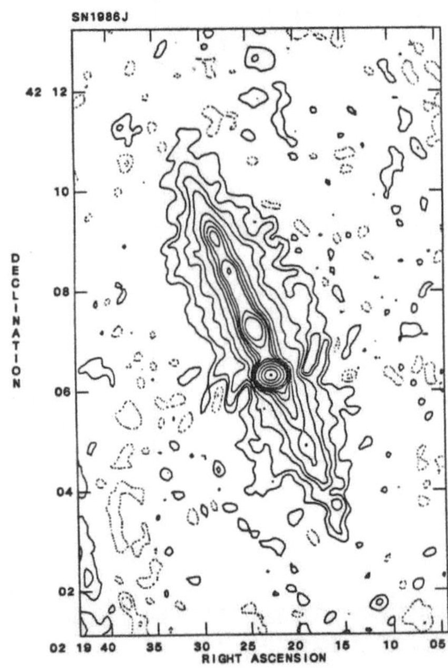

Figure 8. A map of the galaxy NGC891 and its pointlike supernova SN1986J made with the VLA at the frequency and epoch of the third session of the VLBI observations. The contours are –0.32, –0.16, 0.16, 0.32, 0.64, 1, 2, 3, 4, 5, 7, 10, 20, 40, 70, and 110 mJy per beam area. The negative contours are dashed. The coordinates are for epoch B1950.0.

VLBI observations were made at 1.7 GHz on 1986 Sep. 29 (Bartel *et al.* 1987), at 10.7 GHz on 1987 Feb. 23, and at 5.0 GHz on 1987 May 30 (Bartel *et al.* 1988). No significant expansion has yet been observed. With the assumption of uniform expansion, the explosion occurred not later than 1980 (85% confidence) or 1984 (97% confidence). A least-squares fit of an elliptical Gaussian model gives a half-width at half maximum of the major axis of 0.8 ± 0.1 mas with a position angle of $145° \pm 5°$ and a ratio between the minor and the major axis of 0.62 ± 0.05 (Bartel *et al.* 1988).

SN1986J is clearly elongated [see Karovska *et al.* 1988 for a report (announced after this workshop) on recent speckle-interferometric results of SN1987A that provide detection of asymmetry in that supernova]. Deviations from spherical symmetry of a supernova may be caused by a number of processes. First, it is conceivable that the flow pattern of the expanding gas is anisotropic. Rotation of the progenitor star could cause mass ejection predominantly in the equatorial plane (Bodenheimer and Woosley 1983). Further, pulsars have relatively high proper motions of up to $v \sim 400$ km s^{-1} (Lyne, Anderson, and Salter 1982) and a scale height for their galactic z-distribution threefold larger (Manchester and Taylor 1977) than those of OB stars that are the presumed progenitors of pulsars and SNe (O'Connell 1958) and even fourfold larger than those of SNRs (Henning and Wendker 1975). If these properties arise from pulsars receiving a velocity impulse at birth due to an asymmetric explosion (Shklovskii 1970), conservation of momentum would cause anisotropies in the velocity field of the ejected gas. To explain the elongation of SN1986J, the mass of the ejected gas has to be on the order of the mass of the pulsar, provided the pulsar exists (Bartel *et al.* 1988). In this context, it is of interest that the mass of the optically emitting region is $\sim 1\ M_\odot$ if estimated from the observed [OI] luminosity (Hamilton *et al.* 1987). This mass is indeed close to that expected for a pulsar; further, the existence of a pulsar in SN1986J is likely, if SN1986J's optical luminosity and the relative flatness of its light curve are due to a powering source inside the SN nebula (Chevalier 1987).

Second, it is conceivable that the flow pattern of the expanding gas is isotropic but that the density distribution of the circumstellar medium is not. The elongation of SN1986J could be explained on the basis of the circumstellar interaction model, if the combination of the rate of mass loss and wind velocity of the progenitor star varied angularly by at least $\pm 15\%$ (Bartel *et al.* 1988).

Third, the radio emission might not emanate from the shockfront region but rather from the amorphously shaped relativistic particle plasma emanating from the pulsar (Pacini and Salvati 1981). In this case, deviations from spherical symmetry of the brightness distribution of the supernova are a natural consequence. Clearly, an image, and not just a simple model, of the brightness distribution of SN1986J is needed to help distinguish between the suggested causes of the apparent elongation of SN1986J.

e) SN$\sim$1955: t = 20–50 yr

SN$\sim$1955 (SNR41.9+58) in the nearby galaxy M82 is the brightest among more than 20 other hot spots (Kronberg *et al.* 1985), all of which are presumably also SNRs. We assume the galaxy's distance from Earth to be 3.3 Mpc (Tammann and Sandage 1968; but see also Sandage 1984 for a $\sim 50\%$ larger estimate). Combined with estimates of angular radii at epochs 1980 (Wilkinson and de Bruyn 1984) and 1983 (Bartel *et al.* 1987), we find a value for the expansion velocity along the northeast–southwest axis of 6000 ± 3000 km s^{-1} and the date of explosion 1955^{+10}_{-20}. A backwards extrapolation of the SNR's light curve to a time around the date of explosion suggests a maximum flux density at 1.5 GHz of the order of 1 Jy. SN$\sim$1955 is the first SN or young SNR (with an age smaller than, say, 100 yr) for which the brightness distribution could be mapped. Figure 9 shows the map of SN$\sim$1955 (Bartel *et al.* 1987), juxtaposed to an optical image of M82 and a radio image of the central 600 pc of the galaxy. (See also Wilkinson and de Bruyn 1988 for an image of SN$\sim$1955 at 5 GHz.)

The map displays a shell-type morphology. The apparent deformation of the structure, which may originally have been circularly symmetric, may have been caused by relatively large Rayleigh–Taylor instabilities and by the relatively high pressure of the interstellar medium in the nuclear region of M82, which is estimated to be 30–300 times larger than the equivalent pressure in our

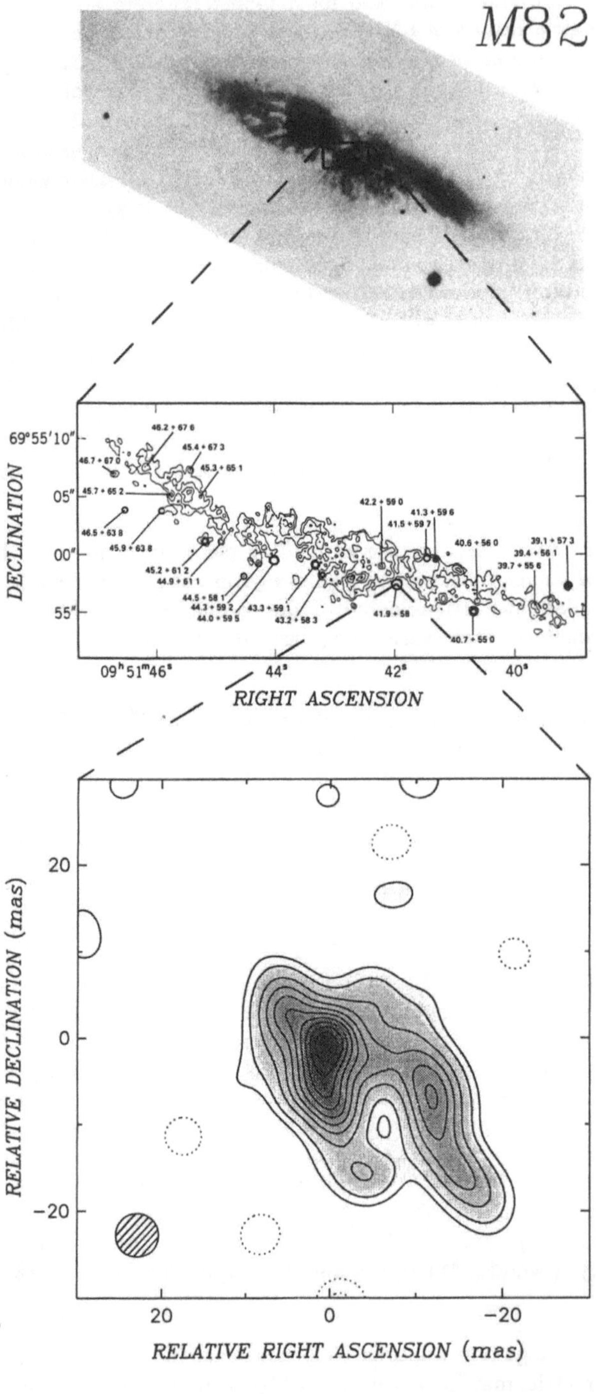

Figure 9. A hybrid map at 2.3 GHz (lower part) of SN~1955 (SNR41.9+58), the strongest compact component in the nuclear region of M82. The contours are at 90, 80, 70, 60, 50, 40, 30, 10, 5, –5% of the peak brightness of 25 mJy per beam area, equivalent to 4.4×10^8 K. The 50% contour of the restoring beam is shown as the striped circle in the lower left corner. The map is shown in relation to a radio map of the inner 600 pc of M82, made by Kronberg *et al.* (1985) with the VLA at a frequency of 4.9 GHz and with an angular resolution of 0.″34. The contours of the VLA map are shown at 40, 30, 20, 10, and 5% of the peak brightness of 102 mJy per beam area. All the sources we observed with VLBI, and analyzed, are labeled. For results from VLBI observations of the sources other than SN~1955 (SNR41.9+58), which are presumably also SNRs, see Bartel *et al.*(1987). The radio map is juxtaposed to an optical image of M82 taken from Sandage's Hubble atlas of galaxies. For each of the three images, north is up and east to the left.

Galaxy (Lugten *et al.* 1986). Further detailed imaging of SN~1955 is of particular interest since it promises to reveal the spectral evolution of each segment of the supernova's shockfront and its expansion into the dense ambient medium. Since the age of SN~1955 is intermediate between that of an extragalactic SN and a galactic SNR, we may perhaps witness the transition from the phase of free expansion of the SN gas to the Sedov phase.

The imaging of SN~1955 and the determination of its expansion rate together with determinations of the sizes, or lower bounds on them, of 13 other SNR candidates, combined with upper bounds on their sizes from VLA observations (Kronberg *et al.*1985) indicate a radio SN rate in the inner 600 pc of M82 of 0.1 yr^{-1} (Bartel *et al.* 1987).

f) SN~1850: $t =\sim 60$ to ~ 200 yr

The supernova remnant SN~1850 in the galaxy NGC4449 (Seaquist and Bignell 1978; Balick and Heckman 1978) at a distance from Earth of a few Mpc shows a declining flux density that was ~ 15 mJy at 1.5 GHz in 1981 (de Bruyn 1983). So far, only an upper bound on the SNR's angular radius of 38 mas (Cas A morphology assumed) has been determined (de Bruyn 1983) with VLBI, but new data may soon allow an image of this source to be made (de Bruyn 1988, priv. communication).

When the upper bound is combined with the galaxy's distance and the SNR's linear expansion velocity (Kirshner and Blair 1980), an upper bound on the SNR's age of ~ 200 yr can be estimated. There is some optical evidence for a lower bound on the age of ~ 60 yr. If this lower bound could be confirmed, the SNR in NGC4449 would be, together with the SNRs in M82, an important element for supernova VLBI research in bridging the gap of ages between the other sources described above with ages between ~ 0.01 and ~ 10 yr and galactic SNRs with ages $\gtrsim 200$ yr (see Reich 1988 for a candidate for the youngest galactic SNR yet).

3. THE EXTRAGALACTIC DISTANCE SCALE

A VLBI determination, at time t after the explosion, of the angular radius, Θ, of a supernova's radiosphere, coupled with an optical-spectroscopic determination of the radial expansion velocity $v_{\rm shock}$, at time t_0 after the explosion, of the supernova's shockfront allows estimates to be made of the distance, D, to the supernova's host galaxy and of Hubble's constant (Bartel 1985; Bartel *et al.* 1985; see also Bartel 1986, for a comparison of this method with other methods (e. g., Branch 1985) that use supernovae as distance indicators):

$$D = \frac{\mu\, \eta\, v_{\rm shock} t}{\kappa \Theta\, m} \left[\frac{t}{t_0}\right]^{m-1} \tag{1}$$

$$H_0 = \frac{v_{\rm redshift} + v_{\rm infall}}{D}\ , \tag{2}$$

The parameter μ describes the effect of any deviation from spherical symmetry on the distance determination. For spherically symmetric SNe, $\mu = 1$. The parameter η is the ratio between the expansion velocity of the radiosphere and that of the shockfront, m a measure of the deceleration (or acceleration) of Θ, with $\Theta \alpha t^m$, and κ the model dependence of Θ. We define the parameter κ so that $\kappa = 1.0$ for a supernova having a shell-like brightness distribution with a shell thickness of $\sim 15\%$. The parameter κ ranges between ~ 1.3 for the brightness distribution being a uniform sphere and ~ 0.8 for the distribution being a ring (see, e.g., Marscher 1985 and also chapter 2 of this paper). For comparison, $\kappa = 0.7$ for the FWHM of a Gaussian. The parameter $v_{\rm redshift}$ denotes the redshift velocity of the host galaxy, and $v_{\rm infall}$ its correction due to local peculiar velocities. For a uniformly expanding, spherically symmetric supernova shell eq. (1) becomes simply: $D = \eta v_{\rm shock} t/\Theta$.

The goal is to determine observationally each of the parameters in eq. (1) and thereby to estimate a host galaxy's distance and H_0. So far, we have obtained such estimates with only two supernovae, SN1979C and SN1980K. For SN1979C, four of the six parameters in eq. (1) have been determined. From VLBI observations, we obtained $\Theta = 0.42 \pm 0.03$ mas at $t = 3.69$ yr and $m = 1.03 \pm 0.15$. From optical spectroscopic observations (Branch *et al.* 1981), the parameter v_{shock} can be estimated. Since the largest unambiguous value for the expansion velocity is $\sim 12000 \pm 200$ km s^{-1}, determined from the blue edge of the absorption trough of the NaD line profile 0.24 yr after the explosion, and since the blue edge of the Hα emission profile, which is equivalent to 10500 km s^{-1}, remains constant until the last date of Branch *et al.*'s spectroscopic observations, 0.65 yr after the explosion, we take $v_{shock} = 12000 \pm 200$ km s^{-1} at $t_0 = 0.65$ yr. A lower bound on the parameter η can be estimated from our VLBI observations of SN1987A, based on the evidence that the physical processes causing the radio emission of SN1987A are indeed comparable with those in SN1979C (see chapter 2a). The lower bounds on the size and velocity of the radio shell relative to, respectively, the radii and velocities inferred from the absorption minimum and the blue edge of the Hα-profile imply that $\eta > 0.9$ for the values for SN1979C. The upper bound on η is model dependent. The circumstellar interaction model predicts $\eta = 1.2$ for $0.9 \lesssim m < 1$ (Chevalier 1985).

Based on our SN1987A observations and on the predictions from the circumstellar interaction model, we assume $\eta = 1.0^{+0.2}_{-0.1}$. Based on observations that many SNRs are approximately circularly shaped (Green 1984), we adopt $\mu = 1.0 \pm 0.15$. Based on observations that $\sim 90\%$ of all known SNRs with discernible morphology have a shell-like structure (Green 1984) and based on the assumption that a ring of emission would be unphysical for SNe and that it therefore can be excluded, we adopt $\kappa = 1.00^{+0.15}_{-0.05}$.

With these values and uncertainties, we get

$$D_{M100} = 22^{+7}_{-6} \text{ Mpc} \quad .$$

For SN1980K, the parameter m could not be determined and the values of the parameters Θ and v_{shock} are more uncertain than the equivalent values for SN1979C. Consequently our estimate of the distance to SN1980K's host galaxy NGC6946 will be only of marginal use. With $\Theta \lesssim 0.5 \pm 0.5$ mas at $t = 2.54$ yr, $v_{shock} = 9500^{+3500}_{-500}$ at $t_0 \sim 0.1$ yr (Barbon *et al.*1982; Uomoto and Kirshner 1986), and $\eta = 1.0^{+0.2}_{-0.1}$, and assuming $m = 1.00 \pm 0.15$, and $\kappa = 1.00^{+0.15}_{-0.05}$, we get $D_{NGC6946} \gtrsim 10^{+\infty}_{-5}$ Mpc , equivalent to

$$D_{NGC6946} > 5 \text{ Mpc} \quad .$$

For estimating H_0 we use only the estimate of the distance to M100. If we take $v_0 + v_{infall} =$ 1250 ± 150 km s^{-1} for the Virgo cluster center (see 12 references in Bartel *et al.* 1985) and assume that the distance to M100 is within 10% of the distance to this center, since M100 is within 5° of the center and has a redshift consistent with this location, we get

$$H_0 = 60 \pm 20 \text{ km s}^{-1} \text{ Mpc} \quad ,$$

with the uncertainties of the estimates representing 1–2σ. A more complete error analysis is pending.

We anticipate making more and more extended VLBI observations of SN1979C. We will model the SN in two dimensions to determine the extent of any deviation from circular symmetry on the plane of the sky and thereby limit the likely uncertainty of the parameter μ. We may also be able to distinguish between the emission region having the morphology of a ring, an optically thin shell, or an optically thin uniform sphere, and thereby also estimate the parameter κ more accurately.

4. CONCLUSIONS

The most important SN VLBI results have been a) the imaging of a young supernova remnant (SN~1955[SNR41.9+58]) and the estimate of the radio SN rate in the central region of its host galaxy (M82), b) the determination of the angular expansion rate, and bounds on any acceleration or deceleration of it, of a supernova (SN1979C), c) the estimate of a useful lower bound on the size of the radio shell of a supernova (SN1987A) relative to the sizes of the line-forming regions, and d) the determinations of the distance to the Virgo cluster of galaxies and H_0.

More observations of the supernovae discussed here and of supernovae yet to be discovered should allow, at least in some cases, investigations of the detailed dynamic and spectral evolutions of different segments of the shockfront expanding into the circumstellar medium, possibly the detection of a pulsar nebula in the center of the radio shell, and the determination of distances to galaxies as far away as $\sim$ 40 Mpc. The VLBA and space-based antennas together with sensitive ground-based antennas will aid considerably in the realization of the projects discussed and may afford us the intriguing opportunity to make a movie of an exploding star.

5. ACKNOWLEDGEMENT

This research was supported in part by the NSF under grant No. AST-8519763.

REFERENCES

Aaronson, N. *et al.* 1982, *Ap. J. Suppl.*, **50**, 241.

Aglietta, M. *et al.* 1987, *Europhys. Lett.*, **3**, 1315.

Balick, B. and Heckmann, T. 1978, *Ap. J. (Letters)*, **226**, L7.

Barbon, R., Ciatti, F., and Rosino, L. 1982, *Astr. Ap.*, **116**, 35.

Bartel, N. 1985, in *Supernovae as Distance Indicators*, Lecture Notes in Physics, ed. N. Bartel (Springer–Verlag, Berlin), **224**, 107.

Bartel, N. 1986, in *Highlights of Astronomy*, ed. J. P. Swings (Reidel, Dordrecht), **7**, 655.

Bartel, N. 1988, in *IAU Symposium 129, The Impact of VLBI on Astrophysics and Geophysics*, eds. M. J. Reid and J. M. Moran (Reidel, Dordrecht), p. 175.

Bartel, N., Rogers, A. E. E., Shapiro, I. I., Gorenstein, M. V., Gwinn, C. R., Marcaide, J. M., and Weiler, K. W. 1985, *Nature*, **318**, 25.

Bartel, N., Rupen, M. R., and Shapiro, I. I. 1988, to be submitted.

Bartel, N. *et al.* 1987, *Ap. J.*, **323**, 505.

Bartel, N. *et al.* 1988, in *Supernova 1987A in the Large Magellanic Cloud*, eds. M. Kafatos and A. Michalitsianos (Cambridge Univ. Press, Cambridge), p. 81.

Bionta, R. M. *et al.* 1987, *Phys. Rev. Lett.*, **58**, 1494.

Blanco, V. M. *et al.* 1987, *Ap. J.*, **320**, 589.

Bodenheimer, P., and Woosley, S. E. 1983, *Ap. J.*, **269**, 281.

Branch, D.. 1985, in *Supernova as Distance Indicators*, Lecture Notes in Physics, ed. N. Bartel (Springer–Verlag, Berlin), **224**, 138.

Branch, D., Falk, S. W., McCall, M. L., Rybski, P., Uomoto, A., and Wills, B. J. 1981, *Ap. J.*, **244**, 780.

Chevalier, R. A. 1982, *Ap. J.*, **259**, 302.

Chevalier, R. A. 1984*a*, *Ap. J. (Letters)*, **285**, L63.

Chevalier, R. A. 1984*b*, *Ann. NY Acad. Sci.*, **422**, 215.

Chevalier, R. A. 1985, in *Supernova as Distance Indicators*, Lecture Notes in Physics, ed. N. Bartel (Springer–Verlag, Berlin), **224**, 123.

Chevalier, R. A. 1987, *Nature*, **329**, 611.

Chevalier, R. A. and Fransson, C. 1987, *Nature*, **328**, 44.

Cowan, J. J., Henry, R. B. C., and Branch, D. *Ap. J.*, **329**, 116.

de Bruyn, A. G. 1983, *Astr. Ap.*, **119**, 301.

Feast, M. W., and Walker, A. R. 1987, *Ann. Rev. Astr. Ap.*, **25**, 345.

Green, D. A. 1984, *M. N. R. A. S.*, **209**, 449.

Hamilton, A. J. S., Vacca, W. D., Pradhan, A. K., Rupen, M. P., Gunn, J. E., and Schneider, D. P. 1988, *IAU Colloq. 101, The Interaction of Supernova Remnants with the Interstellar Medium* (Reidel, Dordrecht), in press.

Hanuschik, R. W., and Dachs, J. 1987, *Astr. Ap. Lett.*, **182**, L29.

Henning, K. and Wendker, H. J. 1975, *Astr. Ap.*, **44**, 91.

Hirata, K. *et al.* 1987, *Phys. Rev. Lett.*, **58**, 1490.

Jauncey, D.L. *et al.* 1988, *Nature*, in press.

Karovska, M., Koechlin, L., Nisenson, P., Papaliolios, C., and Standley, C. 1988, in IAU Circ. 4604.

Kirshner, R. P. and Blair, W. P. 1980, *Ap. J.*, **236**, 135.

Kronberg, P. P., Biermann, P., and Schwab, F. R. 1985, *Ap. J.*, **291**, 693.

Lugten, J. B., Watson, D. M., Crawford, M. K., and Genzel, R. 1986, *Ap. J. (Letters)*, **311**, L51.

Lundqvist, P. and Fransson, C. 1988, *Astr. Ap.*, **192**, 221.

Lyne, A. G., Anderson, B., and Salter, M. J. 1982, *M. N. R. A. S.*, **201**, 503.

Manchester, R. N. and Taylor, J. H. 1977, *Pulsars* (W. H. Freeman and Co., San Francisco).

Marscher, A. P. 1985, in *Supernova as Distance Indicators*, Lecture Notes in Physics, ed. N. Bartel (Springer–Verlag, Berlin), **224**, 130.

Menzies, J. W. *et al.* 1987, *M. N. R. A. S.*, **227**, 39p.

O'Connell, D. J. K., ed. 1958, *Stellar Populations* (Ric. Astr. Specola Vaticana), 5.

Pacini, F. and Salvati, M. 1981, *Ap. J. (Letters)*, **245**, L107.

Preston, R. A., Morabito, D. D., Williams, J. G., Faulkner, J., Jauncey, D. L., and Nicolson, G. D. 1985, *A. J.*, **90**, 1599.

Reich, W. 1988, these proceedings.

Reynolds, J. E. *et al.* 1987, in *Proc. of the ESO workshop on SN1987A*, ed. I. J. Danziger, (ESO Workshop and Conference Proc., Garching), **26**.

Seaquist, E. R. and Bignell, R. C. 1978, *Ap. J. (Letters)*, **226**, L5.

Shapiro, I. I. *et al.* 1988, in *IAU Symposium 129, The Impact of VLBI on Astrophysics and Geophysics*, eds. M. J. Reid and J. M. Moran (Reidel, Dordrecht), p. 185.

Shklovskii, I. S. 1970, *Soviet Astr.*, **13**, 562, transl. from *Astr. Zh.*, **46**, 715.

Storey, M. C., Conrad, G., Cooke, D. J., Troup, E., Wark, R., and Wright, A. E. 1987, IAU Circ. 4432.

Storey, M. C. and Manchester, R. N. 1987, *Nature*, **329**, 421.

Tammann, A. and Sandage, A. R. 1968, *Ap. J.*, **151**, 825.

Turtle, A. J. *et al.* 1987, *Nature*, **327**, 38.

Unger, S. W., Pedlar, A., Axon, D. J., Wilkinson, P. N., and Appleton, P. N. 1984, *M. N. R. A. S.*, **211**, 783.

Weiler, K. W., Sramek, R. A., Panagia, N., van der Hulst, J. M., and Salvati, M. 1986, *Ap. J.*, **301**, 790.

Wilkinson, P. N. and deBruyn, A. G. 1984, *M. N. R. A. S.*, **211**, 593.

Wilkinson, P. N. and deBruyn, A. G. 1988, in *IAU Symposium 129, The Impact of VLBI on Astrophysics and Geophysics*, eds. M. J. Reid and J. M. Moran (Reidel, Dordrecht), p. 187.

THE COMPACT RADIO SOURCES IN THE GALAXY M82
- Supernova-remnants and/or recent supernovae -

Peter L. Biermann
Max-Planck-Institut for Radioastronomie, Bonn

Abstract:

We assemble the radio data for the compact radio sources in the starburst galaxy M82 and compare their properties with the six known radio supernovae. The comparison supports the interpretation of the sources in M82 as old radio supernovae or very young supernova remnants. The spectral index distribution appears to show a bimodal structure, with the spectral index either being close to ~-0.5 or close to ~-1. None of the available theories to explain radio emission has been worked out sufficiently to let us understand this spectral behaviour.

1. Introduction

The starburst galaxy M82 shows a large number of barely resolved and unresolved radio sources at radio wavelengths (Kronberg et al. 1985, Unger et al. 1984). This new radio population is believed to represent recently exploded supernovae or young supernova remnants; this interpretation is supported by VLBI observations (Bartel et al. 1987), and by a discussion of supernova explosions in an interstellar medium of high energy density (Schaaf et al. 1988).

Here we summarize the available data on these sources, discuss their properties, and their relevance for comparisons with the radio emissioh of radio supernovae, among them SN1987A.

2. The data

Fig.1 shows a VLA-radio map of M82 (a modified version of Fig.1 in Kronberg et al. 1985) with the various radio sources numbered; some of these sources are marginal at 5GHz, but stronger at 15GHz.

Table 1 summarizes the pertinent data for those compact sources for which variability information (Glendenning 1985, Kronberg & Sramek 1985), VLBI-information (Bartel et al. 1987), and spectral information exists (Kronberg et al. 1985). Column (1) gives the number of the source in Fig.1, column (2) the name, i.e. the coordinates of the source relative to $09^h51^m00^s$, +69°54'00" (Kronberg et al. 1985), column (3) the integrated flux density at 4.9GHz (Kronberg et al. 1985), column (4) the spectral index between 4.9 and 15GHz, or 1.7 and 4.9GHz (indicated by superscript 1) if a high

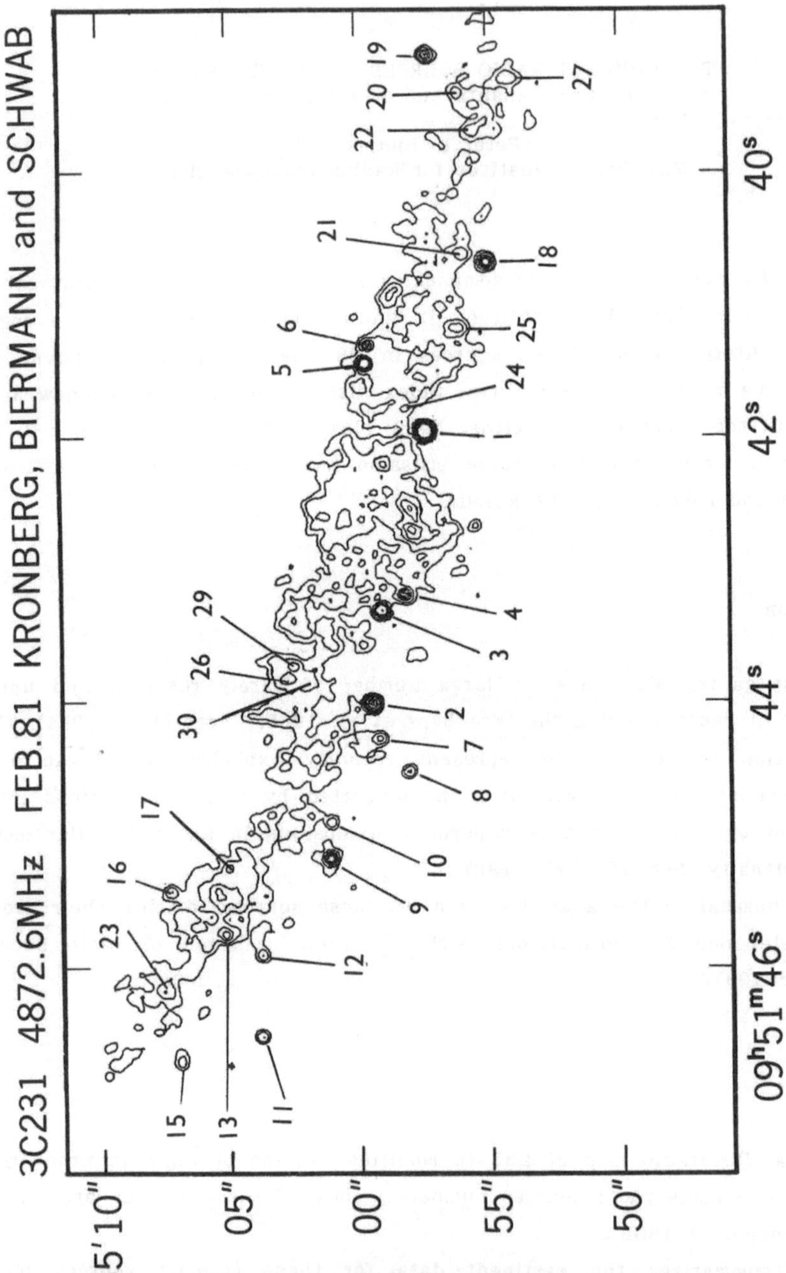

***Fig. 1:** A VLA-radio map of M82, modified from Fig.1 in Kronberg et al. (1985).*

Table 1

Radio sources in M82

Number in Fig.1 (1)	Name (2)	integ.flux density (mJy) (3)	Sp.Index (4)	size (mas) (5)
1	41.9+58	110.5±2.2	-0.93±0.02	24±2x13± 1
2	44.0+59.5	24.5±0.5	-0.46±0.03	≥19± 0
3	43.3+59.1	11.2±0.4	-0.83±0.07	18± 0
4	43.2+58.3	5.5±0.4	-0.62±0.13[1]	≥17± 0
5	41.5+59.7	6.9±1.3	-0.87±0.11	-
6	41.3+59.6	3.2±1.2	+0.01±0.4[1]	≥12± 2
7	44.3+59.2	2.9±0.3	-0.26±0.39[1]	≥11± 2
8	44.5+58.1	2.5±0.4	-0.25±0.49[1]	≥10± 3
9	45.2+61.2	8.0±0.3	-0.48±0.09	60±15
11	46.5+63.8	2.7±0.3	-0.49±0.33[1]	≥85±15
12	45.9+63.8	2.3±0.5	-	≥12± 2
15	46.7+67.0	2.2±0.4	-0.72±0.40[1]	≥60±20
17	45.3+65.1	1.2±0.4	-1.35±0.27[1]	≥70±10
18	40.7+55.0	8.5±0.3	-0.55±0.09	≥75± 5
19	39.1+57.3	5.2±0.2	-0.03±0.27[1]	≥85±10

[1] The spectral index given has been taken from 4.9 and 15GHz data except for the sources marked with "1" for which the two reference frequencies are 1.7 and 4.9GHz.

Table 2

Radio Supernovae

Name	Spectral index	time of radio peak (5GHz) since explosion in days
SN1979C	-0.76±0.08	20
SN1980K	-0.50±0.10	150
SN1981K	-0.91±0.10	<500
SN1983N	-1.03±0.06	15
SN1984L	-1.01±0.17	<40
SN1987A	-0.55±0.15[a]	4[b]

[a] Model-dependent, see Storey & Manchester (1987)
[b] At 843MHz

frequency index is not available (Unger et al.1984, Kronberg et al. 1985, Bartel et al. 1987), and in column (5) the size in milliarcseconds (for 41.9+58 the major and minor axis) from Bartel et al.(1987). We have limited the list to those sources for which either good spectral or size information is available (15 sources).

3. Discussion

Clearly, the sample which we are considering here, is neither homogeneous nor complete. Nevertheless, it is instructuve to see whether the subset of radio sources with good data is consistent with a simple concept of supernova radio source evolution.

The questions, we would like to answer, is a) whether young radio supernovae and the compact radio sources in M82 can be considered to be the same population of sources, and b) whether simple concepts of radio emission from these sources can be induced.

In order to answer these questions, we first limit our M82 source sample to those sources which have a reliable high frequency spectral index. We use only the high frequency index in order to reduce the effects of free-free absorption, recognizable by spectral inversion between low and high frequencies. This results in six sources (40.7+55.0, 41.5+59.7, 41.9+58, 43.3+59.1, 44.0+59.5, 45.2+61.2). For five of these sources size measurements from VLBI exist (Bartel et al. 1987).

Radio supernova data also exist for six sources (Panagia et al. 1986, Weiler et al. 1986, Storey and Manchester 1987: SN1979C, SN1980K, SN1981K, SN1983N, SN1984L, 1987A), listed in Table 2. We note that the spectral index for SN1987A is model-dependent (Turtle et al. 1987, Storey and Manchester 1987).

The distribution of spectral indices for all twelve sources appears to be bimodal; the spectral indices are either close to −0.5 or close to −0.9. Dividing the sample at $\alpha=-0.75$ gives a low spectral index of -0.51 ± 0.04 (five sources) and a high spectral index of -0.91 ± 0.10 (seven sources). There is no obvious difference between radio supernovae and M82 radio sources.

Free-free absorption is necessary to explain the spectral behaviour of the radio supernovae (Weiler et al. 1986), and readily explains the spectral behaviour of the compact sources in M82 as well (Kronberg et al. 1985).

The sizes for the sources in M82 (Bartel et al. 1987) are consistent with an interpretation as moderately young radio supernovae; however, their probable ages are considerably larger than for radio supernovae.

Hence we conclude for our first question, that indeed all reliable data are consistent with radio supernovae and M82 compact sources being the same population.

Theoretical concepts to interpret the data are based either on second order Fermi acceleration (Chevalier 1984), or on first order Fermi acceleration (e.g. Drury 1983). Chevalier's model (1984) can provide an acceptable fit to the data but does not

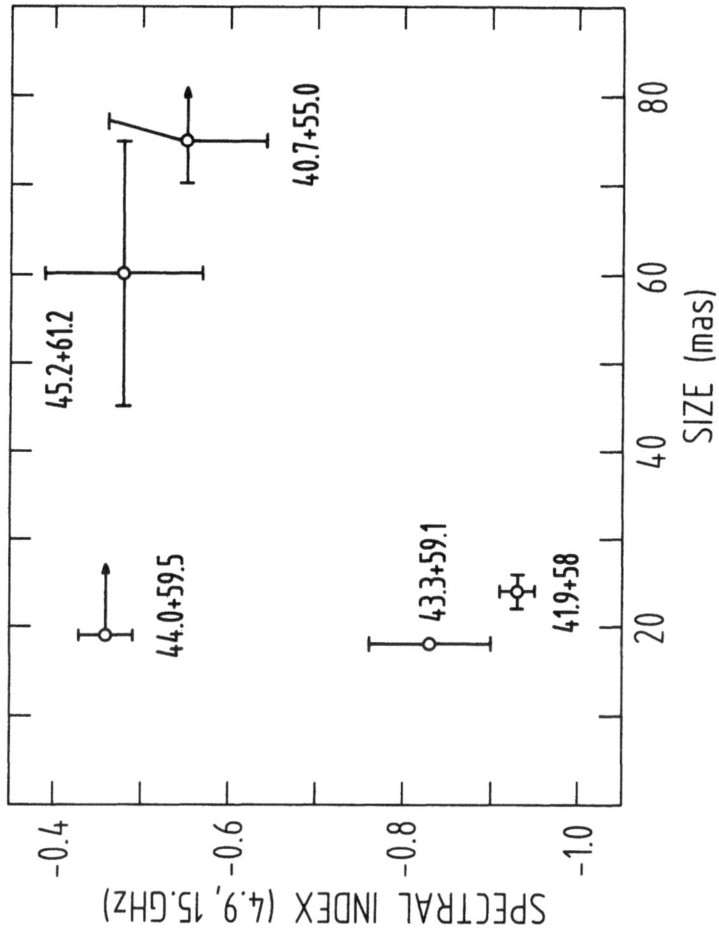

Fig. 2.: *Radio spectral index versus size for those five sources in M82 for which good quality data are available (Kronberg et al. 1985, Bartel et al. 1987).*

provide an explanation for the spectral index observed. First order Fermi acceleration models have not been worked out properly yet for supernova explosions as regards the radio emission. However, in first order Fermi theory strong shocks give a particle spectrum of E^{-2} near to the shock, and approximately $E^{-2.2}$ after integration over the time evolution (Bogdan and Völk 1983), corresponding to a synchrotron spectral index of −0.5 to −0.55. This is clearly **not** consistent with the data. Plotting spectral index versus size (Fig.2) for the M82 sources suggests that especially small (i.e. presumably young) sources show a steep spectral index of ~−0.9. These are just those sources for which the shocks should still be strong, and which are therefore expected to produce a particle spectrum close to E^{-2}.

We thus cannot answer the second question. We do not understand why the M82 sources should be similar to radio supernovae at a much larger scale. A theory is not available which readily explains why radio supernovae − including now the M82 sources − show a tendency for a synchrotron emission spectral index of either ~−0.5 or ~−1.

4. Conclusion

The available data suggest that the compact radio sources in M82 are old radio supernovae. The apparently bimodal distribution of radio spectral indices remains to be explained.

Acknowledgement:

Discussions with Drs. P.P. Kronberg, N. Panagia and H. Völk are gratefully acknowledged.

References:

Bartel, N., Ratner, M.I., Rogers, A.E.E., Shapiro, I.I., Bonometti, R.J., Cohen, N.L., Gorenstein, M.V., Marcaide, J.M., Preston, R.A.: 1987, Ap.J. **323**, 505
Bogdan. T.J., Völk, H.J.: 1983 Astron.&Astrophys. **122**, 129
Chevalier, R.A.: 1984, Ap.J. Letters **285**, L63
Drury, L. O'C.: 1983, Rep. Progress Phys. **46**, 973
Glendenning, B.E.: 1985 Ms.Sc. Thesis, Toronto
Kronberg, P.P., Biermann, P.L., Schwab. F.R.: 1985, Ap.J. **291**, 693
Kronberg, P.P., Sramek, R.A.: 1985, Science **277**, 28
Panagia, N., Sramek, R.A., Weiler, K.W.: 1986, Ap.J. Letters **300**, L55
Schaaf, R., Pietsch, W., Biermann, P.L., Kronberg, P.P., Schmutzler, T.: 1988, Ap.J. (in press)
Storey, M.C.: Manchester, R.N.: 1987, Nature **329**, 421
Turtle, A.J., Campbell-Wilson, D., Bunton, J.D., Jauncey, D.L., Kesteven, M.J., Manchester, R.N., Norris, R.P., Storey, M.C., Reynolds, J.E.: 1987, Nature **327**, 38
Unger, S.W., Pedlar, A., Axon, D.J., Wilkinson, P.N., Appleton, P.N.: 1984, M.N.R.A.S. **211**, 783
Weiler, K.W., Sramek, R.A., Panagia, N., van der Hulst, J.M., Salvati, M.: 1986, Ap.J. **301**, 790

The Supershells of the Large Magellanic Cloud and Their Importance for the Interstellar Medium

J. SPICKER[1] and J.V. FEITZINGER[2,1]

[1] Astronomisches Institut, Ruhr-Universität Bochum
Postfach 10 21 48
D-4630 Bochum, F.R.G.

[2] Sternwarte der Stadt Bochum
Castroper Str. 64
D-4630 Bochum, F.R.G.

Abstract

The first part of this paper describes some observational characteristics of the supershells in the LMC and presents simple estimates of their evolutionary state, based on the supershell theory of MCCRAY ET AL. From the disagreement of theory with observations, possible implications for the phase structure of the ISM in the LMC are inferred.

In the second part, some results of the first application of standard statistical methods to a variety of high-resolution data on the LMC are presented. From an analysis of the two-dimensional autocorrelation and structure functions of HI, HII, radio continuum, IR emission and colour excess, we derive the scale properties of the emission regions and of turbulence in this galaxy. From the appearance of the supershells in the autocorrelation maps we conclude that they are most likely to dynamically influence the ISM in the LMC, in the sense that they produce the observed structures on the largest scales and trigger long term star formation.

The supershells of the LMC: overview of observational results

At least six, possibly nine supershells (or supergiant shells) have been detected on high contrast Hα + [NII] plates of the LMC (MEABURN 1980). These structures are apparently ionized by the tens or hundreds of OB-stars they contain and expand into a neutral ambient medium. They have radii between 300 pc and 700 pc and therefore belong to the largest single structures in the LMC. The smaller shells and bubbles in the LMC have been discussed by BRAUNSFURTH AND FEITZINGER (1983). Figure 1 gives an overview of the approximate form of the supershells and the location of other LMC features in the standard tangential coordinate system. Detection of the shells has also been reported in HI (e.g. ROHLFS ET AL. 1984) and in IR radiation (SCHWERING 1988), but they are also visible in radio continuum and colour excess (see below). Two of the supershells have been investigated in greater detail: LMC2 (CAULET ET AL. 1982) and LMC4 (DOPITA ET AL. 1985), both investigations aimed at clarifying the rôle of the supershells for the ISM in the LMC.

WESTERLUND AND MATHEWSON (1966) already described the supershell LMC4 as a huge HI hole in the LMC, a fact that was confirmed by later studies. A closer look at the ROHLFS ET AL. (1984) maps of integrated HI column density and radial velocity reveals the presence of six supershells in HI (the visibility of the supershells is discussed in Table 1 below).

Two high resolution radio continuum surveys at 21 cm (HAYNES ET AL. 1986) and 13 cm (MOUNTFORT ET AL. 1987) recently became available. They also show some of the supershells, although the resolution of these studies does not permit to detect all of them. SCHWERING (1988) notes the discovery of at least four of the supershells in his detailed IRAS maps of the LMC at all four wavelengths.

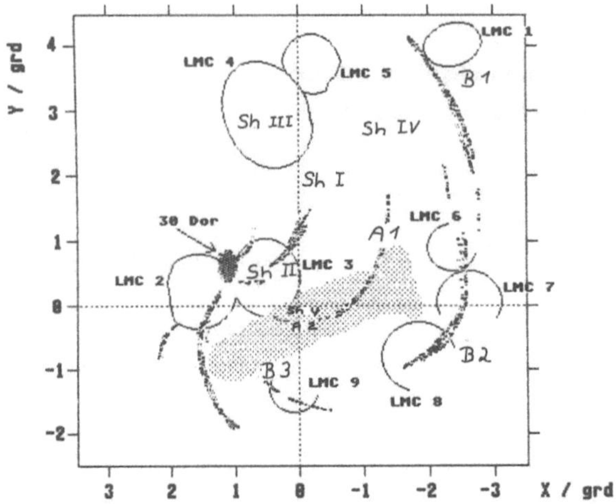

Fig. 1 Synopsis of the main structural characteristics of the LMC in the standard tangential coordinate system. Apart from the supershells LMC 1 to LMC 9 (MEABURN 1980), the main emission regions according to NAIL AND SHAPLEY (1953) and MARTIN ET AL. (1976) are labeled by roman and latin numbers, respectively. The approximate extent of the optical bar is also shown, as well as the locations of some spiral filaments according to FEITZINGER (1980).

We have also analyzed the distribution of colour excess in the LMC (ISSERSTEDT AND KOHL 1984). These data are available for 1508 stars and have to be interpolated to a regular grid in order to obtain contour maps of the mean colour excess. The density of stars in and around LMC4 allows for an interpolation to a dense 0°05 grid based on the data of ten neighbouring stars. LMC4 is clearly visible in this map.

Table 1 summarizes the main observational results for the supershells LMC1 ... LMC8. LMC9 is - apart from the quotation in MEABURN (1980) - not visible in any of the surveys and has therefore not been considered.

Designation	R pc	LH numbers	N_*	n_0 cm^{-3}	age 10^6yr	v_{exp} km/s	visibility
LMC 1	350	15	49	0.5	6 ... 10		Hα, HI, IR
LMC 2	475	101,103-106,108,111	~220	2.5	10...12	30	Hα, HI, IR, cont?
LMC 3	500	57,67,74	~30	0.3	5...10		Hα, HI, (IR)
LMC 4	700	65,72,77	~250	0.5	10...15	36	Hα, HI, IR, cont, ce
LMC 5	400	31	25	0.2	3...6 ?		Hα, HI, IR, cont?
LMC 6	300	4	4	0.7	?		(Hα), (HI), (IR)
LMC 7	400	8,10,21,68	134	0.8	8...10		(Hα), (IR)
LMC 8	450	17	20	0.2	10...15 ?		(Ha), HI, (IR)

Tab. 1: Observational parameters of the supershells LMC 1 to LMC 8. The parameters are:

R	optical radius	LH	Lucke and Hodge regions inside the supershells
N_*	number of blue stars	n_0	ambient HI density
age	from the ages of stars	v_{exp}	expansion velocity

visibility in the various surveys' wavelength

The structure of LMC2 and LMC4

We restrict the following discussion to the supershells LMC 2 and LMC 4, because they are visible in most of the surveys, and because they have been the target of two detailed studies. Fig. 2 displays contour maps of LMC 2 in integrated HI column density and in the 100 μ IR flux. Although this is a very dense region ($n_o \approx 2.5$ cm^{-3}), the shell stands out quite clearly at the left hand boundary of the 30 Dor region.

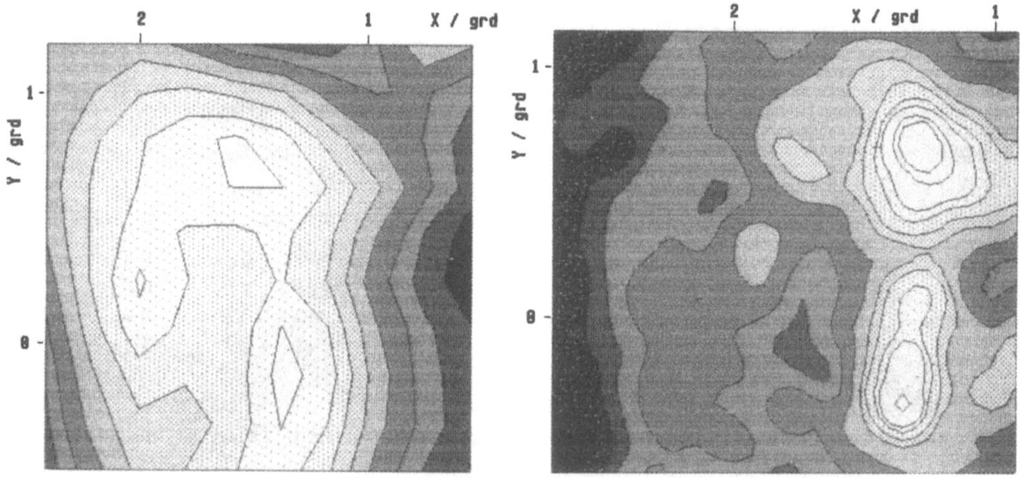

Fig. 2 Contour maps of the supershell LMC 2 in integrated HI (left), and IRAS 100 μ (right)

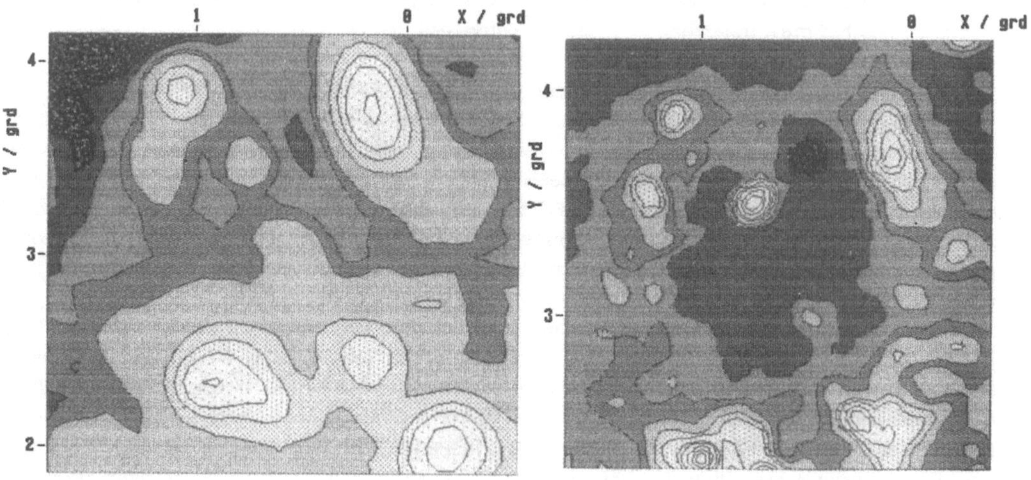

Fig. 3 Contour maps of the supershell LMC 4 in 21 cm radio continuum (left) and IRAS 60 μ (right)

Fig. 3 depicts contour maps of the supershell LMC 4 in 21cm radio continuum and in 60 μ IRAS flux. Again, the huge hole is clearly visible.

MCCRAY AND KAFATOS (1987) and MACLOW AND MCCRAY (1988) discuss the evolution of supershells in numerical models. The combined action of stellar winds and repeated supernova explosions of an OB assocoiation creates a hot pressurized bubble of radius R > 100 pc in the ISM. This model does not incorporate a detailed structure of the ISM, but assumes homogeneous distribution of matter. Relativistic winds, possibly the volume-filling component (KUNDT 1988), are also not included. The bubble is bounded by a thin neutral shell. If the stars of the association obey a typical IMF, the supernova rate will be nearly constant over a period of $t \approx 5 \times 10^7$ yrs. As soon as radiative cooling becomes important, the bubble looses its interior pressure and enters into the snowplow-phase, approximately after some 10^7 yrs at a radius of 100 ... 300pc. Another fate of the bubble could be an outbreak into the galactic halo, if the scale height were less than the bubble radius, or if the OB association were out of the galactic plane. At about the same time, the outer shell becomes gravitationally unstable and is likely to form new molecular complexes, thus giving rise to a new generation of associations.

We have applied this theory to the supershells of the LMC. Table 2 presents the results for LMC 2 and LMC 4.

Shell	R	N_*	E	n_o	v_{exp}	t_c	R_c	v_c	t_i	R_i
	pc		10^{51}erg	cm^{-3}	km/s	10^7 yrs	pc	km/s	10^7 yrs	pc
LMC 2	475	125	0.7	2.5	30	5.4	540	5.9	0.2	220
	475	250	1.5	0.8	30	18.7	1920	6.0	0.4	720
LMC 4	700	159	0.7	0.4	36	21.1	1720	5.2	0.4	640
	700	320	1.0	0.1	36	76.4	6290	5.1	1.1	2150

Tab. 2: Various observational and theoretical parameters of supershells LMC 2 and LMC 4. Two calculations have been made, one with the most probable observed paramers for N_*, E and n_o, the other with assumed extreme values of the parameters. The parameters are:

R	observed optical radius	N_*	numbers of stars	E	Supernova energy
n_o	ISM volume density	v_{exp}	observed expansion velocity		
t_c	start of snowplow-phase	R_c	correspond. radius	v_c	correspond. velocity
t_i	start of grav. instability	R_i	correspond. radius		

The evolution was computed in two possible scenarios: the first calculation was carried out with the observed parameters of the supershells. The numbers of stars were taken from the LH list (LUCKE AND HODGE 1970, LUCKE 1974). The mean energy of the supernovae was taken to be 0.7×10^{51} erg erg (DOPITA 1979). The mean density of the regions containing the supershells was inferred from the HI column density (ROHLFS ET AL. 1984), assuming uniformly filled spherical bubbles. The second calculation assumed extreme values for the parameters, that allow for a maximum of expansion. For LMC 2 and LMC 4, Figs. 4 and 5 show part of the results for the two sets of parameters.

The radii and expansion velocities of the supershells at the critical times, at which either radiative cooling becomes important and the snowplow-phase sets in, or at which the shell becomes subject to gravitational instability, have now to be compared with the corresponding values at the actual ages of the shells. These mean ages (given in Tab. 1) were inferred from the distribution of stars in various age groups as given by ISSERSTEDT (1984). As can be seen from the figures, the supershells are either too young for their observed radii or too old for their observed expansion velocities. Even in the case of the second parameter set, that allows for greater radii and velocities, the observed figures are not matched. According to the observed number of stars and density, LMC 2 could already be in its snowplow-phase, its velocity of expansion is, however, by far too large to fit into the calculations.

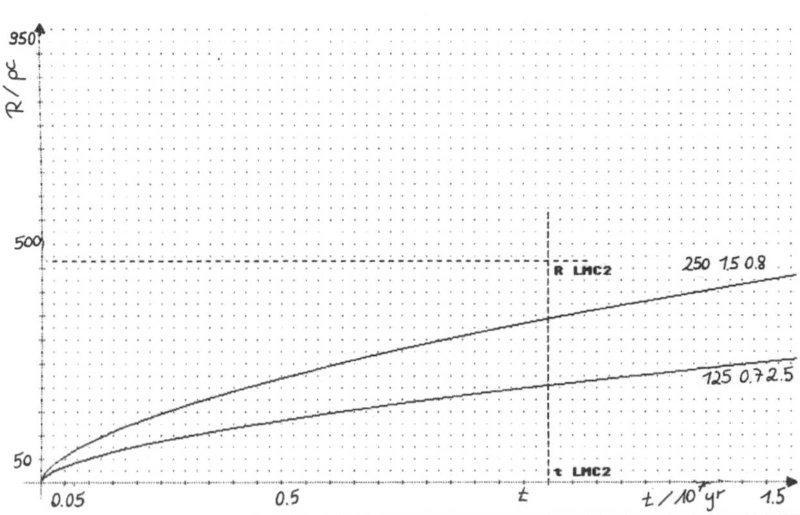

Fig. 4 Evolution of the radius of supershell LMC2 as a function of time for the two sets of parameters. Actual age and corresponding radius are also shown.

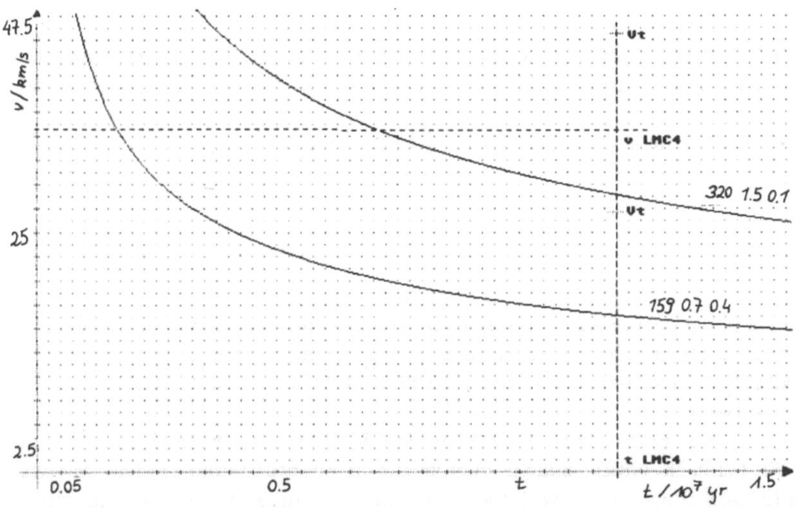

Fig. 5 Evolution of the and expansion velocity of supershell LMC4 as a function of time for the two sets of parameters. Actual age and corresponding velocity are also shown.

These discrepancies call for a closer look at the parameters used in the calculations. These are 1) N_*: the number of stars that became supernovae was taken to be the actual number of blue stars or (for the extreme calculations) its double. The actual number of observed supernova remnants is by far lower than these values, this could be the consequence of severe selection effects (GREEN 1988). We don't see most of the SNRs, because a) the lower brightness ones are still too faint to be detected in the LMC, and b) some of them explode in the rarefied cavities of supershells, where the ambient density is too low ($\lesssim$ 0.1 cm^{-3}) for them to produce an observable remnant. On the other hand, the *inner* parts of the supershells are clearly visible on the UV pictures of the LMC (PAGE AND CARRUTHERS 1981, SMITH ET AL. 1987, VUILLEMIN 1988), so that young objects producing the ionizing radiation *must* be present within the cavities.

2) E: the mean energy a supernova delivers into the ISM. DOPITA (1979) gives a range of 0.2×10^{51} erg to 2×10^{51} erg, with a mean of 0.7×10^{51} erg. The above mentioned selection effects would lead to an overestimate of E.

that either the mean SN energy is lower, that the SN rate is lower, or that thermal evaporation is one of the sinks into which part of the SN energy goes. The blast waves also push ambient cloud medium into previously cleared cavities (this was suggested by CIOFFI 1985 and COX 1986), a fact that is not included into the derivation of the filling factor.

At first sight, there seems to be a contradiction to our calculations of the evolution of the supershells that require *larger* N_* and E and *lower* n_O, whereas HEILES has too much SN energy to offer. On the other hand, the theory of the supershell evolution seems to be far from being settled, as the model of SILICH shows.

Possible solutions to this problem have already been suggested, but none is fully satisfactory:

1) Supershells do not evolve into a uniform medium, but they encounter all the irregularities and clumpiness of the ISM on their way out. Nobody has yet calculated the fate of the clouds entrained by the blast waves, it is not known, how much energy goes into the acceleration of the cloud and intercloud medium. The topic has partly been addressed by CLIFFORD (1984).

2) Condensation of the hot cavity gas onto remaining molecular clouds and evaporation of cold gas from diffuse clouds leaves other sinks for the SN energy.

3) The SN energy could have been largely overestimated and the SN rate could have been underestimated because of selection effects. The minimum mass of stars becoming SN is not that well known for the specific conditions of the LMC. These errors could cancel out, but this is quite uncertain.

4) The effect of type I SN remained unaddressed so far, they are not correlated in space and time, but they are likely to have a larger scale height than have type IIs, so that they could contribute to clear cavities in the lower halo.

None of these explanations alone is very satisfactory (KUNDT, this volume), so that the case is far from being settled.

Statistical methods

We now turn to the statistical properties of the emission regions and supershells of the LMC as inferred from the statistical analysis. The full spectrum of the methods used and the statistical vocabulary has been published in a seperate paper (SPICKER AND FEITZINGER 1988B), we here only briefly summarize the basics.

Throughout our analysis, we assume that the ISM in the LMC is in a general state of turbulence (SPICKER AND FEITZINGER 1988B). Spatial fluctuations of an observable (radial velocity, column density, magnetic field pressure, ...) will then inevitably occur and can be regarded as realizations of a stochastic process. Two-point statistics are the most commonly used tool to describe the behaviour of these fluctuations. Astrophysical data normally consist of observations of an observable N at discrete locations (x, y), only one finite realization of the ensemble of stochastic processes is therefore accessible (i.o.w., the ergodic hypothesis has to be fulfilled).

The two-point covariance tensor defines the correlations of the fluctuations with another variable or itself at a certain displacement (lag) τ. It will reveal to what extent data separated by this lag are correlated. These statistics are only valid for stationary data, so that large scale trends or gradients eventually contained in the data have to be removed in order to extract the stationary components.

An alternative to covariance tensors is given by the two-point structure tensor, which is particularly sensitive to small-scale fluctuations in suppressing large-scale trends. It is also closely related to the Kolmogorov-Obukhov theory of isotropic incompressible subsonic turbulence.

The two-dimensional autocorrelation function (ACF) ρ of an observable N as a function of the discrete lag $(\Delta x, \Delta y)$ is then given by

$$\rho(\Delta x, \Delta y) = \left\langle \left(N(x,y) - \langle N \rangle \right) \left(N(x+\Delta x, y+\Delta y) - \langle N \rangle \right) \right\rangle$$

3) n_o: the mean volume density of the ISM into which the supershells expand. This was taken from the HI column density data assuming simple geometries and a uniform single phase ISM.

The selection effects applicable to 1) and 2) work in opposite directions, so that the errors of these values are likely to cancel out. The most severe error seems to occur through the assumption of a uniform single phase ISM in the LMC. If we assume that the expansion of the supershells takes place into a warm or hot phase - if the latter exists - of the ISM, n_o lowers by factors of 10 or 100, respectively. This seems to be the clue for understanding the evolution of these huge structures.

The multiphase ISM of the LMC

There has been much debate about the phase structure of the ISM in disc galaxies in recent years (see, e.g., SHULL 1987 for an overview). DOPITA (1987) applies this discussion to the LMC. He finds evidence that both a cloud and intercloud medium and a coronal component exist in the LMC. Our calculations of the evolution of the supershells mainly applied to the intercloud medium, for which he gives $n_o = 0.1 \ldots 0.3\ cm^{-3}$ and $T \approx 10^4$ K. As we saw, the observed radii and expansion velocities could not be matched with a medium of that high density. However, as DOPITA claims, the coronal component cannot be maintained at its high temperature by the present supernova rate in the LMC, so that it should have a lower filling factor in the disc. It is therefore not justified to simply assume that the supershells evolve into the coronal component, thus reaching larger radii and velocities.

Ways out of this dilemma have been proposed by a number of authors:

COX (1986) argues that the filling factor of the coronal component of the ISM is not increased by the production of many supershells, but that the remaining cloud and intercloud medium is simply shuffled about. This results in thick and irregular shells, the evolution of which is moderated by the clouds and remnants of earlier shells. There is no creation of really big holes in his model, but merely an agglomeration of interclouds and clouds the blast waves pass through. Thermal evaporation and magnetic fields play a large rôle in that model, which factors have not been included in the picture of MCCRAY ET AL. Parts of these effects have been covered by KREBS AND HILLEBRANDT (1983). CLIFFORD (1984, 1985) has also calculated the interaction of supernova explosions with IS clouds. The explosions are very likely to create the observed mass and velocity spectrum of the smaller clouds, if magnetic fields are taken into account.

These models can be checked against the observations of the supershells in the LMC. The fact that there are young associations and IRAS sources *inside* the shells speaks in favour of this, as does the stellar population study of LMC 4 by REID ET AL. (1987), who find evidence for continuing star formation over most of its lifetime *inside* the shell, which implies the leftover of many star forming clouds *inside* the shell even after the passage of the blast waves. There are also alternative interpretations (see below).

SILICH (1986) also has addressed the evolution of supershells. He uses a thin layer approximation of a strong radiate shock wave and observed IMFs and finds a nearly constant expansion velocity of the supershells in their late evolutionary stage. Application of this model to LMC 2 and LMC 4 gives better agreement with the observed expansion velocities than does the theory of MCCRAY ET AL., at least for those parameter sets that have larger N_* and E and lower n_o.

HEILES (1986, 1987) investigates the effects of the actual distribution of SN of different type on the ISM in the disc and halo of galaxies. For the type II SN, which are correlated in both space and time, he finds a filling factor much larger than unity for the cavities they produce, the disc and halo should be entirely filled with them. There is apparently no galaxy with this property.

That leads him to the conclusion, that SN eject more energy into the ISM and halo than is accounted for by the observations. Among ways out of this dilemma that have been proposed are where the brackets denote the spatial average, and the two-dimensional structure function (STR) by

$$B(\Delta x, \Delta y) = \left\langle \left(N(x,y) - N(x+\Delta x, y+\Delta y) \right) \right\rangle^2$$

Because of the limitations of finite data extent and finite number of realizations of the stochastic process, only estimates of the true ACF and STR can be determined. The most commonly used estimates are the unbiased and the biased estimate (see SPICKER AND FEITZINGER 1988C for a description of their properties), which were both computed.

The basic information contained in these functions is discussed in the papers already mentioned, we just note, that the ACFs of density fluctuations provide the typical sizes and separations of the clumps of radiating matter, whereas the STR is mainly used to calculate the Kolmogorov index of the observable investigated. Typical clump scales are derived from radial cuts through the two-dimensional ACF in different orientations and fitting parabolas to its central parts.

Data and Results

A variety of high-resolution survey data has become available for the LMC in recent years. Most of the surveys are presented in digital form on regular grids, so that they are ideally suited for the statistical analysis outlined in the previous section. Other data (e.g., colour excess) has to be interpolated onto such grids before being subjected to the statistical anlysis.

We have used surveys of HI (providing radial velocity, column density and velocity dispersion information), radio continuum (21cm, 13cm, 74cm), IRAS (100μ, 60μ), Hα, FUV, polarization and colour excess. Some of our results have been published elsewhere, notably on radio continuum and optical polarization (SPICKER AND FEITZINGER 1987), HI (SPICKER AND FEITZINGER 1988B), and IR (SPICKER AND FEITZINGER 1988A).

Three distinct scales of correlation are found, which - as far as resolution permits - occur in all of the observables considered. These scales can be interpreted as scales of a turbulent velocity field (in the case of the radial velocity data), as scales of a turbulent magnetic field (from the radio continuum and polarization data), and as typical sizes of spheres of influence of emission regions. The scale sizes derived from a variety of cuts through the ACF in several orientations give the three distinct scale sizes $\lambda_1 = 200 .. 260$ pc, $\lambda_2 = 550 .. 600$ pc, $\lambda_3 = 900 .. 1000$ pc. λ_1 is typical for the smaller HII regions, λ_2 corresponds to typical spheres of influence of the active star forming regions in the LMC, and λ_3 is the size of the LMC supershells. This is confirmed by the size of the anticorrelations found in their directions.

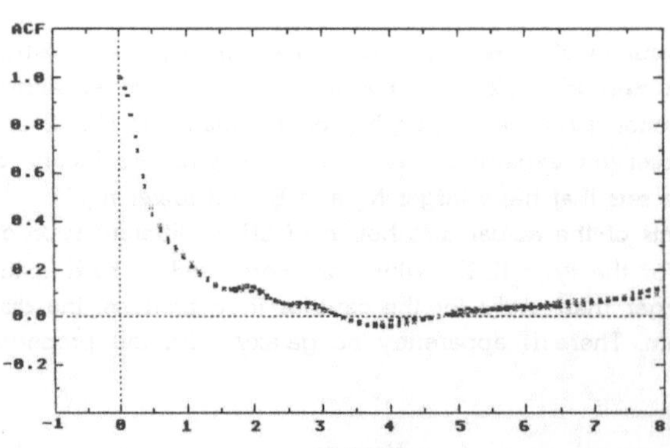

Fig. 6: Radial cut through the two dimensional unbiased estimate of the ACF of the 13 cm radio continuum emission. The value of autocorrelation (normalized to one) is depicted as a function of lag τ in the lagged coordinate system. The parabolic decay around the origin indicates the first scale, the second scale appears as a superposition around ACF = 0.4.

As an example, Fig. 6 shows a radial cut through the two-dimensional unbiased estimate of the ACF of the 13 cm radio continuum emission. There is a double feature of parabolic shape near the origin, which is typical for emission coming from two distinct scale sizes of the emission regions. Other ACFs (in particular those of the IRAS maps) show even three distinct parabolic shaped forms of the ACF near the origin. Other examples can be found in the references quoted above.

Some of the anticorrelations coincide with the positions of the supershells of the LMC. The similar appearance of these structures in maps of radial velocity, emission of various kinds, magnetic field etc. gives strong indication of the importance of these structures for the ISM in the LMC. The supershells turn out to dynamically influence the ISM on large scales, making them most important for the propagation of star formation in that galaxy. They shuffle about the larger components of the ISM (giant molecular clouds, large HI clouds etc.) and inject energy into the ISM on smaller scales in transferring part of their kinetic energy to the small cloud population, which then assumes a state of turbulence. The importance of magnetic fields as stressed by CLIFFORD (1983) is confirmed by the autocorrelation analysis of the nonthermal radio continuum emission (SPICKER AND FEITZINGER 1987), proving that thgis component is also in a state of turbulence. Pronounced deviations from the Kolmogorov law (as inferred from the structure functions not shown here) also indicate the injection of energy into the ISM on all scales.

Summary and Discussion

We have applied common models of the evolution of supershells to the LMC. The theory cannot account for the observed radii and expansion velocities if the ages of the supershells as derived from stellar ages are taken for granted. This fits into the recently discussed problems with SN energies and SN rates and into the uncertainties about the phase structure of the ISM in disc galaxies. A more realistic scenario of supershell evolution will have to take into account a multiphase ISM and the possibilty that these structures do not blow simple empty cavities, but leave apart the remnants of the cloud and intercloud medium existing there before. This picture allows for active star formation to take place not only at the edges of the shells, but also *inside* the shells, a fact that is also observed.

From the statistical analysis of various high resolution data of the LMC, we conclude that turbulence is present in all of the components of the ISM considered, but that its energy spectrum is not compatible with a Kolmogorov type. Energy injection must take place at various scales, not only at the largest ones. Nevertheless, the supershells are - apart from rotation and the bar - the largest structures of dynamical importance in the LMC. They produce turbulence at their outer edges by stirring up the ISM, they produce turbulence within them by shuffling about the cloud and intercloud medium, and they trigger star formation all over the LMC. They are a source of matter rising into the halo and are therefore contributors to the self-regulating mechanism of pattern formation in all three dimensions, which again turns out to be stochastic in nature.

References

Braunsfurth, E., Feitzinger, J.V. 1983, *Astron. Astrophys.* **127**, 113

Caulet, A., Deharveng, L., Georgelin, Y.M., Georgelin, Y.P. 1982, *Astron. Astrophys.* **110**, 185

Cioffi, D. 1985, PhD thesis University of Colorado

Clifford, P. 1984, *Mon. Not. Roy. Astron. Soc.* **211**, 125

Clifford, P. 1985, *Mon. Not. Roy. Astron. Soc.* **216**, 93

Cox, D. 1986, in: Bregman, J.N., Lockman, F.J (Eds), *Gaseous Halos of Galaxies*, NRAO-Workshop #12, Green Bank, p. 239

Dopita, M.A. 1979, *Astrophys. J. Suppl. Ser.* **40**, 455

Dopita, M.A. 1987, in: Peimbert, M., Jugaku, J. (Eds), *Star Forming Regions*, IAU-Symp. #115, Dordrecht: Reidel, p. 501

Dopita, M.A., Mathewson, D.S., Ford, V.L. 1985, *Astrophys. J.* **297**, 599

Feitzinger, J.V. 1980, *Sp. Sci. Rev.* **27**, 35

Feitzinger, J.V., Haynes, R.F., Klein, U., Wielebinski, R., Perschke, M. 1987, *Vistas in Astron.* **30**, 243

Green, D. A. 1988, private communication

Haynes, R.F., Klein, U., Wielebinski, R., Murray, J.D. 1986, *Astron. Astrophys.* **159**, 22

Heiles, C. 1986, in: Bregman, J.N., Lockman, F.J (Eds) 1986, *Gaseous Halos of Galaxies*, NRAO-Workshop #12, Green Bank, p. 79

Heiles, C. 1987, *Astrophys. J.* **315**, 555

Isserstedt, J. 1984, *Astron. Astrophys.* **131**, 347

Isserstedt, J., Kohl, W. 1984, *Astron. Astrophys.* **139**, 115

Krebs, J., Hillebrandt, W. 1983, *Astron. Astrophys.* **128**, 411

Long, K., Helfand, D.J., Grabelsky, D.S. 1981, *Astrophys. J.* **248**, 925

Kundt, W. 1988, private communication

Lucke, P.B. 1974, *Astrophys. J. Suppl. Ser.* **28**, 73

Lucke, P.B., Hodge, P.W. 1970, *Astron. J.* **75**, 171

MacLow, M.-M., McCray, R. 1988, *Astrophys. J.* **324**, 776

Martin, N., Prevot, L., Rebeirot, E., Rousseau, J. 1976, *Astron. Astrophys.* **51**, 345

McCray, R., Kafatos, M. 1987, *Astrophys. J.* **317**, 190

Meaburn, J. 1980, *Mon. Not. Roy. Astron. Soc.* **192**, 365

Mountfort, P.I., Jonas, J.L., de Jager, G., Baart, E.E. 1987, *Mon. Not. Roy. Astron. Soc.* **226**, 917

Nail, V.M., Shapley, H. 1953, *Proc. Nat. Acad. Sci.* **39**, 358

Page, T., Carruthers, G.R. 1981, *Astrophys. J.* **248**, 908

Reid, N., Mould, J., Thompson, I. 1987, *Astrophys. J.* **323**, 433

Rohlfs, K., Kreitschmann, J., Siegman, B.C., Feitzinger, J.V. 1984, *Astron. Astrophys.* **137**, 343

Schwering, P. 1988, *Astron. Astrophys. Suppl. Ser.* in press

Shull, J.M., in: Hollenbach, D.J., Thronson, H.A. (Eds) 1987, *Interstellar Processes*, Dordrecht: Reidel, p. 225

Silich, S.A. 1986, in: de Loore, C.W.H., Willis, A.J., Laskarides, P. (Eds), 1986, *Luminous Stars and Associations in Galaxies*, IAU-Symp. 116, Dordrecht: Reidel , p. 393

Smith, A.M., Cornett, R.H., Hill, R.S. 1987, *Astrophys. J.* **320**, 609

Spicker, J., Feitzinger, J.V. 1987, in: Beck, R., Gräve, R. (Eds), 1987, *Interstellar Magnetic Fields*, Berlin: Springer , p. 82

Spicker, J., Feitzinger, J.V. 1988a, in: Murtagh, F., Heck, A. 1988, *Astronomy from Large Databases*, Proc. ESO , p. 61

Spicker, J., Feitzinger, J.V. 1988b, *Astron. Astrophys.* **191**, 10

Spicker, J., Feitzinger, J.V. 1988c, *Astron. Astrophys.* **191**, 196

Vuillemin, A. 1988, *Astron. Astrophys. Suppl. Ser.* **72**, 249

Westerlund, B.E., Mathewson, D.S. 1966, *Mon. Not. Roy. Astron. Soc.* **131**, 371

This research has been supported by the «Deutsche Forschungsgemeinschaft» under grants Fe 196/3-1 and 196/3-2 and by the «Stiftung Volkswagenwerk». This is most gratefully acknowledged.

Discussion

Biermann: I think it is not that surprising that you do not find a Kolmogorov spectrum, because if you look at the differential equations for turbulence and the injection scale, you never get the inertial range spectrum. You basically demonstrated with your own data, that the injection scale is just the scale you have measured, so you should not get a Kolmogorov spectrum.

Gaskell: Do you get the same inflection of points in the correlation function for different directions and different places?

Spicker: No, the inflection strongly depends on direction and places, which can be traced back into the original maps.

Gaskell: From how many data points do you infer the curvature of the autocorrelation function? Couldn't it be just noise?

Spicker: The actual number of data points depends on what map you are looking at, but the curvature is real, because we are always on the good side of the Nyquist theorem. The scales we derive are normally much larger than the actual resolution of the map. We gain further safety by the calculation of different estimates of the correlation functions.

Extended Onion-Shell Model for Cosmic Ray Spectra Produced by Supernova Remnants

L.A. Zank & H.J. Völk
Max-Planck-Institut für Kernphysik
Postfach 10 39 80, D-6900 Heidelberg 1, FRG

Abstract

A test particle model is used to determine the cosmic ray spectrum produced by supernova remnant (SNR) shock waves. The effects of self-excited upstream Alfvén wave scatterers on the energetic particles and their upper momentum cutoff are included. The number of injected particles is limited to conform with energy flux restrictions and an upper bound on damping effects is obtained. Our work suggests that SNRs remain the most likely source for cosmic rays with energies up to approximately 10^{14} eV/n.

Introduction

Recently much attention has been focused on the concept of cosmic ray acceleration by supernova shock waves in the ISM (see, in particular, Krymsky, 1977; Axford *et al.*, 1977; Bell, 1978a,b; Blandford & Ostriker, 1978). The slope of the particle spectrum obtained by this mechanism corresponds to that of the galactic CR spectrum after adjustment for interstellar propagation effects. (An alternative approach has been advocated by, for example, Kundt, 1983.) We use test particle theory and approximate the particle spectrum at the shock location $R(t)$, for any time t, by employing the steady, plane shock solution, modified to include adiabatic cooling and acceleration time cutoff effects. The cutoff momentum $p_{\rm cut}$ is obtained by equating $\tau(p_{\rm cut}) \propto \kappa(p_{\rm cut})/\dot{R}$, the dynamic time scale of the SNR, with the acceleration time scale $R/\dot{R}$ (Krymsky & Petukhov, 1981; Prishep & Ptuskin, 1981). Previously, the diffusion coefficient κ has been prescribed, rather than determined self-consistently (Blandford & Ostriker, 1980, Bogdan & Völk, 1983, Moraal & Axford, 1983, Jokipii & Ko, 1987). By calculating the overall particle spectrum, including the effect of the Alfvén wave scatterers, we determine κ in a self-consistent manner (at least to first order). For a typical interstellar environment, where the gas pressure is greater than the magnetic pressure, the Alfvén waves experience only non-linear Landau damping. Our calculations provide an upper limit to the detrimental effect of such wave-particle dissipation on shock acceleration. We show that, even by using the unsaturated damping rate of waves propagating down the cosmic ray intensity gradient (Völk & Cesarsky, 1982), the overall particle spectrum produced by a SNR does not differ significantly from the case in which wave damping is ignored. Thus, it appears that SNRs in a homogeneous environment remain prime candidates for the interstellar cosmic ray population below approximately 10^{14} eV/n.

Model and Algorithm

The model described here is an extension of the onion-shell model used by Bogdan & Völk (1983) (see also Moraal & Axford, 1983) to approximate the spatially averaged momentum distribution function for the accelerated particles at the end of the active lifetime of a SNR. The effective diffusion coefficient κ is taken to be that parallel to the magnetic field $\mathbf{B}$. In the onion-shell model, the swept-up matter behind the shock that is propagating into the ISM as a blast wave is separated into a series of shells by imaginary spherical surfaces. Apart from shocked interstellar gas, each shell contains energetic particles, generated by acceleration at the outer shock. The instantaneous Mach number and the compression ratio are taken at the outer shock. It is supposed that, during the lifetime of the SNR, the material within one shell does not mix with that in another shell. Since we are interested only in the overall particle distribution at the end of the lifetime of the SNR, we allow the shells to expand adiabatically back to the ambient ISM pressure and then suppose that the energetic particles escape and mix by diffusion in space. In accordance with test particle theory, the cosmic ray pressure is neglected for the overall SNR dynamics. Under these assumptions, we can determine the final distribution function for each shell and then simply add these contributions to obtain the composite distribution function for the SNR.

Our model differs from the old onion-shell model in that we
(i) determine the diffusion coefficient, κ, and, hence, the upper momentum cutoff, $p_{\rm cut}$, self-consistently for each shell;
(ii) allow the injection momentum $p_{\rm inj}$ to increase with shock velocity $\dot{R}$;
(iii) determine the spatial injection flux density $Q = n_{\rm inj}\dot{R}$ under energy flux constraints;
(iv) include the "sweep-up" phase in our calculations.

In our model, a supernova explosion takes place at time $t = t_0$ and launches a spherical shock into a homogeneous interstellar medium (ISM) of mass density ρ_1 with velocity $\dot{R}_{\rm sw} = 2 \times 10^9$ cm/sec (initially constant). The total blast wave energy is taken to be 4×10^{51} ergs, corresponding to the ejection of 1 solar mass $M_\odot$. When the shock has swept up $5M_\odot$ at the radius $R = R_{\rm sw}$ and time $t = t_{\rm sw}$, a self-similar Sedov solution is assumed to set in, and is approximated by

$$R(t) = (E/\rho_1)^{1/5}\, t^{2/5} \qquad \text{for} \quad t > t_{\rm sw} = R_{\rm sw}/\dot{R}_{\rm sw},$$

The ISM is described by a magnetic field B_1, pressure P_1, mass density ρ_1 and effective sound speed a_1. The shock front Mach number is given by $M = \dot{R}_{\rm sw}/a_1$. The shock continues to expand until the Mach number drops to unity and shock acceleration ends, at which time the internal pressure is approximately equal to P_1 and the energetic particles can escape from the SNR.

In the onion-shell model, the time interval $[t_0, \hat{t}]$ is divided into n subintervals:

$$t_0 < t_1 < t_2 < \ldots < t_{n-1} < t_n \equiv \hat{t},$$

and at each time t_k $(k = 0, 1, 2, \ldots, n)$ an imaginary spherical surface with radius $R(t_k)$ is created. The kth shell contains the material between $R(t_{k-1})$ and $R(t_k)$.

Initially, the accelerated non-thermal component of the kth shell has a spatially averaged isotropic momentum function $f(t_k,p)$, which is approximated by

$$f(t_k,p) \propto p^{-q(t_k)}\{H(\,p(t_k) - p_{\text{inj}}(t_k)\,) - H(\,p(t_k) - p_{\text{cut}}(t_k)\,)\},$$

where $q(t_k) = 3r(t_k)/(r(t_k)-1)$ is the spectral index ($r(t_k)$ is the compression ratio for the kth shell), p_{inj} is the injection momentum and p_{cut} is the upper momentum cutoff.

After adiabatic expansion, the non-thermal distribution function is approximated by

$$\tilde{f}(t_k,p) \propto p^{-q(t_k)}\{H(p(t_k) - \tilde{r}(t_k)^{1/3}p_{\text{inj}}(t_k)) - H(p(t_k) - \tilde{r}(t_k)^{1/3}p_{cut}(t_k))\},$$

where $\tilde{r}(t_k) = \tilde{\rho}_2(t_k)/\rho_2(t_k)$ is the decompression ratio after adiabatic decompression.

It is evident that $\tilde{f}(t_k,p)$ remains a power law, but that the upper and lower momentum cutoffs are shifted downwards by a factor of $\tilde{r}(t_k)^{1/3}$.

The spatially averaged momentum distribution function for the SNR is given by

$$F(p) = \sum_{k=1}^{n} \tilde{f}(t_k,p),$$

from which, on taking the number of shells, n, to be very large and the shells to be infinitesimally thin, we obtain

$$F(p) = \int_{t_0}^{\hat{t}} \tilde{f}(t,p)dt.$$

The actual form of $\tilde{f}(t,p)$ is slightly modified (Drury, private communication) from that of Bogdan & Võlk (1983):

$$\begin{aligned}\tilde{f}(t,p) &= \frac{Q(t)\Delta t R(t)^2(q(t)-3)}{p_{\text{inj}}(t)^3}\tilde{r}(t)^{q(t)/3-1}\left(\frac{p_{\text{inj}}(t)}{p}\right)^{q(t)}\left(\frac{t}{t_0}\right)^{2\nu} \\ &\times \left[H(\,p - p_{\text{inj}}(t)\tilde{r}(t)^{1/3}\,) - H(\,p - p_{\text{cut}}(t)\tilde{r}(t)^{1/3}\,)\right],\end{aligned} \tag{1}$$

where $Q(t)$ is the spatial injection flux density

$$Q(t) = n_{\text{inj}}\dot{R}(t), \qquad n_{\text{inj}} = \delta\rho_1/m_p.$$

We now determine the injection momentum, p_{inj}, upper cutoff momentum, p_{cut} and the unknown parameter δ.

To determine p_{inj} we take the injection energy to be proportional to the downstream thermal energy per particle. We then use the definition of the downstream gas pressure and gas energy flux conservation to obtain

$$p_{\text{inj}}(t) = \sqrt{\frac{3\alpha m_p(\gamma-1)\rho_1\dot{R}(t)^3}{n_2\gamma u_2(t)2}\left\{1 - (\rho_1/\rho_2(t))^2 + \frac{2}{(\gamma-1)M(t)^2}\right\}}.$$

where α (=4) is a constant of proportionality, γ (= 5/3) the adiabatic index, $\rho_2(t)$ the downstream mass density, n_2 the downstream number density and $u_2(t)$ the downstream flow velocity in the shock frame.

The problem of determining the upper momentum cutoff self-consistently is tackled by considering the two cases which arise from the definition of the diffusion coefficient κ (Bell, 1978a)

$$\kappa(p) = \frac{4}{3\pi}\frac{r_g w}{A(k)},$$

where w is the particle speed and $r_g = (pc)/(eB_1)$ denotes the gyroradius of a particle in the interstellar magnetic field (e is the energetic particle charge), $A(k)$ is the wave energy density per unit logarithmic bandwidth and k the wave number (for further details see Völk, Zank & Zank, 1988):

$$\text{(i) } A(k) < 1 \qquad\qquad \text{(ii)} A(k) > 1.$$

The theoretical minimum value of the diffusion mean free path is the gyroradius, in which case the minimum value κ can take is:

$$\kappa_{min}(p) = (4r_g w)/(3\pi). \tag{2}$$

If $A(k) > 1$, then $\kappa(p) < \kappa_{min}(p)$, which is not permissible since other wave dissipation processes prevent it. Hence, for $A(k) \geq 1$, we suppose that $A(k) = 1$ and use the gyroradius limit to determine $p_{\rm cut}$.

By equating τ, the instantaneous acceleration time, with the system age, we obtain for the kth shell:

$$\tau_k \equiv R(t_k)/\dot{R}(t_k) = g(t)\kappa(p_{\rm cut})/\dot{R}(t_k)^2,$$

where $g(t) = 3r(t)(r(t)+1)/(r(t)-1)$. This gives

$$\kappa(p_{\rm cut}) = R(t)\dot{R}(t)/g(t). \tag{3}$$

For $A(k) \geq 1$, we combine equations (2) (with $w = c$) and (3) to obtain:

$$p_{\rm cut}(t) = \frac{R(t)\dot{R}(t)3\pi e B_1}{g(t)4c^2},$$

We note that this expression for $p_{\rm cut}$ is independent of δ.

For $A(k) < 1$, equation (3) is compared with $\kappa(p)$ calculated from unsaturated wave damping:

$$\kappa(p) = \frac{4}{3\pi}\frac{r_g c a_f}{a_F f(t,p)}\left[1 + \frac{V_p c}{3\dot{R}(t)^2\sqrt{2\pi}}\right],$$

where V_p is the proton thermal velocity,

$$a_F = \frac{4}{3\pi} r_g \frac{w}{R(t)}, \qquad a_f = \frac{r_g B_1^2}{8\pi^3 v p^4},$$

$$v = \frac{B_1}{\sqrt{4\pi\rho_1}}, \quad \text{(ISM Alfvén velocity)}$$

to obtain

$$p_{\rm cut}(t) = p_{\rm inj}(t)\left[\frac{\delta R(t)}{D(t)g(t)}\right]^{1/(q(t)-3)},$$

Model	B (Gauss)	a_1 (cm/s)	P_1 (dyne cm^{-2})	n_1 (cm^{-3})	T (K)	v (cm/s)	V_p (cm/s)
Hot	1.4×10^{-6}	1.13×10^{7}	2.9×10^{-13}	3×10^{-3}	7×10^{5}	5.58×10^{6}	7.6×10^{6}
Warm	5×10^{-6}	1.384×10^{6}	2.76×10^{-13}	0.2	1×10^{4}	2.44×10^{6}	9.1×10^{5}

Table 1: Parameters for the hot and warm ISM models

where

$$D(t) = \frac{cB_1 u_2(t)}{6 \times 10^{-3}\pi^2 evR(t)(q(t)-3)}\left[1 + \frac{V_p c}{3\dot{R}(t)^3\sqrt{2\pi}}\right],$$

(see Völk, Zank & Zank, 1988 for a detailed discussion) and δ is determined (iteratively, in the non-gyroradius limit) as follows:

As suggested by observations of interplanetary travelling shocks (Lee, 1983) δ is assigned an initial value of 10^{-3}, with which the cosmic ray pressure, p_{cr}, is determined from

$$p_{cr} = 1/3\int_{p_{inj}}^{p_{cut}} 4\pi p^2 wpf(p)\mathrm{d}p = \int_{p_{inj}}^{p_{cut}} \left(\frac{p}{p_{inj}}\right)^{3-q} \frac{w\hat{\delta}}{3}\mathrm{d}p,$$

where,

$$\hat{\delta} = \frac{\dot{R}n_1(q-3)\delta}{u_2(t)}$$

and n_1 is the upstream number density.

We note that, the determination of p_{cr} in the non-gyroradius limit is a non-linear problem, owing to the dependence on δ.

To ensure that the cosmic ray energy flux does not exceed some fraction of the available kinetic energy flux we require that the cosmic ray pressure satisfy the energy constraint:

$$\frac{\gamma_c}{\gamma_c - 1}u_2(t)p_{cr} \le \epsilon\frac{\rho_1\dot{R}(t)^3}{2}\left[1 - (\rho_1/\rho_2(t))^2 + \frac{2}{(\gamma-1)M(t)^2}\right], \tag{4}$$

where ϵ is a parameter. (We have taken $\epsilon = 1/2$.)

If condition (4) is not satisfied, we reduce δ until the energy condition is satisfied.

Results and Discussion

Table 1 gives the parameters used in our calculations for both the "hot" and the "warm" ISM models. Plots of the resulting number density of particles $F(p)$ in three-dimensional momentum space, spatially integrated over the SNR interior, are shown in Figures 1–3. Our main result, depicted in Figure 1 gives the overall distribution at the end of SNR evolution ($M = 1$) with and without wave damping. Differences occur mainly for $p/mc > 10^2$ in both the hot and the warm models. For the hot model, the spectral slope q for $1 < p/mc < 10^4$ is 4.20 without damping while $q = 4.35$ with unsaturated, i.e. maximum, wave damping. In the warm model, $q = 4.23$

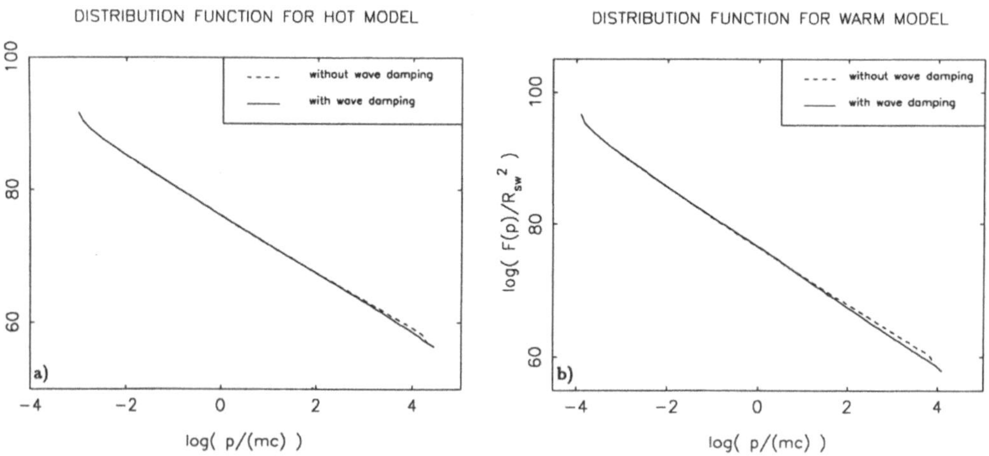

Figure 1: Spatially integrated energetic particle distribution in momentum space $F(p)$ for the hot and warm models.

for $1 < p/mc < 10^4$ without damping and $q = 4.50$ with unsaturated damping. Figure 2 shows separately the spectra during the sweep-up phase and the Sedov phase, with unsaturated wave damping. Clearly, the spectra produced during the high Mach number ($M = 177$) sweep-up phase are very hard $\sim p^{-4}$. The negligible wave-damping in the sweep-up phase offsets the considerable adiabatic decompression and it can be seen that the highest energies come from this phase. The integrated spectrum for the Sedov phase is much steeper. Figure 3 shows intermediate cases, where the evolution was stopped at various minimum shock Mach numbers. The spectrum at progressively higher momenta p/mc is affected only by progressively higher Mach number shocks (*e.g.* $M > 10$ for $p/mc > 10^2$). Figure 4 emphasizes the effect of wave damping on the upper cutoff momentum p_{cut}, given as a function of SNR age t and shock Mach number. With maximal wave damping (lower diagrams in Figures 4a and 4b), p_{cut} drops quite substantially for $M < 10$. For Mach numbers above about 10, the self-excited waves are strong enough to force $\lambda_{mfp} \simeq r_g$ under all conditions. However, SNR's of this small size suffer from adiabatic expansion losses which, for the sweep-up phase $t < t_{sw}$, imply a reduction in p of the order of 10 if no particle escape is allowed until $M = 1$. Fortunately, the drop in p_{cut} with time during the Sedov phase allows some escape of the highest energy particles. For comparison, we show the case without any wave damping in the upper diagrams in Figures 4a and 4b. Here p_{cut} decreases with time as $t^{-1/5}$ during the Sedov phase and drops sharply only near $M = 1$, when the shock degenerates into a sound wave. In the upper diagrams of Figures 4a and 4b, wave dissipation is neglected, whilst unsaturated wave damping is assumed for the lower diagrams. The dotted vertical line in Figures 4a and 4b divides the region where the mean free path λ_{mfp} is given by the Larmor radius limit r_g (early phases, $M > 23$ (hot model), $M > 12$ (warm model) from that where $\lambda_{\mathrm{mfp}} > r_g$ (later phases). The discontinuity at $t \simeq 4 \times 10^{10}$ sec. is due

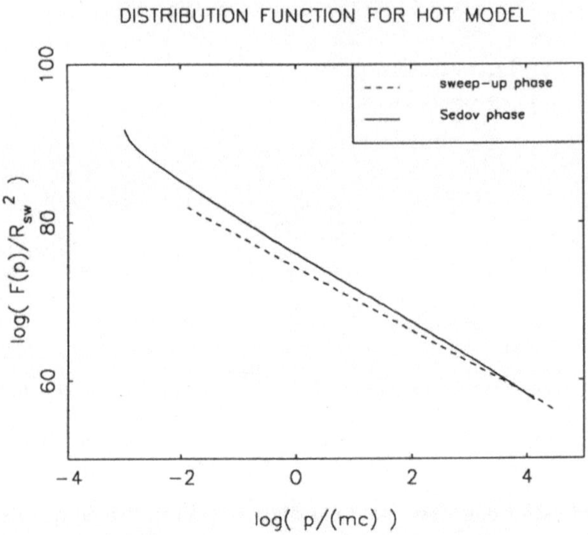

Figure 2: Particle spectrum, showing the contributions from the sweep-up and Sedov phases separately.

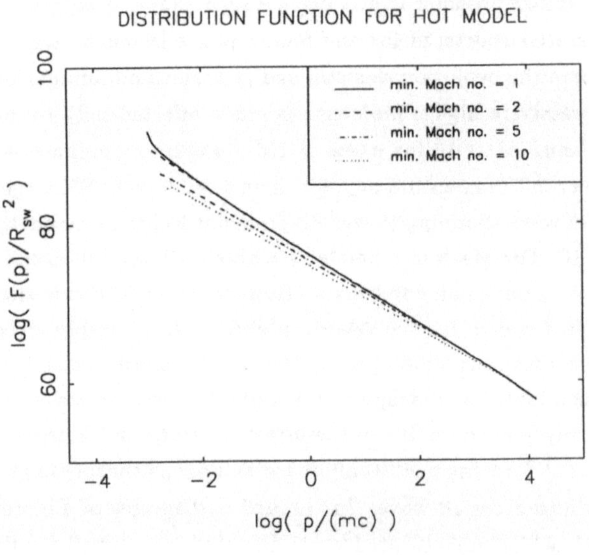

Figure 3: Particle spectrum for various minimum Mach numbers.

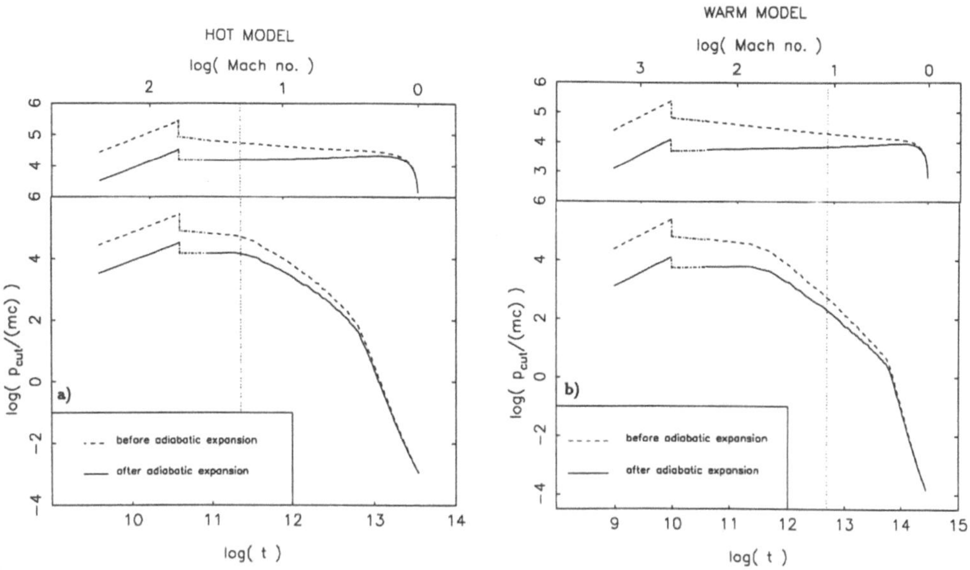

Figure 4: The maximum momentum p_{cut}, as a function of SNR age t (in seconds) and instantaneous Mach number, for the hot and warm models.

to the discontinuous transition from the sweep-up to the Sedov phase; the dash-dot interval corresponds to the first shell in the Sedov phase.

In spite of our having used the maximum wave damping rate, our results show that the overall SNR spectrum is not modified substantially. We have also not attempted to optimize parameters in order to obtain higher energies. Nevertheless, it appears that SNR shock waves in a homogeneous environment may well be the source of interstellar cosmic rays with energies up to approximately 10^{14} eV/n.

Acknowledgements: This research was supported by the Deutsche Sonderforschungsgemeinschaft (SFB 328, Teilprojekt A).

References

Axford, W.I.. Leer, E. & Skadron, G. (1977) *Proc. 15th Int. Cosmic Ray Conf. (Plovdiv)* **11**, 132.
Bell, A.R. (1978a) *M.N.R.A.S.* **182**, 147.
Bell, A.R. (1978b) *M.R.N.A.S.* **182**, 443.
Blandford, R.D. & Ostriker, J.P. (1978) *Ap. J.* **221**, L29.
Blandford, R.D. & Ostriker, J.P. (1980) *Ap. J.* **237**, 793.
Bogdan, T.J. & Völk, H.J. (1983) *Astron. Astrophys.* **122**, 129.
Jokipii, J.R. & Ko, M. (1987) *Proc. 20th Int. Cosmic Ray Conf.* **2**, 179.
Krymsky, G.F. (1977) *Doklady Akad. Nauk. SSSR* **234**, 1306.
Krymsky, G.F. & Petukhov, S.I. (1981) *Proc. 17th Int. Cosmic Ray Conf.* **3**, 503.
Kundt, W. (1983) *Astrophys. Space Sci.* **90**, 59.
Lee, M.A. (1982) *J. Geophys. Res.* **87**, 5063.
Moraal, H. & Axford, W.I. (1983) *Astron. Astrophys.* **125**, 204.
Prishep, V.L. & Ptuskin, V.I. (1981) *Sov. Astr.* **25**, 446.
Völk, H.J. & Cesarsky, C.J. (1982), *Z. Naturforsch.* **37a**, 809.
Völk, H.J., Zank, L.A. & Zank, G.P. (1988) *Astron. Astrophys.* in press.

EPILOGUE

W. Kundt

This volume appears in the Lecture Notes Series; it is meant to be more than a collection of conference reports. So what can be learnt from this book?

Quite often, Conference Proceedings contain opposing views on various interpretations, but it is not always easy to find out about them because minority views tend to be ignored in the reviews, and prefaces tend to leave the reader with the impression that the reviews express the new scientific consensus. As Thomas Kuhn (MIT) says in his 1970 book 'The Structure of Scientific Revolutions': new theories are regularly suppressed. This attitude leads repeatedly to scientific crises, whenever enough details have accumulated that conflict with some paradigm.

As a participant of this workshop, I can see a number of opposing views expressed by the various authors, some of them perhaps leading to another small crisis. They concern the origin, mechanism and energetics of supernova explosions, their influence on the structure of galactic disks and halos, and the way in which Nature accelerates particles to relativistic energies. Minority views can be wrong, but there is no crisis without an initiating minority view. When discussing the highlights of this volume, I shall therefore concentrate on the disagreements between authors of which I am aware, because they are likely at the root of tomorrow's deeper understanding.

<u>SN</u> <u>shells</u> are interpreted as <u>blast</u> <u>waves</u> by most authors because it is thought that the energy of an exploding star is transferred to its surrounding medium like that of a hydrogen bomb in air. The blast wave, or Sedov-Taylor wave, is essentially a thin, hot, spherical mass shell which <u>sweeps</u> the (gaseous) medium in front of it until its momentum is shared by a much larger mass. But there is a second class of explosions - <u>splinter</u> bombs - whose fragments do not sweep because

they have a small volume-filling factor f. Splinters have a much larger range, larger by some factor of order $f^{-1/3}$. For $f = 10^{-3}$, the free-expansion range is some 5 times larger, the energy is shared with roughly 10^2 times the ambient mass, and thermalization leads to temperatures of order 10^7K instead of 10^9K. The two models - the pressure bomb and the splinter bomb - make therefore very different predictions. Which model fits better?

A number of details have been pointed out in the past indicating that SN shells are not well described by the shock-wave model, despite all its seeming success. In this volume, Claas shows that one prediction - the existence of a reverse shock - is at variance with the X-ray data for G 292.0+1.8. Two years ago he had pointed at another difficulty for RCW 86: the X-ray temperature anticorrelates with the expansion velocity for that shell.

In their analysis of the X-ray data for SN 1987A, Bandiera et al conclude that its shell is "heavily fragmented". In my second contribution, I derive this property from quite general assumptions. But a fragmented shell does not sweep and does not form a reverse shock.

Observers often claim that the $r \sim t^{2/5}$ law of a Sedov wave fits the subclass of middle-aged remnants quite well. They confirm this law when they do not draw error bars, like in figure 4 of Reich & Fürst. Strom calls it a "solid physical basis". But Mills et al (1984) have shown that the LMC sample is much better described by $r \sim t$. The splinter model predicts approximately:

$$r \approx \begin{Bmatrix} v_0 t \\ v_0 t_0 \left[1+(t-t_0)/3\, t_0\right]^{2/5} \end{Bmatrix} \quad \text{for} \quad t \begin{Bmatrix} \ll \\ \gtrsim \end{Bmatrix} t_0 \approx 10^{3.5}\text{yr}\ \ \rho_{-25}{}^{1/2}$$

for an initial bulk velocity of $v_0 \lesssim 10^4$ Km s^{-1} and an average ambient mass density ρ in units of 10^{-25} g cm^{-3}, (Kundt & Gopal-Krishna, 1984). Though it converges to a $2/5^{th}$ power law for very old (and faint) stages, it is in good agreement with the Mills et al analysis (without any contrived assumptions on the ISM).

Statistics can be like caoutchouc when too many free parameters are involved. There are, however, predictions which clearly differ for the two models (smooth versus filamentary): a "blowout" - as used in the title of Milne et al - would have to be a structure of the explosion itself, not a consequence of an inhomogeneous CSM, if it involves a significant amount of clumpy ejecta. Similarly, 'barrels' cannot be explained in the same way by the two models, nor would the Penticton source VRO 42.05.01 owe its shape to a hot sheet of the ISM: splinters move like tanks. The splinter model explains bumpy-looking shells, chimneys, and exotics as shaped by the - almost weightless - relativistic plasma.

Another detail which can discriminate between smooth and clumpy ejecta is whether or not the fine structure seen in the shells has the morphology of sheets or ropes. For the Cygnus Loop, Hester (1987) has argued in favour of sheets whereas for S 147, both Kirshner & Arnold (1979) and Fürst & Reich argue in favour of ropes, like for the Crab.

How sharp are the outer edges of SNSs? In my first contribution, indications of large 'halos' around the Crab nebula and Cas A are reviewed, with radii some three times larger than those of the main shells. Forerunner filaments have also been mapped beyond the outer edge of the Cygnus Loop. They are reminiscent of the high velocities signalled by many emission lines from SNe, upto $3 \cdot 10^{4}$ Km s^{-1}, whereas the bulk of the matter moves at $10^{3.8 \pm 0.2}$ Km s^{-1}. Apparently, a small fraction of the ejecta gets accelerated to much higher speeds, perhaps by radiation pressure. In this connection, I am intreagued by the linear-polarization and rotation-measure maps of Milne et al which extend smoothly beyond the outer edges of several shells. If they are confirmed with higher resolution, they demonstrate again that SN shells don't have sharp outer edges.

There is an uncertainty, in the literature, about the ages of SNSs. Shock waves are expected to decelerate and fade on a timescale of $\gtrsim 10^{5}$ years whereas both pulsar associations and statistics favour a maximal SNS age of 10^4 years. A marginal case is CTB 80 whose pulsar has a spindown age of 10^5 yr. Strom and Kundt point at a near-spherical IR and optical shell surrounding the odd-shaped radio source, i.e. we still see a fading shell. Becker's exotic sources, like the flying duck, may likewise be late stages of SN explosions with an active neutron star moving away from the explosion center.

Under what circumstances do we expect to see 'the' central pulsar which Bandiera et al believe to have detected inside SN 1987A? Srinivasan & Bhattacharya demonstrate that in normal shells, only strong pulsars can be detected. Moreover, more than half of all neutron stars may be born as members of binary systems in which they tend to be screened by the companion star's wind; SS 433 is an exception. In most cases, the spindown energy of the binary neutron star may leave the system in the form of cosmic rays (Kundt, 1984). Whether or not a significant fraction of neutron stars have low magnetic dipole moments is not yet decided (Kundt, 1988).

Most workers on cosmic rays consider SN explosions the dominant acceleration sites, whereas direct evidence does not favour this view. At this workshop, Braun inferred a high cosmic-ray ion pressure in the Cygnus Loop from simple assumptions, but Falle and Innes are not convinced. The direct approach by Zank & Völk is at variance with that by Kundt (1984). There is no consensus yet on the origin of the cosmic rays.

There is no consensus either on the mechanism by which a SN ejects its shell. Lattimer discusses the difficulties of the neutrino model. Kundt gives new arguments in support of a magnetic torque. A decaying magnetic spring would convert its energy mainly into relativistic pair plasma, which may help explaining radio SNe (cf. Bartel and Biermann).

On the other hand, Fürst & Reich make the alarming discovery that SN shells, in particular old ones, contain a hard radio component which correlates with the resolved filaments. Are the filaments the sites of in-situ acceleration? Shock-acceleration proponents take this as a direct proof of their approach whereas shock-acceleration sceptics are reminded of field-line reconnections (behind filaments) which are well documented for the Sun and the magnetotail of Earth.

Coming back to SNe again, there is the recent class of type Ib events which correlate with large HII regions. For this class, Van den Bergh determines a progenitor mass of $> 15\ M_\odot$ whereas Panagia concludes at $\lesssim 8\ M_\odot$. The light curves of type Ib events have the same rise and decay times as those of type Ia, which are presently held to derive from exploding white dwarfs. According to theory, this means that their shells involve comparable masses, in conflict with the white-dwarf model. There is an indication that radiation transfer through filamentary shells can explain the exponential tails of SN lightcurves, instead of radioactive heating. If this indication were confirmed, there would be little support left for the exploding-white-dwarf scenario.

An important number for galactic energetics is the birthrate of neutron stars which should be a lower bound on the rate of SNe and SNSs. In her recent review, Trimble (1988) reports a Galactic SN rate of one in (30±10) years, but Gaskell argued in favour of one in $\lesssim 10$ years, and so do Reich & Fürst: $\dot{N}(SNS) = 1/10^{1\pm0.3}$ yr. Blaauw (1985) points out that the progenitor masses of pulsars alone - ignoring binary neutron stars - cannot be much larger than $5\ M_\odot$, simply from counting statistics. Again, a high SN rate is implied.

And what do SN shells tell us about the ISM? Some model builders (Arendt et al) allow themselves density gradients of the ISM whereas others think of discrete components, of temperatures $T \lesssim 10^2$ K (cold), $\approx 10^4$ K (warm), and possibly $\gtrsim 10^5$ K (hot). Of course, density gradients can be realized in a multi-component fashion. But again, a shock wave reacts quite differently from shrapnel, and a secure model should take account of a possible multi-component structure.

Similarly, there is no firm knowledge yet about the medium that blows the LMC supershells (discussed by Spicker & Feitzinger). Is it of local galactic composition, or is it pair plasma?

Unfortunately for this workshop, there was no time left to discuss at any length the best-studied of all SNS-PSR associations, the Crab nebula - on which different scientists have derived different results - nor the best-studied SNS-binary-neutron-star association, W 50. Other meetings will take care of them.

REFERENCES (beyond those found easily in the main text)

Blaauw, A., 1985, in: Birth and Evolution of Massive Stars and Stellar Groups, eds. Boland & van Woerden, Reidel, p. 211

Kirshner, R.P. & Arnold, C.N. 1979: Astrophys.J. 229, 147

Kundt, W., 1984: J.Astrophys.Astron. 5, 277

Kundt, W., 1988: Comments on Astrophysics 12, 113

Kundt, W.& Gopal-Krishna, 1984: Astron.Astrophys. 136, 167

Mills, B.Y., Turtle, A.J. Little, A.G. & Durdin, J.M., 1984: Aust.J.Phys. 37, 321

Trimble, V., 1988: Nature 331, 488

SUBJECT INDEX

LIST OF PARTICIPANTS

Arendt, Richard G.	Dept.of Astronomy, Urbana, IL., USA
Bandiera, Rino	Osservatorio Astrofisica, Firenze, Italy
Bartel, Norbert	Center f.Astrophysics, Cambridge, MASS., USA
Beck, Rainer	MPIfR, Bonn, F.R.G.
Becker, Robert H.	Lawrence Livermore Lab.,U.C.,Livermore,CA.,USA
Biermann, Peter L.	MPIfR, Bonn, F.R.G.
Braun, Robert	VLA/NRAO, Socorro, N.M., USA
Brinkmann, Wolfgang	MPIfA, München, F.R.G.
Claas, Jacques J.	Space Res. Lab., Leiden, Netherlands
Da Costa, Antõnio A.	CEUTL, Lisboa, Portugal
De Boer, Klaas, S.	I.f.A., Bonn University, F.R.G.
Dwek, Eli	NASA/GSFC, Greenbelt, MD., USA
Falle, Samuel A.E.G.	Dept.of.Appl.Math., Univ.of Leeds, England
Fink, Henner H.	MPIfA, München, F.R.G.
Fischer, Daniel	I.f.A., Bonn University, F.R.G.
Fürst, Ernst M.	MPIfR, Bonn, F.R.G.
Gaskell, Martin G.	Astronomy Dept., Ann Arbor, Michigan, USA
Green, Dave A.	Dominion Radio Astron.Observ.,Penticton,Canada
Greidanus, Harm	Leiden Observatory, Leiden, Netherlands
Innes, Davina	MPIfK, Heidelberg, F.R.G.
Hüttemeister, Susanne	I.f.A., Bonn University, F.R.G
Junkes, Norbert	MPIfR, Bonn, F.R.G.
Kundt, Wolfgang	I.f.A., Bonn University, F.R.G.
Lattimer, James	Astronomy Progr., SUNY, Stony Brook, N.Y., USA
Markiewicz, Wojtek J.	MPIfK, Heidelberg, F.R.G.
Meyer, Hinrich	Berg. University, Wuppertal, F.R.G.
Milne, Douglas K.	CSIRO, Epping, NSW, Australia
Panagia, Nino	Istituto di Astronomia, Catania, Italy
Preite-Martinez, Andrea	Istit. di Astrofis. Spaziale, Frascati, Italy
Priester, Wolfgang	I.f.A., Bonn University, F.R.G.
Reich, Patricia	MPIfR, Bonn, F.R.G.
Reich, Wolfgang	MPIfR, Bonn, F.R.G.
Schaaf, Reinhold	I.f.A., Bonn University, F.R.G.
Schaefer, Bradley E.	NASA/GSFC, Greenbelt, MD, USA
Spicker, Jörg	Astronomisches Institut d.Univ., Bochum, F.R.G.
Srinivasan, G.	E.S.O., Garching b. München, F.R.G.
Strom, Richard G.	Radiosterrewacht, Dwingeloo, Netherlands
Trümper, Joachim	MPIfA, München, F.R.G.
Tuffs, Richard	MPIfK, Heidelberg, F.R.G.
Van den Bergh, Sidney	Herzberg Institute/NRC, Victoria, B.C., Canada
Völk, Heinrich J.	MPIfK, Heidelberg, F.R.G.
Wielebinski, Richard	MPIfR, Bonn, F.R.G.
Zank, Lorraine A.	MPIfK, Heidelberg, F.R.G.

Lecture Notes in Mathematics

Lecture Notes in Physics

MIX
Papier aus verantwortungsvollen Quellen
Paper from responsible sources
FSC® C105338

If you have any concerns about our products,
you can contact us on
ProductSafety@springernature.com

In case Publisher is established outside the EU,
the EU authorized representative is:
Springer Nature Customer Service Center GmbH
Europaplatz 3, 69115 Heidelberg, Germany

Printed by Libri Plureos GmbH
in Hamburg, Germany